> 华为ICT认证系列丛书

华为技术认证
HCIA-Datacom 网络技术
实验指南

华为技术有限公司 主编

人民邮电出版社
北京

图书在版编目（CIP）数据

HCIA-Datacom 网络技术实验指南 / 华为技术有限公司主编. -- 北京：人民邮电出版社，2022.4（2024.1重印）
（华为ICT认证系列丛书）
ISBN 978-7-115-58318-5

Ⅰ. ①H… Ⅱ. ①华… Ⅲ. ①计算机网络 Ⅳ. ①TP393

中国版本图书馆CIP数据核字(2021)第261764号

内 容 提 要

本书是《HCIA-Datacom 网络技术学习指南》的配套实验手册，是华为 HCIA-Datacom 认证考试的官方教材。本书从 VRP 系统的基本操作开始介绍，其中包括华为数通设备的文件系统结构及操作方法，并按照具体协议和技术分别设计了多个实验，如静态路由、OSPF 路由协议、以太网基础、生成树、以太网链路聚合、VLAN 间通信、ACL、AAA 配置、NAT、FTP、DHCP、WLAN、IPv6，以及网络编程与自动化。本书最后结合上述多个知识点，以综合实验的形式展示了如何将园区网络的用户需求转化为技术需求，带领读者逐步思考、设计，并搭建出一个完整的园区网络。

本书适合从事网络技术工作的专业人员、正在准备考取华为 HCIA-Datacom 认证的人员阅读，也可作为高等院校相关专业师生的参考图书。

◆ 主　编　华为技术有限公司
　　责任编辑　李　静
　　责任印制　马振武

◆ 人民邮电出版社出版发行　北京市丰台区成寿寺路11号
　邮编　100164　电子邮件　315@ptpress.com.cn
　网址　https://www.ptpress.com.cn
　北京市艺辉印刷有限公司印刷

◆ 开本：787×1092　1/16
　印张：21.25　　　　　　　　　2022年4月第1版
　字数：504千字　　　　　　　　2024年1月北京第13次印刷

定价：139.80元

读者服务热线：(010)81055491　印装质量热线：(010)81055316
反盗版热线：(010)81055315
广告经营许可证：京东市监广登字 20170147 号

编 委 会

主　　任：彭　松

副 主 任：盖　刚

委　　员：孙　刚　邱月峰　史　锐　张　晶　朱殿荣
　　　　　魏　彪　张　博

技术审校：金　珍　郑美霞　袁长龙　陈　睿

主编人员：田　果

参编人员：刘丹宁　韩士良

序　言

乘"数"破浪　智驭未来

当前，数字化、智能化成为经济社会发展的关键驱动力，引领新一轮产业变革。以5G、云、AI为代表的数字技术，不断突破边界，实现跨越式发展，数字化、智能化的世界正在加速到来。

数字化的快速发展，带来了数字化人才需求的激增。《中国ICT人才生态白皮书》预计，到2025年，中国ICT人才缺口将超过2000万人。此外，社会急迫需要大批云计算、人工智能、大数据等领域的新兴技术人才；伴随技术融入场景，兼具ICT技能和行业知识的复合型人才将备受企业追捧。

在日新月异的数字化时代中，技能成为匹配人才与岗位的最基本元素，终身学习逐渐成为全民共识及职场人保持与社会同频共振的必要途径。联合国教科文组织发布的《教育2030行动框架》指出，全球教育需迈向全纳、公平、有质量的教育和终身学习。

如何为大众提供多元化、普适性的数字技术教程，形成方式更灵活、资源更丰富、学习更便捷的终身学习推进机制？如何提升全民的数字素养和ICT从业者的数字能力？这些已成为社会关注的重点。

作为全球ICT领域的领导者，华为积极构建良性的ICT人才生态，将多年来在ICT行业中积累的经验、技术、人才培养标准贡献出来，联合教育主管部门、高等院校、教育机构和合作伙伴等各方生态角色，通过建设人才联盟、融入人才标准、提升人才能力、传播人才价值，构建教师与学生人才生态、终身教育人才生态、行业从业者人才生态，加速数字化人才培养，持续推进数字包容，实现技术普惠，缩小数字鸿沟。

为满足公众终身学习、提升数字化技能的需求，华为推出了"华为职业认证"，这是围绕"云-管-端"协同的新ICT技术架构打造的覆盖ICT领域、符合ICT融合技术发展趋势的人才培养体系和认证标准。目前，华为职业认证内容已融入全国计算机等级考试。

教材是教学内容的主要载体、人才培养的重要保障，华为汇聚技术专家、高校教师、培训名师等，倾心打造"华为 ICT 认证系列丛书"，丛书内容匹配华为相关技术方向认证考试大纲，涵盖云、大数据、5G 等前沿技术方向；包含大量基于真实工作场景的行业案例和实操案例，注重动手能力和实际问题解决能力的培养，实操性强；巧妙串联各知识点，并按照由浅入深的顺序进行知识扩充，使读者思路清晰地掌握知识；配备丰富的学习资源，如 PPT 课件、练习题等，便于读者学习，巩固提升。

在丛书编写过程中，编委会成员、作者、出版社付出了大量心血和智慧，对此表示诚挚的敬意和感谢！

千里之行，始于足下，行胜于言，行而致远。让我们一起从"华为 ICT 认证系列丛书"出发，探索日新月异的 ICT 技术，乘"数"破浪，奔赴前景广阔的美好未来！

华为 ICT 战略与 Marketing 总裁

前 言

随着网络技术的发展，网络世界发生了翻天覆地的变化。对于用户而言，人们或许仅仅感受到服务的部署变得更加灵活和快捷，但是对于网络技术从业人员来说，这种灵活和快捷源于网络世界从基础设施到运维管理都采用了一种全新的架构，或者说是一系列相互紧密关联的新兴技术。伴随 SDN 的标准化进程，大量厂商的控制器产品和平台破茧而出。与此同时，随着虚拟化技术的发展和网络设备软硬件分离理念的推进，网络功能虚拟化随之问世并成为服务提供商的新宠。然而，当逐台设备通过 CLI（Command Line Interface，命令行界面）进行网管的方式难以为继，使用 SNMP（Simple Network Management Protocol，简单网络管理协议）来管理网络同样显得束手束脚时，网络管理和控制方式必然使网络技术的从业者要面对新的挑战。

鉴于网络技术领域的不断发展变化，华为技术有限公司参照行业对从业者的最新需求对大量认证科目进行了重新设计。针对传统的数据通信（数通）方向，华为推出了全新的 Datacom 系列认证。在 HCIA 阶段，华为不仅在考试大纲中增加了 IPv6 的内容，还加入了对 SDN（Software Defined Network，软件定义网络）、自动化和 Python 编程语言的要求，作为传统数通方向的补充。

华为《HCIA-Datacom 网络技术学习指南》是华为 HCIA-Datacom 认证考试的官方教材，本书作为其配套实验手册，由华为技术有限公司联合 YESLAB 培训中心经过精心编写、详细审校，最终创作而成，旨在帮助读者迅速掌握华为 HCIA-Datacom 认证考试所要求的实验知识和技能。本书通过 16 个实验提供了与 HCIA-Datacom 认证考试相关的网络技术的配置示例。同时，通过本书的介绍和演示，读者可以在自己的 PC 上搭建模拟的实验环境并进行实验操作。

在网络技术的学习过程中，实验练习是必不可少的环节。读者不仅要掌握每项网络技术的原理，还要熟悉它在网络设备上的配置方法，进而还应该具备快速定位并解决问题的能力。这种能力与日积月累的经验分不开，读者可以从现在开始，一边学习网络基础知识，一边进行配置练习。

本书主要内容

本书共包含 16 个网络技术基础实验，每个实验都专注于一项网络技术或主题。实验 1 介绍了华为 VRP（Versatile Routing Platform，通用路由平台）系统，为读者后续的实验学习打下基础。

实验 1：华为 VRP 系统的基本操作

华为 VRP 系统是华为数通产品的通用操作系统，本实验将带领读者熟悉 VRP 系统的 CLI 的使用方法，并会介绍各种配置视图、基础命令、快捷键等基础且重要的信息，同时还会通过具体的配置进行直观的演示。

实验 2：构建互联互通的 IP 网络

从这个实验开始，我们正式进入网络技术的练习环节。在这个实验中，我们需要通过静态路由的方式实现主机之间的通信。也就是，管理员需要通过具体的命令指示路由器，针对某个目的地，该往哪里走。

实验 3：OSPF 路由协议基础实验

OSPF（Open Shortest Path First，开放最短路径优先）是适用于各种规模网络环境的动态路由协议，它被广泛应用在各种网络。在这个实验中，我们会解读 OSPF 邻居状态机，展示单区域 OSPF 的配置，通过 Cost 控制 OSPF 的选路，发布默认路由，以及配置 OSPF 认证。

实验 4：构建以太网交换网络

在这个实验中，我们会通过 VLAN（Virtual Local Area Network，虚拟局域网）技术构建简单的以太网交换网络，介绍并演示 VLAN 的效果：VLAN 内的主机之间可以通信，VLAN 间的主机之间不能通信。本实验还展示了划分 VLAN 的几种方法：根据端口、MAC（Media Access Control，媒体访问控制）地址划分 VLAN。实验 4 至实验 7 都属于以太网交换网络实验。

实验 5：生成树基础实验

STP（Spanning Tree Protocol，生成树协议）是以太网络中用来预防环路的机制，它通过一系列比较来为参与以太网交换的端口确定端口角色，并根据端口角色允许端口转发流量，或者阻塞端口转发流量。这个实验将会介绍 STP 的工作原理，以及如何通过各种参数对 STP 的计算结果产生影响。

实验 6：以太网链路聚合实验

以太网链路聚合能够将多条链路"捆绑"为一条链路，使 STP 在进行计算时不会因为这些链路的物理连接（形成环路）而阻塞一些端口。这个实验将会展示如何在华为交换机上配置链路聚合[手动模式和 LACP（Link Aggregation Control Protocol，链路汇聚控制协议）模式]，以及如何在 LACP 模式中对活动链路进行控制。

实验 7：实验 VLAN 间通信实验

这是以太网交换网络部分的最后一个实验，它将实验 5 通过 VLAN 造成的隔离"打破"，通过路由的方式实现 VLAN 间的通信。这个实验通过两种方法实现了 VLAN 间通信：Dot1q 终结子接口和 VLANIF 接口。

实验 8：ACL 配置实验

从这个实验开始，我们会介绍网络安全基础与网络接入相关的技术，涉及 3 个实验：实验 8～实验 10。在实验 8 中，我们首先介绍了 ACL（Access Control Lists，访问控制列表）的工作原理、顺序和类型，接着根据简单的实验要求分别展示了基本 ACL 和高级 ACL 的配置命令和设计思路。

实验 9：本地 AAA 配置实验

AAA 是指认证、授权和计费，这是网络管理中被广泛部署的安全控制机制。这个实验介绍了如何在华为数通设备上实现本地认证和授权，以此提升设备自身的安全性。

实验 10：网络地址转换配置实验

NAT（Network Address Translation，网络地址转换）的安全性只是它提供的功能之一，更为重要的是，它可以使同一个公有地址加上不同端口号给多个私有 IP 地址使用，从而节省了公有 IP 地址的使用。这个实验通过 4 种不同的需求场景展示了 4 种 NAT 配置：静态 NAT、动态 NAT/NAPT、Easy IP 和 NAT Server。

实验 11：FTP 基础配置实验

本实验中，我们将华为路由器配置为 FTP（File Transfer Protocol，文件传输协议）服务器，并以 FTP 的方式从管理员 PC 登录到路由器上对系统中的文件进行操作。在此之前我们还对华为路由器的文件系统的结构及相关配置命令进行了介绍。

实验 12：DHCP 基础配置实验

DHCP（Dynamic Host Configuration Protocol，动态主机配置协议）是广泛部署在不同规模网络中的一项服务，通常终端 PC 会通过 DHCP 获取自己的 IP 地址信息。本实验中，我们先介绍了 DHCP 的工作原理，并通过 3 个场景和需求分别展示了 3 种配置 DHCP 的方法。

实验 13：构建基础 WLAN

在使用华为数通设备部署 WLAN 时，读者需要掌握不同模板和参数之间的嵌套关系。在本实验中，我们先梳理了实验中所使用到的模板和参数，帮助读者更好地理解 WLAN 的配置。接着通过一个简单的实验，使用 AC 对 AP 进行配置和管理，并最终在 STA（Station，站点）上进行测试，成功连接 WLAN。

实验 14：构建简单 IPv6 网络

本实验中，我们首先回顾了 IPv6 地址的书写方法和 IPv6 地址的类型；其次展示了 IPv6 地址的多种配置方式：手动配置、自动配置（有状态配置、无状态配置）；最后介绍了 IPv6 静态路由的配置，完成了一个简单 IPv6 环境的部署。

实验 15：网络编程与自动化基础

本实验中，我们从如何搭建 Python 环境开始介绍，并在本地 Python 环境中通过网络编程来配置和管理网络设备。我们使用了 telnetlib 库和 paramiko 库，paramiko 库超出了 HCIA-Datacom 考试大纲范围，读者可以根据需要进行练习。

实验 16：园区网络项目实战

本实验中，我们着重于展示如何将客户模糊的需求细化为具体的网络技术术语，如何将需求以一种直观的方式进行记录，以及在一个网络设计中需要包含哪些考虑因素。本实验提出的需求仅限于 HCIA-Datacom 考试大纲中包含的知识点，在实际工作中，读者会遇到更多且更复杂的问题，但只要掌握了设计和工作思路，足以应对各种需求。

本书常用图标

本书配套资源可通过扫描封底的"信通社区"二维码，回复数字"583185"进行获取。

关于华为认证的更多精彩内容，请扫码进入华为人才在线官网了解。

华为人才在线

目 录

第1章 华为 VRP 系统的基本操作 ·· 2

    1.1 实验介绍 ··· 4
        1.1.1 关于本实验 ·· 4
        1.1.2 实验目的 ·· 4
        1.1.3 实验组网介绍 ·· 4
        1.1.4 实验任务列表 ·· 4
    1.2 实验配置任务 ··· 5
        1.2.1 连接网络设备的方法 ·· 5
        1.2.2 命令行视图 ·· 7
        1.2.3 命令行的错误提示 ··· 8
        1.2.4 命令行的在线帮助 ··· 10
        1.2.5 命令级别和用户级别 ·· 11
        1.2.6 命令历史 ·· 12
        1.2.7 命令行的快捷键 ··· 12
        1.2.8 网络设备的初始化配置 ·· 13

第2章 构建互联互通的 IP 网络 ··· 18

    2.1 实验介绍 ··· 20
        2.1.1 关于本实验 ·· 20
        2.1.2 实验目的 ·· 20
        2.1.3 实验组网介绍 ·· 20
        2.1.4 实验任务列表 ·· 21
    2.2 实验配置任务 ··· 21
        2.2.1 静态路由 ·· 21
        2.2.2 浮动静态路由 ·· 29
        2.2.3 静态路由负载分担 ··· 35
        2.2.4 默认路由 ·· 36

第3章 OSPF 路由协议基础实验 ·· 38

3.1 实验介绍 ... 41
3.1.1 关于本实验 ... 41
3.1.2 实验目的 ... 41
3.1.3 实验组网介绍 ... 41
3.1.4 实验任务列表 ... 42
3.2 实验配置任务 ... 42
3.2.1 建立单区域 OSPF ... 42
3.2.2 通过开销进行选路 ... 50
3.2.3 通过 OSPF 发布默认路由 ... 54
3.2.4 配置 OSPF 认证 ... 55

第 4 章 构建以太网交换网络 ... 60
4.1 实验介绍 ... 62
4.1.1 关于本实验 ... 62
4.1.2 实验目的 ... 63
4.1.3 实验组网介绍 ... 63
4.1.4 实验任务列表 ... 64
4.2 实验配置任务 ... 64
4.2.1 创建 VLAN ... 64
4.2.2 基于端口划分 VLAN ... 66
4.2.3 基于 MAC 地址划分 VLAN ... 69
4.2.4 验证配置效果 ... 71

第 5 章 生成树基础实验 ... 74
5.1 实验介绍 ... 76
5.1.1 关于本实验 ... 76
5.1.2 实验目的 ... 77
5.1.3 实验组网介绍 ... 77
5.1.4 实验任务列表 ... 77
5.2 实验配置任务 ... 78
5.2.1 搭建拓扑，确认当前环境 ... 78
5.2.2 启用 STP，并将 S1 设置为根交换机 ... 78
5.2.3 启用交换机 S4，并观察 STP 端口状态机 ... 84
5.2.4 调整 STP 参数，影响选举和收敛时间 ... 86
5.2.5 配置 STP 保护参数 ... 91
5.2.6 配置 MSTP ... 96

第6章 以太网链路聚合实验······102

6.1 实验介绍······104
6.1.1 关于本实验······104
6.1.2 实验目的······104
6.1.3 实验组网介绍······105
6.1.4 实验任务列表······105

6.2 实验配置任务······105
6.2.1 配置手动模式······105
6.2.2 配置 LACP 模式······107
6.2.3 设置 LACP 参数······109

第7章 实现 VLAN 间通信实验······116

7.1 实验介绍······119
7.1.1 关于本实验······119
7.1.2 实验目的······119
7.1.3 实验组网介绍······119
7.1.4 实验任务列表······120

7.2 实验配置任务······120
7.2.1 配置 Dot1q 终结子接口，实现 VLAN 间通信······120
7.2.2 配置 VLANIF 接口，实现 VLAN 间通信······126

第8章 ACL 配置实验······128

8.1 实验介绍······130
8.1.1 关于本实验······130
8.1.2 实验目的······130
8.1.3 实验组网介绍······131
8.1.4 实验任务列表······131

8.2 实验配置任务······131
8.2.1 搭建拓扑，并实现全网互通······131
8.2.2 基本 ACL 的配置······135
8.2.3 高级 ACL 的配置······138
8.2.4 ACL 规则的顺序······141

第9章 本地 AAA 配置实验······148

9.1 实验介绍······150
9.1.1 关于本实验······150

9.1.2 实验目的 150
9.1.3 实验组网介绍 150
9.1.4 实验任务列表 151
9.2 实验配置任务 151
9.2.1 配置 hcia-admin 用户 AAA 本地认证 151
9.2.2 配置 hcia-operator 用户 AAA 本地认证 159

第 10 章 网络地址转换配置实验 162

10.1 实验介绍 164
10.1.1 关于本实验 164
10.1.2 实验目的 164
10.1.3 实验组网介绍 164
10.1.4 实验任务列表 165
10.2 实验配置任务 165
10.2.1 静态 NAT 的配置 165
10.2.2 动态 NAT 的配置 168
10.2.3 Easy IP 的配置 171
10.2.4 NAT Server 的配置 173

第 11 章 FTP 基础配置实验 176

11.1 实验介绍 178
11.1.1 关于本实验 178
11.1.2 实验目的 178
11.1.3 实验组网介绍 178
11.1.4 实验任务列表 179
11.2 实验配置任务 179
11.2.1 熟悉文件系统命令 179
11.2.2 FTP 基础配置实验 183

第 12 章 DHCP 基础配置实验 188

12.1 实验介绍 190
12.1.1 关于本实验 190
12.1.2 实验目的 191
12.1.3 实验组网介绍 191
12.1.4 实验任务列表 192
12.2 实验配置任务 192

	12.2.1	配置基于全局地址池的 DHCP 服务器	192
	12.2.2	配置基于接口的 DHCP 服务器	197
	12.2.3	配置 DHCP 中继	200

第13章 构建基础 WLAN ································· 206

13.1 实验介绍 ································· 208
13.1.1 关于本实验 ································· 208
13.1.2 实验目的 ································· 209
13.1.3 实验组网介绍 ································· 210
13.1.4 实验任务列表 ································· 210

13.2 实验配置任务 ································· 210
13.2.1 基本配置 ································· 210
13.2.2 配置 AP 上线 ································· 212
13.2.3 配置 WLAN 业务参数 ································· 215

第14章 构建简单 IPv6 网络 ································· 224

14.1 实验介绍 ································· 226
14.1.1 关于本实验 ································· 226
14.1.2 实验目的 ································· 227
14.1.3 实验组网介绍 ································· 227
14.1.4 实验任务列表 ································· 227

14.2 实验配置任务 ································· 228
14.2.1 手动配置 IPv6 地址 ································· 228
14.2.2 使用 DHCPv6 获得 IPv6 地址 ································· 233
14.2.3 使用 SLAAC 获得 IPv6 地址 ································· 239
14.2.4 配置静态 IPv6 路由 ································· 241

第15章 网络编程与自动化基础 ································· 246

15.1 实验介绍 ································· 248
15.1.1 关于本实验 ································· 248
15.1.2 实验目的 ································· 248
15.1.3 实验组网介绍 ································· 248
15.1.4 实验任务列表 ································· 249

15.2 实验配置任务 ································· 249
15.2.1 搭建 Python 基础环境 ································· 249
15.2.2 通过 Telnet 获取路由器配置 ································· 258

15.2.3 通过 SSH 更改路由器配置 ··· 261

第 16 章 园区网络项目实战 ··· 266

   16.1 实验介绍 ··· 268
      16.1.1 关于本实验 ·· 268
      16.1.2 实验目的 ··· 269
      16.1.3 实验任务列表 ··· 269
   16.2 实验配置任务 ··· 269
      16.2.1 组网方案设计 ··· 269
      16.2.2 二层网络设计 ··· 272
      16.2.3 IP 地址规划 ·· 275
      16.2.4 三层网络设计 ··· 276
      16.2.5 WLAN 设计 ·· 278
      16.2.6 网络设备调试 ··· 279

附录 命令索引 ··· 298

第 1 章
华为 VRP 系统的基本操作

本章主要内容

1.1 实验介绍

1.2 实验配置任务

1.1 实验介绍

1.1.1 关于本实验

本实验中,我们会从华为 CLI(命令行界面)配置界面的结构(配置视图)开始介绍,通过实际配置为读者提供直观演示。读者通过本实验可以熟悉华为 VRP(Versatile Routing Platform,通用路由平台)系统的基本操作。

1.1.2 实验目的

- 理解华为设备命令行视图。
- 掌握配置接口 IP 地址命令的使用方法。
- 掌握路由器配置验证命令的使用方法。
- 掌握删除、保存配置命令的使用方法。
- 了解 CLI 命令行快捷键。

1.1.3 实验组网介绍

VRP 系统的基本操作实验拓扑如图 1-1 所示。

图 1-1 VRP 系统的基本操作实验拓扑

设备连接说明:PC1 通过本地串口连接路由器 AR1 的 Console 口。
本章使用的网络地址见表 1-1。

表 1-1 本章使用的网络地址

设备	接口	IP 地址	子网掩码	默认网关
AR1	G0/0/0	10.1.1.1	255.255.255.0	—
AR2	G0/0/0	10.1.1.2	255.255.255.0	—

1.1.4 实验任务列表

配置任务 1:连接网络设备的方法。
配置任务 2:命令行视图。
配置任务 3:命令行的错误提示。
配置任务 4:命令行的在线帮助。
配置任务 5:命令级别和用户级别。
配置任务 6:命令历史。

配置任务 7：命令行的快捷键。
配置任务 8：网络设备的初始化配置。

1.2 实验配置任务

1.2.1 连接网络设备的方法

一台新设备开箱、上架、接电后，用户需先登录设备。一般设备所支持的首次登录选项中都包含使用 Console 口进行登录，也有越来越多的设备支持使用 MiniUSB 口进行登录。本实验介绍使用 Console 口的方式进行登录。对 MiniUSB 登录方式感兴趣的读者可以查看官方设备配置手册，确认设备支持这种登录方式，并按照指南进行操作。图 1-2 以华为 AR2220 为例展示了网络设备上的 MiniUSB 口和 Console 口，我们从图 1-2 中可以看出 Console 口的物理规格是 RJ45（俗称水晶头）。

图 1-2　MiniUSB 口和 Console 口

步骤 1　物理 Console 线

我们一般使用 Console 线来连接网络设备，传统 Console 线的两端分别为 RJ45 和 DB9，RJ45 端连接网络设备的 Console 口，DB9 端连接用户终端的串口。DB9 是串口的其中一种物理规格。华为网络设备的 Console 线采用 DB9。目前，大多的用户终端已经不再支持物理串口，因此我们需要为 Console 线搭配一个 DB9 转 USB 的转换线（或转换盒），或选用以下类型的 Console 线。

① **Console 线 + DB9 转 USB 转换线**：RJ45 端连接网络设备，DB9（串口）端两两相接，USB 端连接用户终端。

② **USB Console 线**：RJ45 端连接网络设备，USB 端连接用户终端。

③ **蓝牙 Console 线**：RJ45 端连接网络设备，通过蓝牙连接用户终端。

图 1-3 所示为 Console 线 + DB9 转 USB 线示例。图 1-4 所示为 USB Console 线示例。图 1-5 所示为蓝牙 Console 线示例。

图 1-3　Console 线 + DB9 转 USB 线示例

图 1-4 USB Console 线示例

图 1-5 蓝牙 Console 线示例

步骤 2 设置连接参数

无论使用哪种连接方式,网络设备和用户终端都需要使用终端程序与网络设备建立连接并登录网络设备的 CLI。在连接华为网络设备时,读者需要在终端程序中进行参数设置,具体的参数要求见表 1-2。

表 1-2 Console 口参数

参数	值
传输速率/波特率(Speed/baud)	9600 bit/s
流控方式(Flow control)	无
校验方式/奇偶位(Parity)	无
停止位(Stop bits)	1
数据位(Data bits)	8

终端程序有很多种,如 PuTTY,读者可以根据自己的习惯进行选择。图 1-6 所示为使用 PuTTY 进行 Console 连接时的设置页面示例。读者点击"Open"后可登录所连接网络设备的 CLI 命令行。

图 1-6 使用 PuTTY 作为终端程序连接 Console 口的示例

我们对图 1-6 中的 COM 口(本例显示为 COM3)需要进行以下说明。在每次连接不同的网络设备时,本地计算机可能会使用不同的 COM 口,因此读者在连接真实设备进行实验时,在连接 Console 线后,要在安装了终端程序的计算机中通过设备管理器查看当前连接的是几号 COM 口。

实际工作中,管理员登录 Console 进行设备初始化,即配置主机名、设置日期及时间、配置管理 IP 地址、开启 Telnet [实际工作中建议使用 SSH(Secure Shell,安全外壳协议)] 远程登录方式等,为后续配置提供基础环境。

1.2.2 命令行视图

本节主要介绍进入和退出命令行视图的方法。读者登录一台全新的华为路由器时,会在 CLI 中看到<Huawei>,"Huawei"是缺省主机名,这个由"<>"加主机名构成的结构称为命令提示符,它为用户展示了当前所处的命令行视图。

用户可使用对应的命令对华为设备提供的功能进行配置和查询。为了方便用户使用这些命令操作设备,华为按照不同功能将命令分别注册在不同的命令行视图下。在配置某一功能时,用户需要先进入对应的命令行视图,然后执行相应的命令进行配置。表 1-3 中列出了常用的命令行视图以及进入该视图的命令。

表 1-3 常用的命令行视图以及进入该视图的命令

视图名称	命令提示符和命令	描述
用户视图	<Huawei> 用户成功登录设备后就会进入用户视图	在用户视图下,用户可以执行 **display** 命令查看设备和功能的运行状态和统计信息,还可以执行命令来保存设备配置、重启设备等
系统视图	<Huawei>**system-view** Enter system view, return user view with Ctrl+Z. [Huawei] 在用户视图下,执行命令 **system-view** 就可以进入系统视图	在系统视图下,用户可以执行大多数系统配置命令,或进入其他功能配置视图
接口视图	[Huawei]**interface GigabitEthernet 0/0/0** [Huawei-GigabitEthernet0/0/0] 在系统视图下,执行命令 *interface* *interface slot-number* 就可以进入接口视图。 其中 *interface* 参数是接口类型,本例使用了千兆以太网接口 GigabitEthernet;*slot-number* 参数是该接口的编号,本例的 0/0/0 对应的是槽位号/子卡号/接口序号	在接口视图下,用户可以执行与接口相关的配置命令,如 IP 地址、子网掩码等
路由协议视图	[Huawei]**ospf 10** [Huawei-ospf-10] 在系统视图下,执行路由协议命令就可以进入相应的路由协议视图。 本例执行了 OSPF 配置命令并进入了 OSPF 协议视图,在关键字 **ospf** 后面的参数 **10** 是 OSPF 进程号	在路由协议视图下,用户可以执行特定的路由协议命令,来对相关的路由协议进行配置

读者在练习时注意以下两点。首先,在用户视图能够执行的操作很少,以查看操作为主。设备的大多数配置都需要至少进入系统视图,若用户在配置时收到了没有查找到

命令的错误提示（1.2.3 节介绍错误提示），需要首先确定命令行视图是否正确。其次，有些命令既可以在系统视图下执行，也可以在其他视图下执行，因此这条命令所实现的功能与它所处的视图密切相关。以 **lldp enable** 命令为例，该命令在系统视图中被使用时表示全局启用 LLDP 功能，该命令在接口视图下被使用时表示这个接口启用 LLDP 功能，并且接口命令的配置会覆盖全局的设置。

在退出某个视图时，用户可以使用 **quit** 命令退出到上一层视图，也可以使用 **return** 命令或"Ctrl+Z"组合键直接退回到用户视图。例 1-1 为使用命令 **system-view** 从用户视图进入系统视图。示例中会以加粗的形式显示用户输入的字符。

例 1-1　进入系统视图

```
<Huawei>system-view
Enter system view, return user view with Ctrl+Z.
[Huawei]
```

接着用户使用命令 **ospf 10** 进入 OSPF 协议视图。在 OSPF 协议视图中，用户使用命令 **area 0** 进入 OSPF 区域 0 配置视图，具体见例 1-2。

例 1-2　进入 OSPF 协议视图

```
[Huawei]ospf 10
[Huawei-ospf-10]area 0
[Huawei-ospf-10-area-0.0.0.0]
```

此时，用户使用 **quit** 命令可以退到上一层视图，即 OSPF 协议视图，具体见例 1-3。

例 1-3　使用 quit 命令退到上一层视图

```
[Huawei-ospf-10-area-0.0.0.0]quit
[Huawei-ospf-10]
```

如果在 OSPF 区域配置视图中使用 **return** 命令，可以跳过中间的多个视图，直接退回到用户视图，具体见例 1-4。

例 1-4　使用 return 命令直接退回到用户视图

```
[Huawei-ospf-10-area-0.0.0.0]return
<Huawei>
```

1.2.3　命令行的错误提示

用户输入命令并按下回车键后，VRP 会检查命令的语法是否正确，若正确就将命令写入配置中，若不正确 VRP 会向用户提示错误信息。常见的错误信息见表 1-4。

表 1-4　常见的错误信息

错误信息	错误原因
Error:Incomplete command found at '^' position.	"^"处的命令不完整
Error: Unrecognized command found at '^' position.	"^"处的关键字无法识别
Error: Wrong parameter found at '^' position.	"^"处的参数错误
Error:Too many parameters found at '^' position.	"^"处输入的参数过多
Error:Ambiguous command found at '^' position.	"^"处的命令含义不明确

下面我们以几个配置示例进行具体说明。

假设用户想要为路由器配置 IP 地址，但他没有进入具体的接口配置模式，而是在系统模式下输入命令。若用户只输入 **ip**，按下回车键后系统会提示"Error:Incomplete command found at '^' position."，具体见例 1-5。这是因为在系统视图中，存在多个以关键字 **ip** 开头的命令，系统无法确定用户输入的是什么命令，因此就这些系统视图的命令而言，这条命令是不完整的。

例 1-5　错误：命令不完整

```
[Huawei]ip
         ^
Error:Incomplete command found at '^' position.
```

接着用户输入设置 IP 地址的完整命令，收到了另一个错误提示"Error: Unrecognized command found at '^' position."，具体见例 1-6。该命令语法正确，但配置视图选择错误，配置也是无效的。由于系统视图中没有以 **ip address** 开头的命令，因此系统提示无法识别该命令，并使用"^"指示出无法识别的关键字（或参数）。

例 1-6　错误：命令未识别

```
[Huawei]ip address 10.1.1.1 24
        ^
Error: Unrecognized command found at '^' position.
```

用户进入接口视图并打算输入 **ip address 10.1.1.1 255.255.255.0** 命令。若用户只输入 ip address 10.1.1.1 255.255.255.，系统会提示一个错误消息"Error: Wrong parameter found at '^' position."，具体见例 1-7。

例 1-7　错误：参数错误

```
[Huawei-GigabitEthernet0/0/0]ip addres 10.1.1.1 255.255.255.
                                                          ^
Error: Wrong parameter found at '^' position.
```

用户想要为路由器配置主机名"AR1"，但他再次犯了视图错误。他在用户视图下输入简写命令 **sys AR1**，这条命令的完整形式应该是 **sysname AR1**，用来设置设备的主机名（后文会对命令缩写进行介绍）。这时系统会提示错误消息"Error:Too many parameters found at '^' position."，具体见例 1-8。这是因为从用户视图进入系统视图的命令 **system-view** 同样可以简写为 **sys**，并且这条命令的语法中只有这一个关键字，不携带参数。系统认为用户想要输入的命令是 **system-view**，因此给出了参数过多的错误提示。

例 1-8　错误：参数过多

```
<Huawei>sys AR1
            ^
Error:Too many parameters found at '^' position.
```

用户想要使用命令 **display interface** 来查看自己配置的接口，但他忘记了 interface 的拼写，于是输入了 **display in**，按下回车键后系统会提示错误消息"Error:Ambiguous command found at '^' position."，具体见例 1-9。这是因为在关键字 **display** 后面以 **in** 开头的关键字除了 **interface** 之外，还有其他关键字，因此系统无法确定用户要输入的是哪条命令。

例 1-9　错误：命令不明确

```
<Huawei>display in
                ^
Error:Ambiguous command found at '^' position.
```

用户想要配置某项功能，但没有完全记住该命令的语法规则，或者没有记住某个关键字的拼写时，可以求助 VRP 系统（我们将在 1.2.4 节进行具体介绍）。

1.2.4 命令行的在线帮助

用户在使用命令行对设备进行配置时，可以使用在线帮助来获得有关命令语法及其参数解释的实时帮助。

用户可以通过输入"?"来获取帮助信息，系统会根据当前的视图和用户输入的信息来提供帮助信息。命令行的在线帮助分完全帮助和部分帮助，完全帮助适用于忘记了命令语法的情况，部分帮助适用于忘记了关键字拼写的情况。

1. 完全帮助

用户可在任意视图下输入"?"来查看该视图下的所有命令（的开头关键字）及其解释，具体见例 1-10。从本例中我们可以看到，系统回复了很多参数，注意命令输出的最后一行"---- More ----"表示这个列表还有其他内容，此时用户可以按下空格键进行"翻页"，或者按下回车键查看下一行。用户可以按下其他任意键退出列表并返回配置模式。

例 1-10 在用户视图下输入"?"

```
<Huawei>?
User view commands:
  arp-ping                ARP-ping
  autosave                <Group> autosave command group
  backup                  Backup information
  cd                      Change current directory
  clear                   <Group> clear command group
  clock                   Specify the system clock
  cls                     Clear screen
  compare                 Compare configuration file
  copy                    Copy from one file to another
  debugging               <Group> debugging command group
  delete                  Delete a file
  dialer                  Dialer
  dir                     List files on a filesystem
  display                 Display information
  factory-configuration   Factory configuration
  fixdisk                 Try to restory disk
  format                  Format file system
  free                    Release a user terminal interface
  ftp                     Establish an FTP connection
  help                    Description of the interactive help system
  hwtacacs-user           HWTACACS user
  license                 <Group> license command group
  lldp                    Link Layer Discovery Protocol
 ---- More ----
```

从例 1-10 中我们看到，有一条命令是以 **help** 开头的，并且系统提示该命令描述了交互式帮助系统。若用户要再询问这条命令的语法，则可以输入 **"help ?"**，表示想要询问这条命令后续的关键字或参数是什么，此时得到如例 1-11 所示的回复。若系统回复参数"<cr>"，则表示这条命令的语法已经输入完整，用户可按回车键执行这条命令。

例 1-11 输入"help ?"

```
<Huawei>help ?
  <cr>   Please press ENTER to execute command
```

用户输入 **help** 并按下回车键，系统会出现在线帮助系统的解释与用法，具体见例 1-12。

例 1-12　输入命令 help

```
<Huawei>help
Help may be requested at any point in a command by entering
a question mark '?'. If nothing matches, the help list may
be empty
Two styles of help are provided:
  1. Full help is available when you are ready to enter a
     command argument ( e.g. 'display ?' ) and describes each
     possible argument
  2. Partial help is provided when an abbreviated argument
     is entered and you want to know what arguments match
     the input ( e.g. 'display l?' )
```

2．部分帮助

用户想要输入 **display interface**，但只记得第 2 个关键字是以 **in** 开头的，这时就可以使用部分帮助。如例 1-13 所示，用户输入 **display in?**，系统回复了两个参数，这也解释了前文示例中为什么系统认为这条命令不明确。

例 1-13　输入 display in?请求帮助

```
<Huawei>display in?
  info-center   <Group> info-center command group
  interface     <Group> interface command group
```

命令补全功能

从例 1-13 中我们可以看出，有两个参数都是以 **in** 开头的，因此只输入前两个字母会使系统无法区分用户想要输入的命令。但用户再多输入一个字母（即输入 **int**），系统就可以识别用户想要输入的关键字。在排除了歧义的情况下，用户可以使用"Tab"键来进行命令补全，具体见例 1-14。在本例中，用户只输入了第 1 行（**display int**）并按下"Tab"键，系统会自动输出第 2 行（**display interface**），即帮助用户补全了这个关键字。当然，用户也可以在排除歧义后不使用"Tab"键补全命令，而是直接按下回车键执行命令，这就是命令简写。

例 1-14　使用 Tab 进行命令补全

```
<Huawei>display int
<Huawei>display interface
```

当歧义没有被排除时，用户每按一次 Tab 键，系统就会为用户补全关键字列表中的一个关键字，用户可以通过多次按下 Tab 键来选择自己想要使用的关键字。

1.2.5　命令级别和用户级别

在华为网络设备的命令行系统中，对命令和用户进行了分级管理，增强了系统的安全性。每个视图下的每条命令都有默认的级别。用户可以根据实际需要重新设定某条命令的级别，如将高级别命令设置为低级别命令，让低级别用户也可以执行这条命令；或者将低级别命令设置为高级别命令。

命令分 4 个级别，用户分 16 个级别，具体见表 1-5。

表 1-5　用户级别与命令级别

用户级别	命令级别	级别名称	描述
0	0	参观级	对连通性进行初级诊断。典型命令：ping、tracert、telnet、部分 display 等

表 1-5 用户级别与命令级别（续）

用户级别	命令级别	级别名称	描述
1	0~1	监控级	对系统进行维护。典型命令：大多数 display 等
2	0~2	配置级	业务配置命令。包含绝大多数配置命令，如路由配置等
3-15	0~3	管理级	对系统进行高级管理。典型命令：命令级别设置命令、debugging 命令等

1.2.6 命令历史

华为设备会默认记录 10 条用户输入的命令，用户使用 **display history-command** 命令可以查看输入过的 10 条命令，具体见例 1-15。需要注意的是，命令历史是单纯的输入记录。从例 1-15 的命令输出的第 3 条命令我们可以看出，这是前文实验中我们曾输入的一条不完整的命令，当时系统为此弹出了错误提示，但系统仍然把它记录在命令历史中。

例 1-15 查看命令历史

```
<Huawei>display history-command
 help
 quit
 ip add 10.1.1.1 255.255.255.
 interface g0/0/0
 ip address 10.1.1.1 24
 ip
 system-view
 return
 area 1
 area 0
```

用户想要重复上一次输入的命令时，只需要按下"↑"键就可以调出刚输入的命令，再次按下"↑"键，就可以调出上上次输入的命令，以此类推。

1.2.7 命令行的快捷键

华为 VRP 命令行定义了一些系统快捷键，用户熟练使用快捷键可以加速其配置。表 1-6 中列出了常用的系统快捷键。

表 1-6 常用的系统快捷键

功能键	作用
Ctrl+A	移动光标至当前行开头的位置
Ctrl+B 或←	将光标向左移动一个字符
Ctrl+C	停止当前系统正在执行的操作，如 ping 操作等
Ctrl+D	删除当前光标所在位置的字符
Ctrl+E	移动光标至当前行末尾的位置
Ctrl+F 或→	将光标向右移动一个字符
Ctrl+H 或 Backspace	删除光标左侧的一个字符
Ctrl+K	在连接建立阶段中止呼出的连接
Ctrl+N	显示命令历史中的后一条命令
Ctrl+P	显示命令历史中的前一条命令

表 1-6 常用的系统快捷键（续）

功能键	作用
Ctrl+T	输入问号（?）
Ctrl+W	删除光标左侧的一个字符串
Ctrl+X	删除光标左侧所有字符
Ctrl+Y	删除光标所在位置及其右侧所有字符
Ctrl+Z	返回到用户视图
Ctrl+]	终止呼入的连接或重定向连接
Esc+B	将光标向左侧移动一个字符串
Esc+D	删除光标右侧的一个字符串
Esc+F	将光标向右侧移动一个字符串

1.2.8 网络设备的初始化配置

通过 Console 首次连接网络设备后，用户需要使用设备缺省（默认）的用户名/密码进行登录。读者可以在华为官网搜索"缺省账号与密码"。例 1-16 中展示了首次登录后的"自动配置"选项。我们可输入"**y**"停止自动配置。

例 1-16　首次登录路由器

```
<Huawei>
 Warning: Auto-Config is working. Before configuring the device, stop Auto-
Config. If you perform configurations when Auto-Config is running, the DHCP,
routing, DNS, and VTY configurations will be lost. Do you want to stop Auto-
Config? [y/n]:y
 Info: Auto-Config has been stopped.
<Huawei>
```

步骤 1　设置日期和时间

网络设备上的时间保持一致至关重要，在工作环境中，我们会使用 NTP 进行自动且精确的时钟同步。本例使用手动配置的方式调整路由器的日期和时间。命令语法如下所示。

① **clock timezone** *time-zone-name* {**add** | **minus**} *offset*：用户视图命令，配置时区。*time-zone-name* 参数用来指定时区的名称。可选关键字 **add** 和 **minus**（以及关键字后面的 *offset* 参数）分别指定了该时区相比于 UTC（Universal Time Coordinated，协调世界时）的偏移量，其中 **add** 表示+，**minus** 表示-。*offset* 参数的格式为 *HH:MM:SS*，用户设置北京时间（UTC +8）时，需要将 *offset* 参数设置为 08:00:00。

② **clock datetime** *HH:MM:SS YYYY-MM-DD*：用户视图命令，配置设备本地时间。

例 1-17　设置时区和日期及时间

```
<Huawei>clock timezone CST add 08:00:00
<Huawei>clock datetime 14:58:00 2021-03-08
<Huawei>display clock
2021-03-08 14:58:25
Monday
Time Zone(CST) : UTC+08:00
```

例 1-17 中的第 1 条命令将路由器的时区取名为 CST（中国标准时间）北京时间（UTC+8）。第 2 条命令配置了日期和时间，第 3 条命令查看了时区和日期时间设置。需要注意的是，设置时区和日期及时间的命令是用户视图下的命令，在配置时要看清配置视图。

步骤 2 设置用户名和 IP 地址

本例使用的命令语法如下所示，例 1-18 中展示了具体的配置。

① **system-view**：用户视图命令，用来进入系统视图。

② **sysname** *host-name*：系统视图命令，配置主机名。

③ **interface** *interface-type interface-number*：系统视图命令，进入接口视图。

④ **ip address** *ip-address* {*mask* | *mask-length*}：接口视图命令，配置 IP 地址。

⑤ **display this**：这条命令会根据用户所在的视图显示出相关的配置信息。

例 1-18 设置用户名和 IP 地址

```
<Huawei>system-view
Enter system view, return user view with Ctrl+Z.
[Huawei]sysname AR1
[AR1]interface GigabitEthernet 0/0/0
[AR1-GigabitEthernet0/0/0]ip address 10.1.1.1 255.255.255.0
[AR1-GigabitEthernet0/0/0]display this
[V200R003C00]
#
interface GigabitEthernet0/0/0
 ip address 10.1.1.1 255.255.255.0
#
return
```

例 1-18 所示命令在设置 G0/0/0 接口的 IP 地址时，子网掩码使用的是 255.255.255.0 的形式，也可以使用掩码长度为"24"的进行配置。

步骤 3 设置 Console 密码

实际工作中，为了保障网络设备的安全性，相关人员需要为 Console 口设置登录密码。本例将 Console 口的密码设置为 Huawei@123，同时为通过 Console 登录的用户设置用户等级为 15 级。本例使用的命令语法如下，例 1-19 中展示了具体的配置。

① **user-interface console 0**：进入 Console 口。

② **authentication-mode password**：设置 Console 口密码。

③ **user privilege level** *level*：为通过 Console 登录的用户设置用户等级。

例 1-19 设置 Console 密码

```
[AR1]user-interface console 0
[AR1-ui-console0]authentication-mode password
Please configure the login password (maximum length 16):Huawei@123
[AR1-ui-console0]user privilege level 15
```

在为 Console 口设置了密码后，我们可以从 PC1 再次连接 AR1 的 Console 口。这次我们会看到命令行中弹出了需要输入密码的提示（Password:），输入密码"Huawei@123"后，用户能够成功登录到 AR1 中，并看到命令提示符<AR1>。

步骤 4 设置 Telnet

实际工作中，为了进行远程设备管理，管理员会为设备配置远程登录策略，出于安全性的考虑，建议使用 SSH，不建议使用 Telnet。SSH 超出了 HCIA-Datacom 的范围，因此本书只演示 Telnet 的配置。

首先我们在 AR1 上配置本地用户，用户名为 admin-user，密码为 Huawei@123，用户等级为 15 级，具体见例 1-20。本例使用的命令语法如下。

① **aaa**：进入 AAA 视图，进行用户接入的安全配置，如创建用户、设定用户级别等。

② **local-user** *user-name* **password cipher** *password*：配置本地用户的登录密码。

③ **local-user** *user-name* **privilege level** *level*：配置本地用户的用户级别。

④ **local-user** *user-name* **service-type** {**telnet**}：配置本地用户的接入方式，只有用户接入方式与这条命令的设置匹配时，才允许用户接入。该命令中省略了一些选项。

例 1-20　配置本地用户

```
[AR1]aaa
[AR1-aaa]local-user admin-user password cipher Huawei@123
Info: Add a new user.
[AR1-aaa]local-user admin-user privilege level 15
[AR1-aaa]local-user admin-user service-type telnet
```

其次在 AR1 上启用 Telnet，并让其使用本地用户进行认证，具体见例 1-21。本例使用的命令语法如下：

① **telnet server enable**：启用 Telnet 服务器。

② **user-interface** *ui-type first-ui-number* [*last-ui-number*]：进入 VTY 视图。

③ **protocol inbound** {**all** | **ssh** | **telnet**}：指定 VTY 用户界面所支持的协议。

④ **user privilege level** *level*：指定通过 VTY 登录的用户级别。

例 1-21　启用 Telnet

```
[AR1]telnet server enable
[AR1]user-interface vty 0 4
[AR1-ui-vty0-4]protocol inbound telnet
[AR1-ui-vty0-4]authentication-mode aaa
[AR1-ui-vty0-4]user privilege level 15
```

此时 AR1 已完成被 Telnet 连接前的准备。读者可以按照上述步骤为 AR2 进行初始化，在此期间至少要为 AR2 配置 IP 地址，因为我们将通过 AR2 测试 AR1 的 Telnet 设置。在例 1-22 中，用户从 AR2 使用命令 **telnet 10.1.1.1**，通过用户名 admin-user 和密码 Huawei@123（CLI 中不会显示输入的密码）登录到 AR1。这不是用户第一次 Telnet 登录，因此登录后自动显示了上一次的登录信息。

例 1-22　测试 AR1 的 Telnet 设置

```
<AR2>telnet 10.1.1.1
  Press CTRL_] to quit telnet mode
  Trying 10.1.1.1 ...
  Connected to 10.1.1.1 ...

Login authentication


Username:admin-user
Password:
  ------------------------------------------------------------------------------
  User last login information:
  ------------------------------------------------------------------------------
  Access Type: Telnet
  IP-Address : 10.1.1.2
  Time       : 2021-03-08 18:15:40+08:00
  ------------------------------------------------------------------------------
<AR1>
```

登录后用户可以通过命令 **display users** 来查看当前设备的用户信息，具体见例 1-23。

例 1-23 查看登录设备的用户信息

```
<AR1>display users
  User-Intf    Delay     Type    Network Address      AuthenStatus      AuthorcmdFlag
+ 129 VTY 0   00:00:00   TEL     10.1.1.2             pass
  Username : admin-user
```

第 2 章
构建互联互通的 IP 网络

本章主要内容

2.1　实验介绍

2.2　实验配置任务

在本章的实验中，我们使用静态路由来实现主机之间的通信。静态路由是指用户手动配置在路由器上的路由，可使路由器根据目的 IP 地址转发数据包。相比于动态路由协议，静态路由的配置简单，但灵活性差。我们通常用它来实现稳定性较强的默认路由的指定。

2.1 实验介绍

2.1.1 关于本实验

在本实验中，我们要实现 PC1 与 PC2 之间的通信。我们通过以下配置任务带领读者完成单条静态路由、备份路由、负载分担和默认路由的配置。

2.1.2 实验目的

- 理解数据包转发规则。
- 理解双向通信的概念。
- 掌握静态路由的配置方法。
- 掌握 ping 测试方法。
- 掌握 tracert 测试方法。

2.1.3 实验组网介绍

本章的实验拓扑如图 2-1 所示。

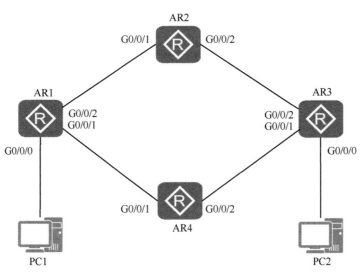

图 2-1 本章的实验拓扑

本章使用的网络地址见表 2-1。

表 2-1　本章使用的网络地址

设备	接口	IP 地址	子网掩码	默认网关
AR1	G0/0/0	172.16.10.1	255.255.255.0	—
AR1	G0/0/1	10.10.14.1	255.255.255.0	—
AR1	G0/0/2	10.10.12.1	255.255.255.0	—
AR2	G0/0/1	10.10.12.2	255.255.255.0	—
AR2	G0/0/2	10.10.23.2	255.255.255.0	—
AR3	G0/0/0	172.16.30.3	255.255.255.0	—
AR3	G0/0/1	10.10.34.3	255.255.255.0	—
AR3	G0/0/2	10.10.23.3	255.255.255.0	—
AR4	G0/0/1	10.10.14.4	255.255.255.0	—
AR4	G0/0/2	10.10.34.4	255.255.255.0	—
PC1	E0/0/1	172.16.10.10	255.255.255.0	172.16.10.1
PC2	E0/0/1	172.16.30.10	255.255.255.0	172.16.30.3

2.1.4　实验任务列表

配置任务 1：静态路由（PC1 与 PC2 之间的通信使用 AR1-AR2-AR3 链路）。

配置任务 2：浮动静态路由（PC1 与 PC2 之间的通信使用主链路 AR1-AR2-AR3，备份链路 AR1-AR4-AR3）。

配置任务 3：静态路由负载分担（使用两条路径实现 PC1 与 PC2 之间的通信）。

配置任务 4：默认路由。

2.2　实验配置任务

2.2.1　静态路由

本实验的目的是通过 AR1-AR2-AR3 的路径实现 PC1 与 PC2 之间的通信。

步骤 1　搭建拓扑并对设备进行初始化配置

在本实验中，我们首先进行基础拓扑的搭建，并按照拓扑的连线和 IP 地址规划，实现两两设备之间的直连通信并进行测试。用户可按照表 2-1 中的信息为两台 PC（Personal Computer，个人计算机）配置 IP 地址、子网掩码和网关。

我们为 AR1、AR2、AR3 和 AR4 进行初始化配置，即设置主机名和 IP 地址。这些命令在上一实验中已经进行过具体的介绍和演示，因此我们这里只为读者列出这 4 台设备上的配置命令作为参考，不对具体的命令进行冗余解释。AR1、AR2、AR3 和 AR4 上的初始化命令见例 2-1～例 2-4。

例 2-1　AR1 的初始化配置

```
<Huawei>system-view
Enter system view, return user view with Ctrl+Z.
[Huawei]sysname AR1
[AR1]interface GigabitEthernet 0/0/0
[AR1-GigabitEthernet0/0/0]ip address 172.16.10.1 24
```

```
[AR1-GigabitEthernet0/0/0]quit
[AR1]interface GigabitEthernet 0/0/1
[AR1-GigabitEthernet0/0/1]ip address 10.10.14.1 24
[AR1-GigabitEthernet0/0/1]quit
[AR1]interface GigabitEthernet 0/0/2
[AR1-GigabitEthernet0/0/2]ip address 10.10.12.1 24
```

例 2-2　AR2 的初始化配置

```
<Huawei>system-view
Enter system view, return user view with Ctrl+Z.
[Huawei]sysname AR2
[AR2]interface GigabitEthernet 0/0/1
[AR2-GigabitEthernet0/0/1]ip address 10.10.12.2 24
[AR2-GigabitEthernet0/0/1]quit
[AR2]interface GigabitEthernet 0/0/2
[AR2-GigabitEthernet0/0/2]ip address 10.10.23.2 24
```

例 2-3　AR3 的初始化配置

```
<Huawei>system-view
Enter system view, return user view with Ctrl+Z.
[Huawei]sysname AR3
[AR3]interface GigabitEthernet 0/0/0
[AR3-GigabitEthernet0/0/0]ip address 172.16.30.3 24
[AR3-GigabitEthernet0/0/0]quit
[AR3]interface GigabitEthernet 0/0/1
[AR3-GigabitEthernet0/0/1]ip address 10.10.34.3 24
[AR3-GigabitEthernet0/0/1]quit
[AR3]interface GigabitEthernet 0/0/2
[AR3-GigabitEthernet0/0/2]ip address 10.10.23.3 24
```

例 2-4　AR4 的初始化配置

```
<Huawei>system-view
Enter system view, return user view with Ctrl+Z.
[Huawei]sysname AR4
[AR4]interface GigabitEthernet 0/0/1
[AR4-GigabitEthernet0/0/1]ip address 10.10.14.4 24
[AR4-GigabitEthernet0/0/1]quit
[AR4]interface GigabitEthernet 0/0/2
[AR4-GigabitEthernet0/0/2]ip address 10.10.34.4 24
```

每一部分的配置结束后我们都要进行验证，以确保我们所做的配置是正确且达到实验目的的。我们可以通过查看路由器上的配置来进行验证，例 2-5～例 2-8 展示了 AR1、AR2、AR3 和 AR4 上的接口配置，示例中使用的查看接口命令为 **display ip interface brief**，这条命令可查看设备上所有接口的简要配置。我们使用这条命令可快速查看该接口的物理状态和协议状态，以及接口的 IP 地址设置。有些命令输出中会展示大量信息，因此我们会在示例中以阴影的形式标注需要读者格外关注的信息。

例 2-5　查看 AR1 的接口配置

```
<AR1>display ip interface brief
*down: administratively down
^down: standby
(l): loopback
(s): spoofing
The number of interface that is UP in Physical is 4
The number of interface that is DOWN in Physical is 0
The number of interface that is UP in Protocol is 4
The number of interface that is DOWN in Protocol is 0

Interface                 IP Address/Mask      Physical    Protocol
GigabitEthernet0/0/0      172.16.10.1/24       up          up
GigabitEthernet0/0/1      10.10.14.1/24        up          up
GigabitEthernet0/0/2      10.10.12.1/24        up          up
NULL0                     unassigned           up          up(s)
```

例 2-6　查看 AR2 的接口配置

```
<AR2>display ip interface brief
*down: administratively down
^down: standby
(l): loopback
(s): spoofing
The number of interface that is UP in Physical is 3
The number of interface that is DOWN in Physical is 1
The number of interface that is UP in Protocol is 3
The number of interface that is DOWN in Protocol is 1

Interface                     IP Address/Mask      Physical    Protocol
GigabitEthernet0/0/0          unassigned           down        down
GigabitEthernet0/0/1          10.10.12.2/24        up          up
GigabitEthernet0/0/2          10.10.23.2/24        up          up
NULL0                         unassigned           up          up(s)
```

例 2-7　查看 AR3 的接口配置

```
<AR3>display ip interface brief
*down: administratively down
^down: standby
(l): loopback
(s): spoofing
The number of interface that is UP in Physical is 4
The number of interface that is DOWN in Physical is 0
The number of interface that is UP in Protocol is 4
The number of interface that is DOWN in Protocol is 0

Interface                     IP Address/Mask      Physical    Protocol
GigabitEthernet0/0/0          172.16.30.3/24       up          up
GigabitEthernet0/0/1          10.10.34.3/24        up          up
GigabitEthernet0/0/2          10.10.23.3/24        up          up
NULL0                         unassigned           up          up(s)
```

例 2-8　查看 AR4 的接口配置

```
<AR4>display ip interface brief
*down: administratively down
^down: standby
(l): loopback
(s): spoofing
The number of interface that is UP in Physical is 3
The number of interface that is DOWN in Physical is 1
The number of interface that is UP in Protocol is 3
The number of interface that is DOWN in Protocol is 1

Interface                     IP Address/Mask      Physical    Protocol
GigabitEthernet0/0/0          unassigned           down        down
GigabitEthernet0/0/1          10.10.14.4/24        up          up
GigabitEthernet0/0/2          10.10.34.4/24        up          up
NULL0                         unassigned           up          up(s)
```

其实，确认配置是否正确最好的做法是以效果进行验证：测试连通性。一方面因为实现连通性是我们执行配置的最终目的，另一方面是因为配置正确并不代表具有连通性。这时我们需要登录每台设备并发出 **ping** 命令。例 2-9 和例 2-10 分别以 AR1 和 AR3 为例，测试了这两台路由器上 3 个直连连接的连通性。

例 2-9　测试 AR1 的直连连通性

```
<AR1>ping 172.16.10.10
  PING 172.16.10.10: 56  data bytes, press CTRL_C to break
    Reply from 172.16.10.10: bytes=56 Sequence=1 ttl=128 time=360 ms
    Reply from 172.16.10.10: bytes=56 Sequence=2 ttl=128 time=20 ms
    Reply from 172.16.10.10: bytes=56 Sequence=3 ttl=128 time=10 ms
    Reply from 172.16.10.10: bytes=56 Sequence=4 ttl=128 time=10 ms
    Reply from 172.16.10.10: bytes=56 Sequence=5 ttl=128 time=10 ms

  --- 172.16.10.10 ping statistics ---
    5 packet(s) transmitted
    5 packet(s) received
    0.00% packet loss
```

```
    round-trip min/avg/max = 10/82/360 ms

<AR1>ping 10.10.12.2
  PING 10.10.12.2: 56  data bytes, press CTRL_C to break
    Reply from 10.10.12.2: bytes=56 Sequence=1 ttl=255 time=100 ms
    Reply from 10.10.12.2: bytes=56 Sequence=2 ttl=255 time=10 ms
    Reply from 10.10.12.2: bytes=56 Sequence=3 ttl=255 time=30 ms
    Reply from 10.10.12.2: bytes=56 Sequence=4 ttl=255 time=30 ms
    Reply from 10.10.12.2: bytes=56 Sequence=5 ttl=255 time=20 ms

  --- 10.10.12.2 ping statistics ---
    5 packet(s) transmitted
    5 packet(s) received
    0.00% packet loss
    round-trip min/avg/max = 10/38/100 ms

<AR1>ping 10.10.14.4
  PING 10.10.14.4: 56  data bytes, press CTRL_C to break
    Reply from 10.10.14.4: bytes=56 Sequence=1 ttl=255 time=80 ms
    Reply from 10.10.14.4: bytes=56 Sequence=2 ttl=255 time=20 ms
    Reply from 10.10.14.4: bytes=56 Sequence=3 ttl=255 time=40 ms
    Reply from 10.10.14.4: bytes=56 Sequence=4 ttl=255 time=20 ms
    Reply from 10.10.14.4: bytes=56 Sequence=5 ttl=255 time=20 ms

  --- 10.10.14.4 ping statistics ---
    5 packet(s) transmitted
    5 packet(s) received
    0.00% packet loss
    round-trip min/avg/max = 20/36/80 ms
```

例 2-10 测试 AR3 的直连连通性

```
<AR3>ping 172.16.30.10
  PING 172.16.30.10: 56  data bytes, press CTRL_C to break
    Reply from 172.16.30.10: bytes=56 Sequence=1 ttl=128 time=10 ms
    Reply from 172.16.30.10: bytes=56 Sequence=2 ttl=128 time=20 ms
    Reply from 172.16.30.10: bytes=56 Sequence=3 ttl=128 time=20 ms
    Reply from 172.16.30.10: bytes=56 Sequence=4 ttl=128 time=20 ms
    Reply from 172.16.30.10: bytes=56 Sequence=5 ttl=128 time=20 ms

  --- 172.16.30.10 ping statistics ---
    5 packet(s) transmitted
    5 packet(s) received
    0.00% packet loss
    round-trip min/avg/max = 10/18/20 ms

<AR3>ping 10.10.23.2
  PING 10.10.23.2: 56  data bytes, press CTRL_C to break
    Reply from 10.10.23.2: bytes=56 Sequence=1 ttl=255 time=50 ms
    Reply from 10.10.23.2: bytes=56 Sequence=2 ttl=255 time=30 ms
    Reply from 10.10.23.2: bytes=56 Sequence=3 ttl=255 time=20 ms
    Reply from 10.10.23.2: bytes=56 Sequence=4 ttl=255 time=20 ms
    Reply from 10.10.23.2: bytes=56 Sequence=5 ttl=255 time=40 ms

  --- 10.10.23.2 ping statistics ---
    5 packet(s) transmitted
    5 packet(s) received
    0.00% packet loss
    round-trip min/avg/max = 20/32/50 ms

<AR3>ping 10.10.34.4
  PING 10.10.34.4: 56  data bytes, press CTRL_C to break
    Reply from 10.10.34.4: bytes=56 Sequence=1 ttl=255 time=60 ms
    Reply from 10.10.34.4: bytes=56 Sequence=2 ttl=255 time=30 ms
    Reply from 10.10.34.4: bytes=56 Sequence=3 ttl=255 time=20 ms
    Reply from 10.10.34.4: bytes=56 Sequence=4 ttl=255 time=20 ms
    Reply from 10.10.34.4: bytes=56 Sequence=5 ttl=255 time=30 ms

  --- 10.10.34.4 ping statistics ---
    5 packet(s) transmitted
    5 packet(s) received
    0.00% packet loss
    round-trip min/avg/max = 20/32/60 ms
```

步骤 2 配置静态路由

实现了直连链路的连通性后,我们需要进行下一步,通过 AR1-AR2-AR3 的路径实现 PC1 与 PC2 之间的通信。首先我们要确认当前 PC1 与 PC2 的通信情况。我们在 PC1 上执行 **ping 172.16.30.10** 命令(结果是我们已经知道的),以及 **tracert 172.16.30.10** 命令。tracert 172.16.30.10 命令是路由追踪命令,它利用 ICMP(Internet Control Message Protocol,控制报文协议)使转发路径中的每一跳设备都进行回复,以此描绘出数据包的转发路径。在本实验中,我们用它来确认路径是在哪里中断的。

例 2-11 为在 PC1 上追踪路由。

例 2-11 在 PC1 上追踪路由

```
PC1>ping 172.16.30.10

Ping 172.16.30.10: 32 data bytes, Press Ctrl_C to break
Request timeout!
Request timeout!
Request timeout!
Request timeout!
Request timeout!

--- 172.16.30.10 ping statistics ---
  5 packet(s) transmitted
  0 packet(s) received
  100.00% packet loss

PC1>tracert 172.16.30.10

traceroute to 172.16.30.10, 8 hops max
(ICMP), press Ctrl+C to stop
 1   *   *   *
 2   *   *   *
 3

PC1>
```

如例 2-11 的路由追踪命令所示,*表示请求超时,读者可以通过使用快捷键"Ctrl+C"中断测试。我们按照拓扑中数据包应该使用的路径方向进行排查,从 AR1 开始。

我们先通过 AR1 的路由表确认 AR1 上的路由。例 2-12 中展示了 AR1 的路由表,本例使用的命令是 **display ip routing-table**。本书会对 **display** 这类查看命令的输出信息进行解释,但不会对所有参数进行解释,而是依照当前实验的主题,选取与之最相关的参数,突出重点进行介绍。

例 2-12 查看 AR1 的路由表

```
<AR1>display ip routing-table
Route Flags: R - relay, D - download to fib
------------------------------------------------------------------------------
Routing Tables: Public
         Destinations : 13       Routes : 13

Destination/Mask    Proto   Pre  Cost        Flags NextHop         Interface

     10.10.12.0/24  Direct  0    0             D   10.10.12.1      GigabitEthernet0/0/2
     10.10.12.1/32  Direct  0    0             D   127.0.0.1       GigabitEthernet0/0/2
   10.10.12.255/32  Direct  0    0             D   127.0.0.1       GigabitEthernet0/0/2
     10.10.14.0/24  Direct  0    0             D   10.10.14.1      GigabitEthernet0/0/1
     10.10.14.1/32  Direct  0    0             D   127.0.0.1       GigabitEthernet0/0/1
   10.10.14.255/32  Direct  0    0             D   127.0.0.1       GigabitEthernet0/0/1
      127.0.0.0/8   Direct  0    0             D   127.0.0.1       InLoopBack0
      127.0.0.1/32  Direct  0    0             D   127.0.0.1       InLoopBack0
127.255.255.255/32  Direct  0    0             D   127.0.0.1       InLoopBack0
     172.16.10.0/24 Direct  0    0             D   172.16.10.1     GigabitEthernet0/0/0
     172.16.10.1/32 Direct  0    0             D   127.0.0.1       GigabitEthernet0/0/0
   172.16.10.255/32 Direct  0    0             D   127.0.0.1       GigabitEthernet0/0/0
255.255.255.255/32  Direct  0    0             D   127.0.0.1       InLoopBack0
```

从例 2-12 中的 AR1 路由表可以看到，本实验使用的 6 个地址段中的 3 个（10.10.12.0/24、10.10.14.0/24 和 172.16.10.0/24），也是 AR1 所直连的 3 个地址段。这是因为当我们在路由器的接口上配置了一个 IP 地址和子网掩码后，它会自动将其放入本地路由表中。"Proto"列中的"Direct"就说明这条路由是直连路由，来自于管理员在接口上的 IP 地址配置。

路由器等网络设备是根据 IP 路由表来进行数据包路由的。因此，当 AR1 的路由表中没有 PC2 所在的网段（172.16.30.0/24）时，它就无法对去往 PC2 的数据包进行路由。解决方法是通过某种方法将去往该网段的路由添加到 AR1 的路由表中，有手动添加和动态添加两种方式，即静态路由和动态路由协议。本实验将演示如何手动添加静态路由，后面的实验中会对具体的动态路由协议配置进行演示。

本实验使用的配置静态路由的命令语法如下。为了突出重点，本书所展示的一些命令语法并不是完整的语法，省略了实验中没有用到的可选关键字和可选参数，想要了解完整命令语法的读者可以参考华为命令手册。

ip route-static *ip-address mask-length nexthop-address*：配置 IPv4 静态路由，其中包含 3 个参数，第 1 个参数是 IP 地址。第 2 个参数是子网掩码，与设置 IP 地址的命令相同，子网掩码可以由两种形式表示，本例使用子网掩码长度这种简洁的形式。第 3 个参数是下一跳，本例会使用下一跳 IP 地址作为下一跳参数。

例 2-13 中展示了使用下一跳地址配置的结果。

例 2-13 使用下一跳地址参数配置静态路由

```
[AR1]ip route-static 172.16.30.0 24 10.10.12.2
[AR1]display ip routing-table
Route Flags: R - relay, D - download to fib
------------------------------------------------------------------------
Routing Tables: Public
        Destinations : 14       Routes : 14

Destination/Mask    Proto   Pre  Cost       Flags NextHop         Interface

      10.10.12.0/24  Direct  0    0           D   10.10.12.1      GigabitEthernet0/0/2
      10.10.12.1/32  Direct  0    0           D   127.0.0.1       GigabitEthernet0/0/2
    10.10.12.255/32  Direct  0    0           D   127.0.0.1       GigabitEthernet0/0/2
      10.10.14.0/24  Direct  0    0           D   10.10.14.1      GigabitEthernet0/0/1
      10.10.14.1/32  Direct  0    0           D   127.0.0.1       GigabitEthernet0/0/1
    10.10.14.255/32  Direct  0    0           D   127.0.0.1       GigabitEthernet0/0/1
        127.0.0.0/8  Direct  0    0           D   127.0.0.1       InLoopBack0
        127.0.0.1/32 Direct  0    0           D   127.0.0.1       InLoopBack0
  127.255.255.255/32 Direct  0    0           D   127.0.0.1       InLoopBack0
      172.16.10.0/24 Direct  0    0           D   172.16.10.1     GigabitEthernet0/0/0
      172.16.10.1/32 Direct  0    0           D   127.0.0.1       GigabitEthernet0/0/0
    172.16.10.255/32 Direct  0    0           D   127.0.0.1       GigabitEthernet0/0/0
      172.16.30.0/24 Static  60   0           RD  10.10.12.2      GigabitEthernet0/0/2
  255.255.255.255/32 Direct  0    0           D   127.0.0.1       InLoopBack0
```

如例 2-13 所示，使用下一跳地址参数配置静态路由后，AR1 的路由表中（倒数第 2 行）出现了这个新网段的路由。与其他路由条目相比，这条路由有两点明显区别："Pre"列为 60、"Flags"列为 RD。

"Pre"列表示路由优先级，数字越小优先级越高，从本例中我们可以看出直连路由的优先级为 0（最高），静态路由的优先级为 60，管理员可以手动修改路由的优先级，本章浮动路由的实验中会演示它的用法。

"Flags"列表示路由标记，读者从 **display ip routing-table** 命令输出的第 1 行可以看到路由标记（Route Flags）的解释：D 表示该路由已下载到 FIB（转发信息表）中，可直接使用；R 表示迭代路由，路由器需要根据下一跳 IP 地址计算出本地的出接口。本例 AR1 计算出的出接口为 G0/0/2。

有读者会提出，既然路由器还需要做进一步的判断才能确定出接口，管理员在静态路由的配置中指明出接口不是更方便的做法吗？实际上，VRP 系统可支持以出接口作为下一跳参数，但这种做法并不适用于以太网环境。若管理员只指定了出接口，但在可能存在多台网络设备的以太网环境中，仍无法确认应该把数据包转发给哪台设备。因此对于以太网接口来说，我们可以指定下一跳 IP 地址，或者同时指定出接口和下一跳 IP 地址。另外，这里说的出接口是指配置路由的那台路由器本地的接口，而下一跳是指对端路由设备的 IP 地址。在本例中，出接口指的是 AR1 的 G0/0/2 接口，下一跳地址指的是 AR2 的 G0/0/1 接口 IP 地址（10.10.12.2/24）。

现在 AR1 已经知道了该如何去往 PC2，但路径中的 AR2 还不知道该如何去往 PC2。例 2-14 在 AR2 上添加了去往 PC2 的静态路由，还查看了路由表中的静态路由条目。

例 2-14　在 AR2 上添加去往 PC2 的静态路由

```
[AR2]display ip routing-table
Route Flags: R - relay, D - download to fib
------------------------------------------------------------------------------
Routing Tables: Public
         Destinations : 11       Routes : 11

Destination/Mask    Proto   Pre  Cost      Flags NextHop         Interface

      10.10.12.0/24  Direct  0    0          D   10.10.12.2      GigabitEthernet0/0/1
      10.10.12.2/32  Direct  0    0          D   127.0.0.1       GigabitEthernet0/0/1
    10.10.12.255/32  Direct  0    0          D   127.0.0.1       GigabitEthernet0/0/1
      10.10.23.0/24  Direct  0    0          D   10.10.23.2      GigabitEthernet0/0/2
      10.10.23.2/32  Direct  0    0          D   127.0.0.1       GigabitEthernet0/0/2
    10.10.23.255/32  Direct  0    0          D   127.0.0.1       GigabitEthernet0/0/2
         127.0.0.0/8 Direct  0    0          D   127.0.0.1       InLoopBack0
       127.0.0.1/32  Direct  0    0          D   127.0.0.1       InLoopBack0
 127.255.255.255/32  Direct  0    0          D   127.0.0.1       InLoopBack0
      172.16.30.0/24 Static  60   0          RD  10.10.23.3      GigabitEthernet0/0/2
 255.255.255.255/32  Direct  0    0          D   127.0.0.1       InLoopBack0
```

此时，在 PC1 去往 PC2 的路径中，所有路由器都知道该如何转发去往 PC2 的数据包了，但 PC1 还无法与 PC2 进行通信，读者可以在自己的实验环境中进行验证。这是因为无论是 ping（ICMP）还是其他操作，要想实现访问都是需要有来有回的。至此，PC1 向 PC2 发起 ping 测试，数据包的走向如下。

① PC1 会将数据包发给自己的默认网关 AR1。
② AR1 会按照刚配置的静态路由，将去往 PC2 的数据包转发给 AR2。
③ AR2 会按照刚配置的静态路由，将去往 PC2 的数据包转发给 AR3。
④ AR3 会按照直连路由将数据包转发给 PC2。
⑤ PC2 会处理并将返回的数据包发给自己的默认网关 AR3。
⑥ AR3 在路由表中找不到去往 PC1（172.16.10.0/24）网段的路由，因此丢弃数据包。

如何证明 PC2 已经收到并返回了消息呢？我们可以在 AR3 的 G0/0/0 和 G0/0/2 接口上同时开启抓包。AR3 G0/0/0 接口上的抓包信息如图 2-2 所示。

我们可以看到，对于每个 ping 请求（ICMP Echo request），AR3 都从 PC2 那里收到了相应的应答（ICMP Echo reply）。AR3 G0/0/2 接口上的抓包信息如图 2-3 所示。

从 AR3 G0/0/2 接口的抓包信息中我们可以看出，AR3 没有转发任何应答消息。现在我们在 AR3 上补充 PC1 网段的路由，具体见例 2-15。

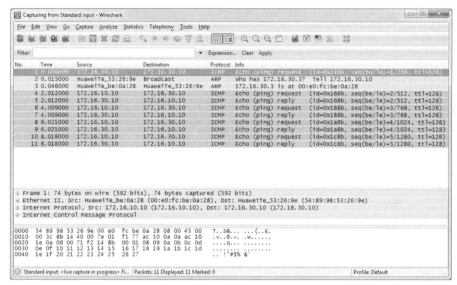

图 2-2　AR3 G0/0/0 接口上的抓包信息

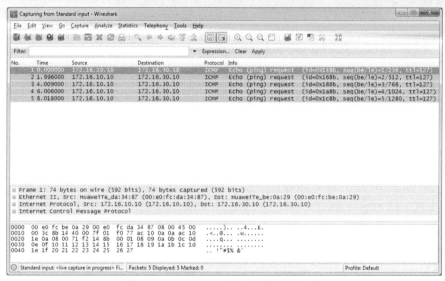

图 2-3　AR3 G0/0/2 接口上的抓包信息

例 2-15　在 AR3 上添加 PC1 网段的静态路由

```
[AR3]ip route-static 172.16.10.0 24 10.10.23.2
```

前文数据包转发流程中的第 6 步已经修复，现在的流程如下。

① PC1 会将数据包发给自己的默认网关 AR1。
② AR1 会按照静态路由将去往 PC2 的数据包转发给 AR2。
③ AR2 会按照静态路由将去往 PC2 的数据包转发给 AR3。
④ AR3 会按照直连路由将数据包转发给 PC2。
⑤ PC2 会处理并将返回的数据包发给自己的默认网关 AR3。
⑥ AR3 会按照刚配置的静态路由，将去往 PC1 的数据包转发给 AR2。

⑦ AR2 在路由表中找不到去往 PC1（172.16.10.0/24）网段的路由，因此丢弃数据包。

在从 PC2 返回 PC1 的路径中，AR2 也需要知道该如何去往 PC1，因此例 2-16 在 AR2 上添加了 PC1 网段的静态路由。

例 2-16 在 AR2 上添加 PC1 网段的静态路由

```
[AR2]ip route-static 172.16.10.0 24 10.10.12.1
```

例 2-16 中的命令解决了上述第 7 步中的问题，现在流程变得完整，具体如下。

① PC1 会将数据包发给自己的默认网关 AR1。
② AR1 会按照静态路由将去往 PC2 的数据包转发给 AR2。
③ AR2 会按照静态路由将去往 PC2 的数据包转发给 AR3。
④ AR3 会按照直连路由将数据包转发给 PC2。
⑤ PC2 会处理并将返回的数据包发给自己的默认网关 AR3。
⑥ AR3 会按照静态路由将去往 PC1 的数据包转发给 AR2。
⑦ AR2 会按照刚配置的静态路由，将去往 PC1 的数据包转发给 AR1。
⑧ AR1 会按照直连路由将数据包转发给 PC1。
⑨ PC1 收到返回的数据包。

例 2-17 中展示了从 PC1 向 PC2 再次发起 ping 测试的示例，这次 ping 成功了。

例 2-17 PC1 向 PC2 发起 ping 测试

```
PC1>ping 172.16.30.10

Ping 172.16.30.10: 32 data bytes, Press Ctrl_C to break
From 172.16.30.10: bytes=32 seq=1 ttl=125 time=46 ms
From 172.16.30.10: bytes=32 seq=2 ttl=125 time=31 ms
From 172.16.30.10: bytes=32 seq=3 ttl=125 time=47 ms
From 172.16.30.10: bytes=32 seq=4 ttl=125 time=47 ms
From 172.16.30.10: bytes=32 seq=5 ttl=125 time=125 ms

--- 172.16.30.10 ping statistics ---
  5 packet(s) transmitted
  5 packet(s) received
  0.00% packet loss
  round-trip min/avg/max = 31/31/32 ms
```

2.2.2 浮动静态路由

为了使 PC1 与 PC2 之间的通信更为可靠，我们可以在拓扑中增加冗余性，即在它们之间启用另一条备份路径。如图 2-4 所示，我们可以选用的路径是 AR1-AR4-AR3。在这个实验中，我们要通过在 AR1、AR4 和 AR3 上各添加一些路由来实现这个目标。

步骤 1 AR1 上的配置

我们按照从 PC1 到 PC2 的通信路径来逐一研究每台设备上应有的配置。目前 AR1 已经知道了要使用本地接口 G0/0/2 通过 AR2 去往 PC2，我们现在要让它在 G0/0/2 接口或其所连接的网络出现问题时，选用本地接口 G0/0/1 通过 AR4 去往 PC2。我们可以使用浮动静态路由来配置这种备用路径。浮动静态路由的命令语法如下，在静态路由条目后添加 **preference** 关键字并设置路由优先级。例 2-18 中展示了 AR1 上的浮动静态路由配置。

ip route-static *ip-address mask-length nexthop-address* **preference** *preference*：在这个命令语法中，关键字 **preference** 及其后面的优先级参数是新知识，也是配置浮动静态路

由的关键。从前面的实验中我们知道静态路由的路由优先级是 60（可修改），并且优先级值越小，路由的优先级越高。因此我们可以把这条备用路由的路由优先级设置得比 60 大，从而实现备份功能。动态路由协议也有各自默认的路由优先级，因此我们在改变或设置静态路由的路由优先级时，也要考虑动态路由协议的路由优先级。

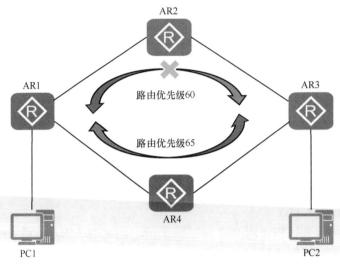

图 2-4　路由备份

例 2-18　AR1 上的浮动静态路由配置

```
[AR1]ip route-static 172.16.30.0 24 10.10.14.4 preference 65
```

接下来我们查看 AR1 的路由表。例 2-19 中展示了 AR1 上的路由表。

例 2-19　查看 AR1 的路由表

```
[AR1]display ip routing-table
Route Flags: R - relay, D - download to fib
------------------------------------------------------------------------------
Routing Tables: Public
         Destinations : 14      Routes : 14

Destination/Mask       Proto   Pre  Cost     Flags NextHop        Interface

     10.10.12.0/24     Direct  0    0          D   10.10.12.1     GigabitEthernet0/0/2
     10.10.12.1/32     Direct  0    0          D   127.0.0.1      GigabitEthernet0/0/2
   10.10.12.255/32     Direct  0    0          D   127.0.0.1      GigabitEthernet0/0/2
     10.10.14.0/24     Direct  0    0          D   10.10.14.1     GigabitEthernet0/0/1
     10.10.14.1/32     Direct  0    0          D   127.0.0.1      GigabitEthernet0/0/1
   10.10.14.255/32     Direct  0    0          D   127.0.0.1      GigabitEthernet0/0/1
       127.0.0.0/8     Direct  0    0          D   127.0.0.1      InLoopBack0
       127.0.0.1/32    Direct  0    0          D   127.0.0.1      InLoopBack0
 127.255.255.255/32    Direct  0    0          D   127.0.0.1      InLoopBack0
     172.16.10.0/24    Direct  0    0          D   172.16.10.1    GigabitEthernet0/0/0
     172.16.10.1/32    Direct  0    0          D   127.0.0.1      GigabitEthernet0/0/0
   172.16.10.255/32    Direct  0    0          D   127.0.0.1      GigabitEthernet0/0/0
     172.16.30.0/24    Static  60   0          RD  10.10.12.2     GigabitEthernet0/0/2
 255.255.255.255/32    Direct  0    0          D   127.0.0.1      InLoopBack0
```

从 AR1 的路由表中我们可以确认，刚才配置的浮动静态路由没有出现在这里。这是因为能够被放入路由表的路由都是最优路由，是用来指导数据包转发的路由。因此去往相同目的地的多条路由，路由优先级较高的会被放入路由表中，若多条路由的路由优先级相同，则会同时被放入路由表中，从而形成静态路由负载分担。下一个实验中我们会介绍静态路由负载分担。

为了让这条备用路由"浮现"在路由表中,我们手动关闭 AR1 的 G0/0/2 接口来模拟这条线路的故障。例 2-20 中展示了关闭接口的配置,以及关闭接口后 AR1 的路由表。用来关闭接口的命令是接口视图的命令 **shutdown**,开启接口的命令是 **undo shutdown**。

例 2-20 关闭接口后再次查看 AR1 路由表

```
[AR1]interface GigabitEthernet 0/0/2
[AR1-GigabitEthernet0/0/2]shutdown
[AR1-GigabitEthernet0/0/2]quit
[AR1]display ip routing-table
Route Flags: R - relay, D - download to fib
------------------------------------------------------------------------------
Routing Tables: Public
         Destinations : 11       Routes : 11

Destination/Mask    Proto   Pre  Cost      Flags NextHop         Interface

      10.10.14.0/24  Direct  0    0           D  10.10.14.1      GigabitEthernet0/0/1
      10.10.14.1/32  Direct  0    0           D  127.0.0.1       GigabitEthernet0/0/1
    10.10.14.255/32  Direct  0    0           D  127.0.0.1       GigabitEthernet0/0/1
       127.0.0.0/8   Direct  0    0           D  127.0.0.1       InLoopBack0
       127.0.0.1/32  Direct  0    0           D  127.0.0.1       InLoopBack0
 127.255.255.255/32  Direct  0    0           D  127.0.0.1       InLoopBack0
      172.16.10.0/24 Direct  0    0           D  172.16.10.1     GigabitEthernet0/0/0
      172.16.10.1/32 Direct  0    0           D  127.0.0.1       GigabitEthernet0/0/0
    172.16.10.255/32 Direct  0    0           D  127.0.0.1       GigabitEthernet0/0/0
      172.16.30.0/24 Static  65   0          RD  10.10.14.4      GigabitEthernet0/0/1
 255.255.255.255/32  Direct  0    0           D  127.0.0.1       InLoopBack0
```

这时我们可以在 AR1 的路由表中看到这条优先级为 65 的浮动静态路由了。它是一条递归路由,且下一跳指向了 AR4(10.10.14.4)。接下来我们登录 AR4 进行配置。

步骤 2　AR4 上的配置

在 AR4 中,为了能够帮助 PC1 和 PC2 转发数据包,它需要分别知道这两个网段的路由。因此我们会在这里使用静态路由命令为它配置这两个网段的路由。例 2-21 中展示了 AR4 上的静态路由配置。

例 2-21 在 AR4 上配置静态路由

```
[AR4]ip route-static 172.16.10.0 24 10.10.14.1
[AR4]ip route-static 172.16.30.0 24 10.10.34.3
```

现在我们通过 AR4 的路由表来查看这两条静态路由,具体见例 2-22。现在 AR4 已经准备好为 PC1 和 PC2 转发数据包了。

例 2-22 AR4 路由表中的静态路由

```
[AR4]display ip routing-table
Route Flags: R - relay, D - download to fib
------------------------------------------------------------------------------
Routing Tables: Public
         Destinations : 12       Routes : 12

Destination/Mask    Proto   Pre  Cost      Flags NextHop         Interface

      10.10.14.0/24  Direct  0    0           D  10.10.14.4      GigabitEthernet0/0/1
      10.10.14.4/32  Direct  0    0           D  127.0.0.1       GigabitEthernet0/0/1
    10.10.14.255/32  Direct  0    0           D  127.0.0.1       GigabitEthernet0/0/1
      10.10.34.0/24  Direct  0    0           D  10.10.34.4      GigabitEthernet0/0/2
      10.10.34.4/32  Direct  0    0           D  127.0.0.1       GigabitEthernet0/0/2
    10.10.34.255/32  Direct  0    0           D  127.0.0.1       GigabitEthernet0/0/2
       127.0.0.0/8   Direct  0    0           D  127.0.0.1       InLoopBack0
       127.0.0.1/32  Direct  0    0           D  127.0.0.1       InLoopBack0
 127.255.255.255/32  Direct  0    0           D  127.0.0.1       InLoopBack0
      172.16.10.0/24 Static  60   0          RD  10.10.14.1      GigabitEthernet0/0/1
      172.16.30.0/24 Static  60   0          RD  10.10.34.3      GigabitEthernet0/0/2
 255.255.255.255/32  Direct  0    0           D  127.0.0.1       InLoopBack0
```

步骤 3　AR3 上的配置

AR3 与 AR1 上的配置相同，都是使用静态浮动路由配置一条备用路由，将其指向 AR4。例 2-23 中展示了 AR3 上的静态浮动路由配置。

例 2-23　在 AR3 上配置静态浮动路由

```
[AR3]ip route-static 172.16.10.0 24 10.10.34.4 preference 65
```

现在我们查看 AR3 的路由表。例 2-24 中展示了 AR3 的路由表。

例 2-24　查看 AR3 的路由表

```
[AR3]display ip routing-table
Route Flags: R - relay, D - download to fib
------------------------------------------------------------------------------
Routing Tables: Public
         Destinations : 14      Routes : 14

Destination/Mask     Proto   Pre   Cost       Flags NextHop         Interface

     10.10.23.0/24   Direct  0     0            D   10.10.23.3      GigabitEthernet0/0/2
     10.10.23.3/32   Direct  0     0            D   127.0.0.1       GigabitEthernet0/0/2
   10.10.23.255/32   Direct  0     0            D   127.0.0.1       GigabitEthernet0/0/2
     10.10.34.0/24   Direct  0     0            D   10.10.34.3      GigabitEthernet0/0/1
     10.10.34.3/32   Direct  0     0            D   127.0.0.1       GigabitEthernet0/0/1
   10.10.34.255/32   Direct  0     0            D   127.0.0.1       GigabitEthernet0/0/1
        127.0.0.0/8  Direct  0     0            D   127.0.0.1       InLoopBack0
       127.0.0.1/32  Direct  0     0            D   127.0.0.1       InLoopBack0
   127.255.255.255/32 Direct 0     0            D   127.0.0.1       InLoopBack0
     172.16.10.0/24  Static  60    0            RD  10.10.23.2      GigabitEthernet0/0/2
     172.16.30.0/24  Direct  0     0            D   172.16.30.3     GigabitEthernet0/0/0
     172.16.30.3/32  Direct  0     0            D   127.0.0.1       GigabitEthernet0/0/0
   172.16.30.255/32  Direct  0     0            D   127.0.0.1       GigabitEthernet0/0/0
   255.255.255.255/32 Direct 0     0            D   127.0.0.1       InLoopBack0
```

我们发现途经 AR2 去往 PC1 的那条路由优先级为 65 的路由并没有出现在路由表中，这是因为 AR3 并不能感知到"远端" AR1 与 AR2 之间的链路故障。在静态路由环境中，若管理员不进行干预，PC1 与 PC2 之间就会失去连通性。此时 AR3 仍会选择 AR2 为去往 PC1 的下一跳路由器，但这条路已经走不通了。

我们再次从 PC1 向 PC2 发起 ping 测试，结果是 100%丢包。同时我们在 AR3 的 G0/0/1 和 G0/0/2 接口分别开启抓包，从图 2-5 和图 2-6 中我们可以看出，ping 请求的转发路径是 AR1-AR4-AR3，ping 应答的转发路径是 AR3-AR2，到 AR2 就会断开。

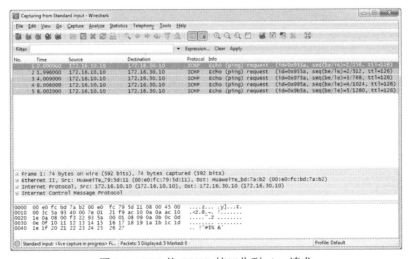

图 2-5　AR3 从 G0/0/1 接口收到 ping 请求

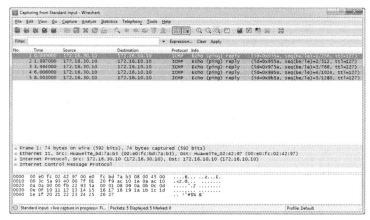

图 2-6　AR3 从 G0/0/2 接口转发 ping 应答

本实验中，这条浮动静态路由没有出现在例 2-24 所示的路由表中，这是因为命令 **display ip routing-table** 输出的路由表是"最优路由表"，其实浮动静态路由器已经存在于路由表中，我们可以通过命令 **display ip routing-table protocol static** 进行查看，具体见例 2-25。

例 2-25　在 AR3 上查看不活跃的静态路由

```
[AR3]display ip routing-table protocol static
Route Flags: R - relay, D - download to fib
------------------------------------------------------------------------------
Public routing table : Static
         Destinations : 1        Routes : 2       Configured Routes : 2

Static routing table status : <Active>
         Destinations : 1        Routes : 1

Destination/Mask    Proto   Pre   Cost      Flags  NextHop         Interface

    172.16.10.0/24  Static  60    0           RD   10.10.23.2      GigabitEthernet0/0/2

Static routing table status : <Inactive>
         Destinations : 1        Routes : 1

Destination/Mask    Proto   Pre   Cost      Flags  NextHop         Interface

    172.16.10.0/24  Static  65    0           R    10.10.34.4      GigabitEthernet0/0/1
```

例 2-25 中命令输出的最后一行为我们配置的浮动静态路由，它当前的状态为 Inactive（不活跃）。本例手动配置的静态路由环境中，我们只有再关闭 AR3 的 G0/0/2 接口，这条路由才"浮现"出来。

例 2-26 为关闭 AR3 的 G0/0/2 接口后，再次查看 AR3 的路由表。

例 2-26　再次查看 AR3 的路由表

```
[AR3]interface GigabitEthernet 0/0/2
[AR3-GigabitEthernet0/0/2]shutdown
[AR3-GigabitEthernet0/0/2]quit
[AR3]display ip routing-table
Route Flags: R - relay, D - download to fib
------------------------------------------------------------------------------
Routing Tables: Public
         Destinations : 11       Routes : 11

Destination/Mask    Proto   Pre   Cost      Flags  NextHop         Interface

      10.10.34.0/24  Direct  0     0           D    10.10.34.3      GigabitEthernet0/0/1
      10.10.34.3/32  Direct  0     0           D    127.0.0.1       GigabitEthernet0/0/1
    10.10.34.255/32  Direct  0     0           D    127.0.0.1       GigabitEthernet0/0/1
```

```
       127.0.0.0/8    Direct  0    0            D   127.0.0.1      InLoopBack0
       127.0.0.1/32   Direct  0    0            D   127.0.0.1      InLoopBack0
 127.255.255.255/32   Direct  0    0            D   127.0.0.1      InLoopBack0
      172.16.10.0/24  Static  65   0            RD  10.10.34.4     GigabitEthernet0/0/1
      172.16.30.0/24  Direct  0    0            D   172.16.30.3    GigabitEthernet0/0/0
      172.16.30.3/32  Direct  0    0            D   127.0.0.1      GigabitEthernet0/0/0
    172.16.30.255/32  Direct  0    0            D   127.0.0.1      GigabitEthernet0/0/0
  255.255.255.255/32  Direct  0    0            D   127.0.0.1      InLoopBack0
```

步骤 4 验证浮动静态路由的效果

至此我们完成了实验中的配置部分,现在通过在 PC1 上向 PC2 发起 ping 测试来验证效果。例 2-27 中展示了 ping 测试的结果。

例 2-27 在 PC1 上向 PC2 发起 ping 测试

```
PC1>ping 172.16.30.10

Ping 172.16.30.10: 32 data bytes, Press Ctrl_C to break
Request timeout!
Request timeout!
Request timeout!
From 172.16.30.10: bytes=32 seq=4 ttl=125 time=31 ms
From 172.16.30.10: bytes=32 seq=5 ttl=125 time=32 ms

--- 172.16.30.10 ping statistics ---
  5 packet(s) transmitted
  2 packet(s) received
  60.00% packet loss
  round-trip min/avg/max = 0/31/32 ms
```

我们看到这个 ping 测试最终成功了,但前 3 次测试没有成功。这是因为在构成 ICMP(Internet Control Message Protocol,控制报文协议)(ping)请求包时,设备需要知道下一跳设备的 MAC 地址才能封装这个 ICMP 请求包并发送出去。无论是路由器上的静态路由、浮动路由配置,还是在 PC 上的 IP 地址和默认网关配置,都解决了 IP 层面的路由问题。但设备不仅需要知道下一跳 IP 地址,还需要知道其 MAC 地址,这是通过 ARP 解析实现的。前 3 次丢包是因为在 ARP 查找的过程中耽误了时间,导致 ICMP 等待超时,因此返回了超时消息。对此感兴趣的读者可以"沿途"开启抓包并查看 ARP 消息。

现在 PC1 和 PC2 之间可以通过 AR1-AR4-AR3 实现通信了。当 AR1 与 AR2 之间的链路恢复后,AR1 会自动将去往 PC2 的路径恢复为通过 AR2 进行转发。例 2-28 展示了我们把 AR1 的 G0/0/2 接口重新开启,并查看了 AR1 的路由表。

例 2-28 重新打开 AR1 的 G0/0/2 接口并查看路由表

```
[AR1]interface GigabitEthernet 0/0/2
[AR1-GigabitEthernet0/0/2]undo shutdown
[AR1-GigabitEthernet0/0/2]quit
[AR1]display ip routing-table
Route Flags: R - relay, D - download to fib
------------------------------------------------------------------------------
Routing Tables: Public
         Destinations : 14       Routes : 14

Destination/Mask      Proto   Pre  Cost         Flags NextHop        Interface

      10.10.12.0/24   Direct  0    0            D     10.10.12.1     GigabitEthernet0/0/2
      10.10.12.1/32   Direct  0    0            D     127.0.0.1      GigabitEthernet0/0/2
    10.10.12.255/32   Direct  0    0            D     127.0.0.1      GigabitEthernet0/0/2
      10.10.14.0/24   Direct  0    0            D     10.10.14.1     GigabitEthernet0/0/1
      10.10.14.1/32   Direct  0    0            D     127.0.0.1      GigabitEthernet0/0/1
    10.10.14.255/32   Direct  0    0            D     127.0.0.1      GigabitEthernet0/0/1
       127.0.0.0/8    Direct  0    0            D     127.0.0.1      InLoopBack0
       127.0.0.1/32   Direct  0    0            D     127.0.0.1      InLoopBack0
 127.255.255.255/32   Direct  0    0            D     127.0.0.1      InLoopBack0
      172.16.10.0/24  Direct  0    0            D     172.16.10.1    GigabitEthernet0/0/0
      172.16.10.1/32  Direct  0    0            D     127.0.0.1      GigabitEthernet0/0/0
    172.16.10.255/32  Direct  0    0            D     127.0.0.1      GigabitEthernet0/0/0
      172.16.30.0/24  Static  60   0            RD    10.10.12.2     GigabitEthernet0/0/2
  255.255.255.255/32  Direct  0    0            D     127.0.0.1      InLoopBack0
```

从 AR1 的路由表中我们可以看出，优先级为 60 且通过 AR2（10.10.12.2）的路由重新出现了，优先级为 65 且通过 AR4 的备用路由"浮动"了下去。为了恢复 PC1 与 PC2 之间的通信，用户要开启 AR3 的 G0/0/2 接口。此处不再展示 AR3 上的配置和最后的验证，读者可以在自己的实验环境中开启 AR3 的 G0/0/2 接口并再次测试 PC1 与 PC2 之间的连通性。

2.2.3 静态路由负载分担

前面的实验中，同一时间 PC1 与 PC2 之间只有一条路径：主用路径 AR1-AR2-AR3 或备用路径 AR1-AR4-AR3。本实验将同时使用这两条路径来实现 PC1 与 PC2 之间的通信。图 2-7 为路由负载分担拓扑。

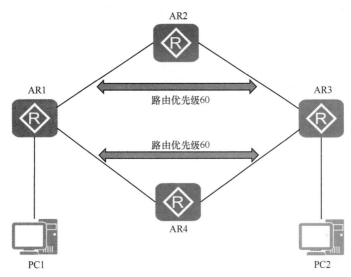

图 2-7　路由负载分担

我们先要删除上一个实验中配置的两条浮动静态路由，并以普通静态路由的方式再次配置通过 AR4 进行数据包转发的路由。例 2-29 和例 2-30 中展示了 AR1 和 AR3 上的相应配置。

例 2-29　AR1 上的配置变更

```
[AR1]undo ip route-static 172.16.30.0 255.255.255.0 10.10.14.4 preference 65
[AR1]ip route-static 172.16.30.0 24 10.10.14.4
```

例 2-30　AR3 上的配置变更

```
[AR3]undo ip route-static 172.16.10.0 255.255.255.0 10.10.34.4 preference 65
[AR3]ip route-static 172.16.10.0 24 10.10.34.4
```

此时我们查看 AR1 的路由表。如例 2-31 所示，读者可以在 AR1 的路由表中看到两条去往 172.16.30.0/24 的路由。

例 2-31　在 AR1 中查看负载分担的路由

```
[AR1]display ip routing-table
Route Flags: R - relay, D - download to fib
------------------------------------------------------------------------------
Routing Tables: Public
         Destinations : 14       Routes : 15
```

```
Destination/Mask      Proto   Pre  Cost       Flags  NextHop         Interface
      10.10.12.0/24   Direct  0    0          D      10.10.12.1      GigabitEthernet0/0/2
      10.10.12.1/32   Direct  0    0          D      127.0.0.1       GigabitEthernet0/0/2
      10.10.12.255/32 Direct  0    0          D      127.0.0.1       GigabitEthernet0/0/2
      10.10.14.0/24   Direct  0    0          D      10.10.14.1      GigabitEthernet0/0/1
      10.10.14.1/32   Direct  0    0          D      127.0.0.1       GigabitEthernet0/0/1
      10.10.14.255/32 Direct  0    0          D      127.0.0.1       GigabitEthernet0/0/1
        127.0.0.0/8   Direct  0    0          D      127.0.0.1       InLoopBack0
        127.0.0.1/32  Direct  0    0          D      127.0.0.1       InLoopBack0
    127.255.255.255/32 Direct 0    0          D      127.0.0.1       InLoopBack0
      172.16.10.0/24  Direct  0    0          D      172.16.10.1     GigabitEthernet0/0/0
      172.16.10.1/32  Direct  0    0          D      127.0.0.1       GigabitEthernet0/0/0
      172.16.10.255/32 Direct 0    0          D      127.0.0.1       GigabitEthernet0/0/0
      172.16.30.0/24  Static  60   0          RD     10.10.12.2      GigabitEthernet0/0/2
                      Static  60   0          RD     10.10.14.4      GigabitEthernet0/0/1
    255.255.255.255/32 Direct 0    0          D      127.0.0.1       InLoopBack0
```

读者可以从 PC1 向 PC2 发起追踪路由（tracert）测试，来验证实验结果，具体见例 2-32。

例 2-32 验证实验结果

```
PC1>tracert 172.16.30.10

traceroute to 172.16.30.10, 8 hops max
(ICMP), press Ctrl+C to stop
 1  172.16.10.1    15 ms  16 ms  15 ms
 2  10.10.14.4     32 ms  15 ms  16 ms
 3  10.10.34.3     31 ms  31 ms  16 ms
 4  172.16.30.10   31 ms  16 ms  16 ms
```

我们从实验结果可以看出，现在从 PC1 去往 PC2 的流量交替通过两条路径 AR1-AR2-AR3 和 AR1-AR4-AR3 进行传递。

2.2.4 默认路由

默认路由是指当路由表中没有更精确的路由匹配条目时所匹配的路由。本实验中，我们要把 AR1 和 AR3 上的两条静态路由删除，并通过默认路由实现 PC1 与 PC2 之间的通信，选用路径 AR1-AR2-AR3。

例 2-33 展示了 AR1 上的配置变更。我们需要在 AR1 上删除两条静态路由，这两条静态路由去往相同目的地，因此可以使用一条命令将其全部删除（我们可以使用 **undo ip route-static 172.16.30.0 24** 命令）。在执行了这条命令后，我们再次通过命令 **display ip routing-table protocol static** 查看当前 AR1 上配置的静态路由，会发现路由器没有返回任何信息。

例 2-33 AR1 上的配置变更

```
[AR1]undo ip route-static 172.16.30.0 24
[AR1]display ip routing-table protocol static
[AR1]
[AR1]ip route-static 0.0.0.0 0 10.10.12.2
```

读者可以按照 AR1 上的配置，在 AR3 上执行配置变更，将去往 PC1 的默认路由指向 AR2 的 G0/0/2 接口。例 2-34 中展示了 AR3 配置变更后的路由表。

例 2-34 AR3 路由表中的默认路由

```
[AR3]display ip routing-table protocol static
Route Flags: R - relay, D - download to fib
------------------------------------------------------------------------------
Public routing table : Static
        Destinations : 1        Routes : 1        Configured Routes : 1

Static routing table status : <Active>
        Destinations : 1        Routes : 1

Destination/Mask    Proto   Pre   Cost       Flags NextHop         Interface

     0.0.0.0/0     Static   60    0            RD  10.10.23.2      GigabitEthernet0/0/2

Static routing table status : <Inactive>
        Destinations : 0        Routes : 0
```

配置完成后，读者可以通过 ping 来测试配置结果，具体见例 2-35。

例 2-35 测试实验结果

```
PC1>ping 172.16.30.10

Ping 172.16.30.10: 32 data bytes, Press Ctrl_C to break
From 172.16.30.10: bytes=32 seq=1 ttl=125 time=47 ms
From 172.16.30.10: bytes=32 seq=2 ttl=125 time=63 ms
From 172.16.30.10: bytes=32 seq=3 ttl=125 time=46 ms
From 172.16.30.10: bytes=32 seq=4 ttl=125 time=47 ms
From 172.16.30.10: bytes=32 seq=5 ttl=125 time=78 ms

--- 172.16.30.10 ping statistics ---
  5 packet(s) transmitted
  5 packet(s) received
  0.00% packet loss
  round-trip min/avg/max = 46/56/78 ms
```

从本实验中我们可以看出，默认静态路由是一种特殊的静态路由，因此读者可以按照静态路由负载分担的方法，将 AR1-AR4-AR3 路径设置为默认路由，实现负载分担。本书将不展示这部分配置，感兴趣的读者可以在自己的实验环境中自行尝试进行配置和验证。

第 3 章
OSPF 路由协议基础实验

本章主要内容

3.1 实验介绍

3.2 实验配置任务

OSPF 协议是一种链路状态路由协议，适用于中大型网络环境。它使用开销（Cost）作为度量参数，每台参与 OSPF 路由的路由器都会通过 Dijkstra 算法计算出去往各个远端网段的最优路径（最短路径树）。计算出的结果保存在 LSDB（Link State Database，链路状态数据库）中，在同一个 OSPF 域中，同一个区域中的 LSDB 都是相同的。在 OSPF 的区域（Area）概念中，每个 OSPF 域中都有一个骨干区域（Area 0），并且其他区域都会与 Area 0 直接相连（否则需要额外的配置）。

OSPF 是在实践中广泛使用的 IGP（内部网关协议），在 OSPF 路由器之间能够相互交换路由信息前，它们需要先建立邻接关系。为了建立稳定状态的完全邻接关系，邻接的两台路由器上须匹配如下参数：

- Hello 时间间隔；
- Dead 时间间隔；
- 区域编号；
- 认证（若启用）；
- 链路 MTU（Maximum Transmission Unit，最大传输单元）；
- 子网号；
- 子网掩码。

在建立邻接关系的过程中，两台邻接路由器之间会经历以下状态。

① 失效状态（Down）：这是初始状态，表示还未从邻居路由器那里收到 Hello 包。

② 尝试状态（Attempt）：只适用于 NBMA（Non-Broadcast Multiple Access，非广播多路访问）链路，表示暂未从邻居路由器那里收到 Hello 包，但会以特定间隔向邻居发送 Hello 包进行再次尝试。

③ 初始状态（Init）：收到了邻居发来的 Hello 包，但未建立双向通信，即邻居发来的 Hello 包中不包含自己的路由器 ID。

④ 双向通信状态（2-Way）：双方已经建立了双向通信，在广播链路中，2-Way 状态下会进行 DR 和 BDR 的选举。

⑤ 信息交换初始状态（ExStart）：这是建立邻接关系的第一步，在此阶段会确定后续信息交换过程中的主/从关系，以及确定初始 DD（Database description，数据库描述）数据包的序列号。

⑥ 信息交换状态（ExChange）：在这个过程中路由器会向邻居发送 DD 数据包来描述自己完整的链路状态数据库。

⑦ 信息加载状态（Loading）：在这个状态中，路由器会向邻居发送 LSQ（Link State Request，链路状态请求）数据包，来请求上一步发现的最新 LSA（Link State Advertisements，链路状态通告）。

⑧ 完全邻接状态（Full）：这时邻居路由器形成了完全邻接关系。

本实验会通过单区域 OSPF 的配置，为读者介绍与 OSPF 相关的基础知识。

第 3 章 OSPF 路由协议基础实验

3.1 实验介绍

3.1.1 关于本实验

这个实验使用与静态路由实验相同的拓扑，以此突显动态路由协议的灵活性，帮助读者理解 OSPF 的基础概念、配置和排错。

3.1.2 实验目的

- 理解 OSPF 的基本概念。
- 掌握单区域 OSPF 的配置。
- 掌握 OSPF 邻居状态的解读。
- 掌握通过 Cost 控制 OSPF 选路的方法。
- 掌握 OSPF 默认路由发布的方法。
- 掌握 OSPF 认证的配置方法。

3.1.3 实验组网介绍

图 3-1 为单区域 OSPF 实验拓扑。

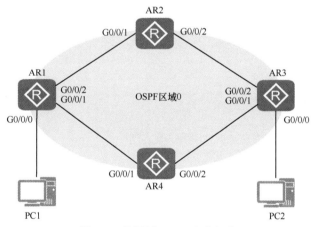

图 3-1 单区域 OSPF 实验拓扑

在图 3-1 所示的本实验拓扑中，4 台路由器上参与 OSPF 路由的接口都属于骨干区域（区域 0）。每台路由器上除了连接其他路由器的物理接口外，还各自建立了一个环回接口，环回接口的子网掩码为 32 位。本章使用的网络地址见表 3-1。

表 3-1 本章使用的网络地址

设备	接口	IP 地址	子网掩码	默认网关
AR1	G0/0/0	172.16.10.1	255.255.255.0	—
	G0/0/1	10.10.14.1	255.255.255.0	—
	G0/0/2	10.10.12.1	255.255.255.0	—
	Lo0	10.10.0.1	255.255.255.255	—

表 3-1 本章网络地址（续）

设备	接口	IP 地址	子网掩码	默认网关
AR2	G0/0/1	10.10.12.2	255.255.255.0	—
	G0/0/2	10.10.23.2	255.255.255.0	—
	Lo0	10.10.0.2	255.255.255.255	—
AR3	G0/0/0	172.16.30.3	255.255.255.0	—
	G0/0/1	10.10.34.3	255.255.255.0	—
	G0/0/2	10.10.23.3	255.255.255.0	—
	Lo0	10.10.0.3	255.255.255.255	—
AR4	G0/0/1	10.10.14.4	255.255.255.0	—
	G0/0/2	10.10.34.4	255.255.255.0	—
	Lo0	10.10.0.4	255.255.255.255	—
PC1	E0/0/1	172.16.10.10	255.255.255.0	172.16.10.1
PC2	E0/0/1	172.16.30.10	255.255.255.0	172.16.30.3

3.1.4 实验任务列表

配置任务 1：建立单区域 OSPF。

配置任务 2：通过开销进行选路。

配置任务 3：通过 OSPF 发布默认路由。

配置任务 4：配置 OSPF 认证。

3.2 实验配置任务

3.2.1 建立单区域 OSPF

本实验中，我们需要通过 OSPF 协议，让 4 台路由器分别学习两台 PC（Personal Computer，个人计算机）的网段，以及每台路由器的环回接口地址，实现全网 IP 互通。

步骤 1 构建直连连通性

读者先要构建如图 3-1 所示拓扑，然后按照表 3-1 中的地址，配置路由器的物理接口并实现直连互通，最后配置环回接口待后续使用。

例 3-1～例 3-4 展示了路由器 AR1、AR2、AR3 和 AR4 上的基础 IP 地址配置。

例 3-1 AR1 上的基础配置

```
<Huawei>system-view
Enter system view, return user view with Ctrl+Z.
[Huawei]sysname AR1
[AR1]interface GigabitEthernet 0/0/0
[AR1-GigabitEthernet0/0/0]ip address 172.16.10.1 24
[AR1-GigabitEthernet0/0/0]quit
[AR1]interface GigabitEthernet 0/0/1
[AR1-GigabitEthernet0/0/1]ip address 10.10.14.1 24
[AR1-GigabitEthernet0/0/1]quit
[AR1]interface GigabitEthernet 0/0/2
[AR1-GigabitEthernet0/0/2]ip address 10.10.12.1 24
[AR1-GigabitEthernet0/0/2]quit
[AR1]interface LoopBack 0
[AR1-LoopBack0]ip address 10.10.0.1 32
```

例 3-2　AR2 上的基础配置

```
<Huawei>system-view
Enter system view, return user view with Ctrl+Z.
[Huawei]sysname AR2
[AR2]interface GigabitEthernet 0/0/1
[AR2-GigabitEthernet0/0/1]ip address 10.10.12.2 24
[AR2-GigabitEthernet0/0/1]quit
[AR2]interface GigabitEthernet 0/0/2
[AR2-GigabitEthernet0/0/2]ip address 10.10.23.2 24
[AR2-GigabitEthernet0/0/2]quit
[AR2]interface LoopBack 0
[AR2-LoopBack0]ip address 10.10.0.2 32
```

例 3-3　AR3 上的基础配置

```
<Huawei>system-view
Enter system view, return user view with Ctrl+Z.
[Huawei]sysname AR3
[AR3]interface GigabitEthernet 0/0/0
[AR3-GigabitEthernet0/0/0]ip address 172.16.30.3 24
[AR3-GigabitEthernet0/0/0]quit
[AR3]interface GigabitEthernet 0/0/1
[AR3-GigabitEthernet0/0/1]ip address 10.10.34.3 24
[AR3-GigabitEthernet0/0/1]quit
[AR3]interface GigabitEthernet 0/0/2
[AR3-GigabitEthernet0/0/2]ip address 10.10.23.3 24
[AR3-GigabitEthernet0/0/2]quit
[AR3]interface LoopBack 0
[AR3-LoopBack0]ip address 10.10.0.3 32
```

例 3-4　AR4 上的基础配置

```
<Huawei>system-view
Enter system view, return user view with Ctrl+Z.
[Huawei]sysname AR4
[AR4]interface GigabitEthernet 0/0/1
[AR4-GigabitEthernet0/0/1]ip address 10.10.14.4 24
[AR4-GigabitEthernet0/0/1]quit
[AR4]interface GigabitEthernet 0/0/2
[AR4-GigabitEthernet0/0/2]ip address 10.10.34.4 24
[AR4-GigabitEthernet0/0/2]quit
[AR4]interface LoopBack 0
[AR4-LoopBack0]ip address 10.10.0.4 32
```

配置完成后，读者可以按照前文中介绍的 ping 测试方法来测试直连连接是否已经建立。测试前要在 PC1 和 PC2 上配置 IP 地址、子网掩码和网关。

环回接口是虚拟接口，自属于一个网段，在其他路由器上需要配置静态路由或动态路由才能访问环回接口。环回接口有多种用途，常见用途之一就是我们在实验中会用到的以环回接口作为 OSPF 路由器 ID。

我们以 AR2 为例查看当前 AR2 上的路由表，具体见例 3-5。当前路由器上既没有配置静态路由，也没有配置动态路由协议，因此路由表中只包含直连路由。

例 3-5　AR2 的路由表（只包含直连条目）

```
[AR2]display ip routing-table
Route Flags: R - relay, D - download to fib
------------------------------------------------------------------------------
Routing Tables: Public
         Destinations : 11       Routes : 11
Destination/Mask    Proto   Pre  Cost      Flags NextHop         Interface
      10.10.0.2/32  Direct  0    0           D   127.0.0.1       LoopBack0
     10.10.12.0/24  Direct  0    0           D   10.10.12.2      GigabitEthernet0/0/1
     10.10.12.2/32  Direct  0    0           D   127.0.0.1       GigabitEthernet0/0/1
   10.10.12.255/32  Direct  0    0           D   127.0.0.1       GigabitEthernet0/0/1
     10.10.23.0/24  Direct  0    0           D   10.10.23.2      GigabitEthernet0/0/2
     10.10.23.2/32  Direct  0    0           D   127.0.0.1       GigabitEthernet0/0/2
   10.10.23.255/32  Direct  0    0           D   127.0.0.1       GigabitEthernet0/0/2
      127.0.0.0/8   Direct  0    0           D   127.0.0.1       InLoopBack0
      127.0.0.1/32  Direct  0    0           D   127.0.0.1       InLoopBack0
  127.255.255.255/32 Direct 0    0           D   127.0.0.1       InLoopBack0
  255.255.255.255/32 Direct 0    0           D   127.0.0.1       InLoopBack0
```

步骤 2 让 AR1、AR2 和 AR3 加入 OSPF 域

我们先在 AR1、AR2 和 AR3 这 3 台路由器上启用 OSPF 进程，以此实现 PC1 与 PC2 之间的通信。配置 OSPF 的步骤如下。

① 创建并进入 OSPF 进程：使用系统视图命令 **ospf** [*process-id* | **router-id** *router-id*]。

② 创建并进入 OSPF 区域：使用 OSPF 协议视图命令 **area** *area-id*。

③ 将接口加入 OSPF 区域：可以使用 OSPF 区域视图命令 **network** *ip-address wildcard-mask* 将属于配置子网中的接口加入这个 OSPF 区域中，或者使用接口视图命令 **ospf enable** [*process-id*] **area** *area-id* 将接口加入 OSPF 区域中。

本实验会以 3 种不同的方式在 AR1、AR2 和 AR3 上启用 OSPF 协议，使其之间建立 OSPF 邻居关系并实现 PC1 与 PC2 之间的相互通信。

我们先在 AR1 上配置 OSPF，具体见例 3-6。

例 3-6 在 AR1 上配置 OSPF

```
[AR1]ospf 1 router-id 10.10.0.1
[AR1-ospf-1]area 0
[AR1-ospf-1-area-0.0.0.0]network 172.16.10.0 0.0.0.255
[AR1-ospf-1-area-0.0.0.0]network 10.10.12.0 0.0.0.255
[AR1-ospf-1-area-0.0.0.0]network 10.10.14.0 0.0.0.255
[AR1-ospf-1-area-0.0.0.0]network 10.10.0.1 0.0.0.0
```

在 AR1 上，用户通过系统视图命令 **ospf 1 router-id 10.10.0.1** 配置了 OSPF 进程号 1，并将路由器 ID 设置为 AR1 的环回接口地址。OSPF 进程号只具有本地意义，这说明了以下两点。第一，同一个 OSPF 域中的多台路由器上可以使用不同的 OSPF 进程号。本实验所使用的 OSPF 进程号选择与路由器相同的编号，即 AR1 的 OSPF 进程号为 1，AR2 的 OSPF 进程号为 2，以此类推。本例只为展示说明 OSPF 进程号仅为本地有效，并不是推荐配置，在实际工作中，仍建议读者在同一个 OSPF 域中使用相同的 OSPF 进程号。第二，同一台路由器上可以同时运行多个 OSPF 进程，即同时参与多个 OSPF 域的工作。

我们再使用 OSPF 协议视图的命令 **area 0** 进入 OSPF 区域视图。区域编号有两种配置方法，一种是本例中使用的十进制，另一种是点分十进制，也就是与 IP 地址的格式相同。区域 0 可以表示为区域 0.0.0.0。

在 OSPF 区域视图中，用户可以使用 **network** 命令在接口上启用 OSPF，或者与例 3-7 中 AR2 的配置一样，在接口视图中进行启用。对于 **network** 命令，我们需要注意它的语法和取值。就语法而言，IP 子网后面是通配符掩码，而不是子网掩码。就取值而言，接口 IP 地址的掩码长度要大于等于 **network** 命令中使用的掩码长度。

另外，对于环回接口，OSPF 默认会以 32 位主机路由的方式对外发布，与环回接口上配置的子网掩码无关。

接下来用户在 AR2 上通过接口模式启用 OSPF，具体见例 3-7。

例 3-7 在 AR2 上配置 OSPF

```
[AR2]ospf 2 router-id 10.10.0.2
[AR2-ospf-2]area 0
[AR2-ospf-2-area-0.0.0.0]quit
[AR2-ospf-2]quit
[AR2]interface GigabitEthernet 0/0/1
[AR2-GigabitEthernet0/0/1]ospf enable 2 area 0
[AR2-GigabitEthernet0/0/1]quit
[AR2]interface GigabitEthernet 0/0/2
[AR2-GigabitEthernet0/0/2]ospf enable 2 area 0
[AR2-GigabitEthernet0/0/2]quit
[AR2]interface LoopBack 0
[AR2-LoopBack0]ospf enable 2 area 0
```

例 3-7 中展示了 AR2 中的完整配置，从中我们可以看出，用户使用了 OSPF 进程号 2，并且在要启用 OSPF 的接口视图中，用户使用命令 **ospf enable** 对其进行启用。一台路由器上可以运行多个 OSPF 进程，并且一个 OSPF 进程中可能会有多个区域，因此在这条命令中，用户需要明确指定进程号和区域号。

事实上，AR1 的 G0/0/2 接口上已经启用了 OSPF，因此当 AR2 上与 AR1 直连的 G0/0/1 接口启用了 OSPF 后，AR1 和 AR2 之间就会启动建立 OSPF 邻居关系的流程，具体见例 3-8。这个提示输出中包含较多信息，我们着重关注最后两个参数，以第一个和最后一个消息为例，第一个消息的最后两个参数表示邻居关系的前一个状态为 Down，当前状态为 Init，且 OSPF 的邻居关系建立过程已经开始。最后一个消息表示前一个状态为 Loading，当前状态为 Full，邻居关系建立完成。

例 3-8　AR2 与 AR1 协商建立 OSPF 邻居关系的过程

```
[AR2]
Mar 18 2021 02:32:06-08:00 AR2 %%01OSPF/4/NBR_CHANGE_E(l)[2]:Neighbor changes ev
ent: neighbor status changed. (ProcessId=512, NeighborAddress=1.12.10.10, Neighb
orEvent=HelloReceived, NeighborPreviousState=Down, NeighborCurrentState=Init)
[AR2]
Mar 18 2021 02:32:06-08:00 AR2 %%01OSPF/4/NBR_CHANGE_E(l)[3]:Neighbor changes ev
ent: neighbor status changed. (ProcessId=512, NeighborAddress=1.12.10.10, Neighb
orEvent=2WayReceived, NeighborPreviousState=Init, NeighborCurrentState=2Way)
[AR2]
Mar 18 2021 02:32:06-08:00 AR2 %%01OSPF/4/NBR_CHANGE_E(l)[4]:Neighbor changes ev
ent: neighbor status changed. (ProcessId=512, NeighborAddress=1.12.10.10, Neighb
orEvent=AdjOk?, NeighborPreviousState=2Way, NeighborCurrentState=ExStart)
[AR2]
Mar 18 2021 02:32:06-08:00 AR2 %%01OSPF/4/NBR_CHANGE_E(l)[5]:Neighbor changes ev
ent: neighbor status changed. (ProcessId=512, NeighborAddress=1.12.10.10, Neighb
orEvent=NegotiationDone, NeighborPreviousState=ExStart, NeighborCurrentState=Exc
hange)
[AR2]
Mar 18 2021 02:32:06-08:00 AR2 %%01OSPF/4/NBR_CHANGE_E(l)[6]:Neighbor changes ev
ent: neighbor status changed. (ProcessId=512, NeighborAddress=1.12.10.10, Neighb
orEvent=ExchangeDone, NeighborPreviousState=Exchange, NeighborCurrentState=Loadi
ng)
[AR2]
Mar 18 2021 02:32:06-08:00 AR2 %%01OSPF/4/NBR_CHANGE_E(l)[7]:Neighbor changes ev
ent: neighbor status changed. (ProcessId=512, NeighborAddress=1.12.10.10, Neighb
orEvent=LoadingDone, NeighborPreviousState=Loading, NeighborCurrentState=Full)
[AR2]
```

如例 3-9 所示，我们在 AR2 上使用 **display ospf 2 peer brief** 查看 AR2 上当前的 OSPF 邻居。从这条命令中我们可以看到，AR2 所使用的路由器 ID（10.10.0.2）、AR2 通过哪个本地接口（G0/0/1）与邻居建立的关系、邻居是以邻居的路由器 ID 表示的（10.10.0.1），以及与这个邻居之间的状态。

例 3-9　在 AR2 上查看 OSPF 邻居 AR1

```
[AR2]display ospf 2 peer brief

         OSPF Process 2 with Router ID 10.10.0.2
                 Peer Statistic Information
 ----------------------------------------------------------------------------
 Area Id         Interface                    Neighbor id      State
 0.0.0.0         GigabitEthernet0/0/1         10.10.0.1        Full
 ----------------------------------------------------------------------------
```

如例 3-10 所示，用户在 AR3 上配置 OSPF，在本例中使用了一种偷懒的方法在接口上启用 OSPF：OSPF 区域视图中，在 **network** 后以 0.0.0.0 0.0.0.0 匹配所有 IP 子网，因此可以把路由器上的所有接口都启用 OSPF。这种方法虽然简单，但在实际工作中，管理员需要根据实际情况谨慎选择。

例 3-10 在 AR3 上配置 OSPF

```
[AR3]ospf 3 router-id 10.10.0.3
[AR3-ospf-3]area 0
[AR3-ospf-3-area-0.0.0.0]network 0.0.0.0 0.0.0.0
[AR3-ospf-3-area-0.0.0.0]return
<AR3>
Mar 18 2021 04:38:58-08:00 AR3 %%01OSPF/4/NBR_CHANGE_E(l)[3]:Neighbor changes ev
ent: neighbor status changed. (ProcessId=768, NeighborAddress=2.23.10.10, Neighb
orEvent=HelloReceived, NeighborPreviousState=Down, NeighborCurrentState=Init)
<AR3>
Mar 18 2021 04:38:58-08:00 AR3 %%01OSPF/4/NBR_CHANGE_E(l)[4]:Neighbor changes ev
ent: neighbor status changed. (ProcessId=768, NeighborAddress=2.23.10.10, Neighb
orEvent=2WayReceived, NeighborPreviousState=Init, NeighborCurrentState=2Way)
<AR3>
Mar 18 2021 04:38:58-08:00 AR3 %%01OSPF/4/NBR_CHANGE_E(l)[5]:Neighbor changes ev
ent: neighbor status changed. (ProcessId=768, NeighborAddress=2.23.10.10, Neighb
orEvent=AdjOk?, NeighborPreviousState=2Way, NeighborCurrentState=ExStart)
<AR3>
Mar 18 2021 04:38:58-08:00 AR3 %%01OSPF/4/NBR_CHANGE_E(l)[6]:Neighbor changes ev
ent: neighbor status changed. (ProcessId=768, NeighborAddress=2.23.10.10, Neighb
orEvent=NegotiationDone, NeighborPreviousState=ExStart, NeighborCurrentState=Exc
hange)
<AR3>
Mar 18 2021 04:38:58-08:00 AR3 %%01OSPF/4/NBR_CHANGE_E(l)[7]:Neighbor changes ev
ent: neighbor status changed. (ProcessId=768, NeighborAddress=2.23.10.10, Neighb
orEvent=ExchangeDone, NeighborPreviousState=Exchange, NeighborCurrentState=Loadi
ng)
<AR3>
Mar 18 2021 04:38:58-08:00 AR3 %%01OSPF/4/NBR_CHANGE_E(l)[8]:Neighbor changes ev
ent: neighbor status changed. (ProcessId=768, NeighborAddress=2.23.10.10, Neighb
orEvent=LoadingDone, NeighborPreviousState=Loading, NeighborCurrentState=Full)
<AR3>
```

例 3-10 中还展示了 AR3 与 AR2 之间 OSPF 的建立过程，由此我们可以知道 AR3 的 G0/0/2 已经成功启用了 OSPF，要想查看其他接口的状态，用户输入了命令 **display ospf 3 interface all**，查看路由器上所有启用了 OSPF 的接口，具体见例 3-11。

例 3-11 查看 AR3 上启用了 OSPF 的接口

```
[AR3]display ospf 3 interface all
        OSPF Process 3 with Router ID 10.10.0.3
            Interfaces

 Area: 0.0.0.0         (MPLS TE not enabled)
 Interface: 172.16.30.3 (GigabitEthernet0/0/0)
 Cost: 1       State: DR      Type: Broadcast    MTU: 1500
 Priority: 1
 Designated Router: 172.16.30.3
 Backup Designated Router: 0.0.0.0
 Timers: Hello 10 , Dead 40 , Poll  120 , Retransmit 5 , Transmit Delay 1

 Interface: 10.10.34.3 (GigabitEthernet0/0/1)
 Cost: 1       State: DR      Type: Broadcast    MTU: 1500
 Priority: 1
 Designated Router: 10.10.34.3
 Backup Designated Router: 0.0.0.0
 Timers: Hello 10 , Dead 40 , Poll  120 , Retransmit 5 , Transmit Delay 1

 Interface: 10.10.23.3 (GigabitEthernet0/0/2)
 Cost: 1       State: BDR     Type: Broadcast    MTU: 1500
 Priority: 1
 Designated Router: 10.10.23.2
 Backup Designated Router: 10.10.23.3
 Timers: Hello 10 , Dead 40 , Poll  120 , Retransmit 5 , Transmit Delay 1

 Interface: 10.10.0.3 (LoopBack0)
 Cost: 0       State: P-2-P   Type: P2P          MTU: 1500
 Timers: Hello 10 , Dead 40 , Poll  120 , Retransmit 5 , Transmit Delay 1
```

我们以例 3-11 中 AR3 的 G0/0/2 接口（连接 AR2）为例，逐一解释每个参数。每个接口显示区域的第一行以 IP 地址和接口编号开始，参数如下。

① **Cost**（开销）：OSPF 会根据接口的带宽自动计算开销值（接口开销 = 带宽参考值/接口带宽），并取计算结果的整数部分作为接口开销值（当结果小于 1 时取 1）。用户可以通过命令改变某个接口的开销，也可以通过改变带宽参考值来进行统一更改。但要注意，用户在更改带宽参考值时，要保证在 OSPF 域中的所有 OSPF 路由器上都进行相同的更改。后面会通过实验介绍更改接口开销的命令。

② **State**（状态）：AR3 是当前这个网段中的 BDR（Backup Designated Router，备份指定路由器）。DR（Designated Router，指定路由器）/BDR 的选举规则如下。① 接口的 DR 优先级值高者赢得选举；② 若 DR 优先级相同，则路由器 ID 值高者赢得选举。一个 OSPF 域中路由器 ID 是唯一的，因此一定会有选举结果。

本例中，AR3 的路由器 ID 10.10.0.3 大于 AR2 的路由器 ID 10.10.0.2，但 AR2 是第一台存在于这个广播网段中的 OSPF 路由器，因此它便优先占据了 DR 的角色。后续即使上线了路由器 ID 值更高的路由器（如 AR3），也只能成为 BDR。这是因为为了维持网络的稳定，DR 和 BDR 都不具有抢占特性。若当前 BDR 下线，网络中会进行 BDR 选举；若当前 DR 下线，当前 BDR 会成为 DR，网络中会进行 BDR 的选举。为了使 DR/BDR 的选举更具可控性，管理员可以通过指定 DR 优先级影响选举结果。

③ **Type**（类型）：链路层协议为以太网时，OSPF 网络类型为广播（Broadcast）；当链路层协议是 PPP、HDLC 和 LAPB 时，OSPF 网络类型为点到点（P-2-P），从命令输出中我们也可以看出，环回接口同样为点到点；其他类型还包括 NBMA 和点到多点。当链路层协议是帧中继、X.25 时，OSPF 网络类型为 NBMA；点到多点只能是管理员手动更改的结果，通常会根据需要把非全连接的 NBMA 网络类型更改为点到多点。

④ **MTU**：链路 MTU，前文中已介绍过它属于建立 OSPF 邻居关系必须匹配的参数之一。若该参数不匹配，OSPF 邻居关系会停留在 ExStart 状态，无法进入后续阶段并形成完全邻接关系。

⑤ **Priority**（优先级）：接口的 DR 优先级，管理员可以通过接口命令 **ospf dr-priority** *priority* 进行更改，默认值为 1，值越大优先级越高。

⑥ **Designated Router**（指定路由器）和 **Backup Designated Router**（备份指定路由器）：显示出这个广播网络中 DR 和 BDR 的接口 IP 地址。

⑦ **Timers**（计时器）：Hello 计时器、Dead 计时器、轮询计时器、重传计时器、传输延迟。这里读者需要着重关注前两个计时器，要想建立稳定的邻居关系，邻居之间的 Hello 计时器和 Dead 计时器要相同。

至此我们已经完成了部分 OSPF 配置，例 3-9 中展示了如何快速查看 OSPF 邻居汇总信息，用户还可以使用 **display ospf peer** 命令来查看邻居详细信息，具体见例 3-12。

例 3-12 在 AR3 上查看 OSPF 邻居详细信息

```
[AR3]display ospf peer
         OSPF Process 3 with Router ID 10.10.0.3
                 Neighbors

 Area 0.0.0.0 interface 10.10.23.3(GigabitEthernet0/0/2)'s neighbors
 Router ID: 10.10.0.2        Address: 10.10.23.2
   State: Full  Mode:Nbr is  Slave  Priority: 1
```

```
DR: 10.10.23.2  BDR: 10.10.23.3  MTU: 0
Dead timer due in 35  sec
Retrans timer interval: 5
Neighbor is up for 00:00:19
Authentication Sequence: [ 0 ]
```

从例 3-12 中我们可以看出，AR3 上目前只有一个 OSPF 邻居，这条命令的输出内容中有一些是与命令 **display ospf 3 interface all** 的输出内容相同的。需要注意的是，这条命令中的 Dead 计时器不是默认的 40 秒，这是因为它是活动的倒计时值，若 35 秒内还没有收到下一个 Hello 数据包，OSPF 就会认为邻居已失效。另外从这条命令中还可以看出 OSPF 邻居已经建立了多长时间，本例为 19 秒。

接下来我们查看路由。PC1 与 PC2 之间应该可以通过 AR1-AR2-AR3 路径进行通信。我们先在路径中的 AR2 上查看路由表中的 OSPF 路由，具体见例 3-13。

例 3-13 在 AR2 上查看路由表中的 OSPF 路由

```
[AR2]display ip routing-table protocol ospf
Route Flags: R - relay, D - download to fib
------------------------------------------------------------------------
Public routing table : OSPF
        Destinations : 6        Routes : 6

OSPF routing table status : <Active>
        Destinations : 6        Routes : 6

Destination/Mask    Proto   Pre  Cost        Flags NextHop         Interface

      10.10.0.1/32  OSPF    10   1             D   10.10.12.1      GigabitEthernet0/0/1
      10.10.0.3/32  OSPF    10   1             D   10.10.23.3      GigabitEthernet0/0/2
     10.10.14.0/24  OSPF    10   2             D   10.10.12.1      GigabitEthernet0/0/1
     10.10.34.0/24  OSPF    10   2             D   10.10.23.3      GigabitEthernet0/0/2
     172.16.10.0/24 OSPF    10   2             D   10.10.12.1      GigabitEthernet0/0/1
     172.16.30.0/24 OSPF    10   2             D   10.10.23.3      GigabitEthernet0/0/2

OSPF routing table status : <Inactive>
        Destinations : 0        Routes : 0
```

例 3-13 中使用命令 **display ip routing-table protocol ospf** 筛选出了通过 OSPF 学到的路由。我们可以发现，AR2 已经通过 OSPF 学习到了所有非直连的网段（不包括与 AR4 相关网段），这包括 AR1 和 AR3 的环回接口、AR1 和 AR3 与 AR4 之间连接的链路，以及 AR1 和 AR3 与 PC1 和 PC2 相连的网段。现在我们在 PC1 上对 PC2 发起 ping 命令进行测试，具体见例 3-14。

例 3-14 从 PC1 向 PC2 发起 ping 测试

```
PC1>ping 172.16.30.10

Ping 172.16.30.10: 32 data bytes, Press Ctrl_C to break
From 172.16.30.10: bytes=32 seq=1 ttl=125 time=46 ms
From 172.16.30.10: bytes=32 seq=2 ttl=125 time=62 ms
From 172.16.30.10: bytes=32 seq=3 ttl=125 time=31 ms
From 172.16.30.10: bytes=32 seq=4 ttl=125 time=31 ms
From 172.16.30.10: bytes=32 seq=5 ttl=125 time=47 ms

--- 172.16.30.10 ping statistics ---
  5 packet(s) transmitted
  5 packet(s) received
  0.00% packet loss
  round-trip min/avg/max = 31/43/62 ms
```

步骤 3 将 AR4 加入 OSPF 域

我们通过这一步着重观察 OSPF 中的等价路径。首先在 AR4 中配置 OSPF，具体见例 3-15。

例 3-15 在 AR4 上配置 OSPF

```
[AR4]ospf 4 router-id 10.10.0.4
[AR4-ospf-4]area 0
[AR4-ospf-4-area-0.0.0.0]network 10.10.14.0 0.0.0.255
[AR4-ospf-4-area-0.0.0.0]network 10.10.34.0 0.0.0.255
[AR4-ospf-4-area-0.0.0.0]network 10.10.0.4 0.0.0.0
```

验证 AR4 上的 OSPF 邻居关系，具体见例 3-16。AR4 已经与两个邻居建立了完全的 OSPF 邻接关系。

例 3-16 查看 AR4 上的 OSPF 邻居关系

```
[AR4]display ospf peer brief
         OSPF Process 4 with Router ID 10.10.0.4
                Peer Statistic Information
 ----------------------------------------------------------------
 Area Id          Interface                    Neighbor id       State
 0.0.0.0          GigabitEthernet0/0/1         10.10.0.1         Full
 0.0.0.0          GigabitEthernet0/0/2         10.10.0.3         Full
 ----------------------------------------------------------------
```

从图 3-1 中我们可以看出，PC1 去往 PC2 可以采用两条路径。我们通过 AR1 和 AR3 的路由表进行验证。例 3-17 和例 3-18 分别展示了 AR1 和 AR3 路由表中的 OSPF 路由。

例 3-17 查看 AR1 路由表中的 OSPF 路由

```
[AR1]display ip routing-table protocol ospf
Route Flags: R - relay, D - download to fib
----------------------------------------------------------------
Public routing table : OSPF
        Destinations : 6        Routes : 8

OSPF routing table status : <Active>
        Destinations : 6        Routes : 8

Destination/Mask    Proto   Pre  Cost      Flags NextHop        Interface

    10.10.0.2/32    OSPF    10   1            D  10.10.12.2     GigabitEthernet0/0/2
    10.10.0.3/32    OSPF    10   2            D  10.10.12.2     GigabitEthernet0/0/2
                    OSPF    10   2            D  10.10.14.4     GigabitEthernet0/0/1
    10.10.0.4/32    OSPF    10   1            D  10.10.14.4     GigabitEthernet0/0/1
    10.10.23.0/24   OSPF    10   2            D  10.10.12.2     GigabitEthernet0/0/2
    10.10.34.0/24   OSPF    10   2            D  10.10.14.4     GigabitEthernet0/0/1
    172.16.30.0/24  OSPF    10   3            D  10.10.12.2     GigabitEthernet0/0/2
                    OSPF    10   3            D  10.10.14.4     GigabitEthernet0/0/1

OSPF routing table status : <Inactive>
        Destinations : 0        Routes : 0
```

例 3-18 查看 AR3 路由表中的 OSPF 路由

```
[AR3]display ip routing-table protocol ospf
Route Flags: R - relay, D - download to fib
----------------------------------------------------------------
Public routing table : OSPF
        Destinations : 6        Routes : 8

OSPF routing table status : <Active>
        Destinations : 6        Routes : 8

Destination/Mask    Proto   Pre  Cost      Flags NextHop        Interface

    10.10.0.1/32    OSPF    10   2            D  10.10.34.4     GigabitEthernet0/0/1
                    OSPF    10   2            D  10.10.23.2     GigabitEthernet0/0/2
    10.10.0.2/32    OSPF    10   1            D  10.10.23.2     GigabitEthernet0/0/2
    10.10.0.4/32    OSPF    10   1            D  10.10.34.4     GigabitEthernet0/0/1
    10.10.12.0/24   OSPF    10   2            D  10.10.23.2     GigabitEthernet0/0/2
    10.10.14.0/24   OSPF    10   2            D  10.10.34.4     GigabitEthernet0/0/1
    172.16.10.0/24  OSPF    10   3            D  10.10.23.2     GigabitEthernet0/0/2
                    OSPF    10   3            D  10.10.34.4     GigabitEthernet0/0/1

OSPF routing table status : <Inactive>
        Destinations : 0        Routes : 0
```

我们通过例 3-18 对静态路由和 OSPF 路由进行比较。首先读者可以注意到 Pre（优先级）一列，OSPF 的路由优先级默认是 10，静态路由的路由优先级默认是 60，优先级值越小，路由优先级越高，直连路由的路由优先级为 0。其次，OSPF 路由比静态路由多了 Cost 值，这个值是这条路径上所有出接口开销的总和。例如，经过计算，GigabitEthernet 接口的 Cost 值为 1，本例的路由器使用的都是 GigabitEthernet 接口，因此本例中每个接口的 OSPF Cost 值都为 1。以 10.10.12.0/24 这条路由为例，AR3 在向这个网段转发数据包时，数据包会经过 AR3 的 G0/0/2 出接口、AR2 的 G0/0/1 出接口，因此这两个接口的开销值相加就是这条路由的开销，即 1+1=2。

另外，OSPF 通过计算发现去往同一目的地的两条路由的开销值相等时，OSPF 会自动实现负载分担。在下一个实验中，管理员要通过手动调整接口的 OSPF 开销，来影响路由器的选路，使两条负载分担路由区分为一条主用路由、一条备用路由。

至此单区域 OSPF 就配置完成，接下来我们通过实验查看 OSPF 的一些特性和配置选项。

3.2.2　通过开销进行选路

本实验中，管理员要通过 AR1 上的配置，使 AR1 优先使用 AR1-AR2-AR3 路径转发去往 PC2 的数据包，将 AR1-AR4-AR3 路径作为去往 PC2 的备用路径，当 AR1 与 AR2 之间的链路断开后，网络中的路由器会自动启用备用路径，如图 3-2 所示。

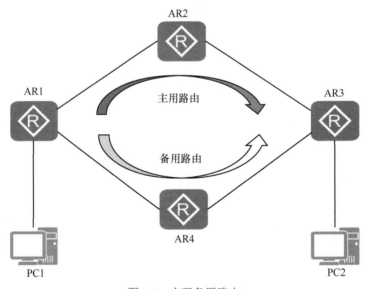

图 3-2　实现备用路由

步骤 1　更改接口的 OSPF 开销

为了影响 AR1 的 OSPF 选路，使 AR1-AR4-AR3 路径成为备用路由，用户可以增加 AR1 连接 AR4 的接口（G0/0/1）的 OSPF 开销，使 AR1-AR2-AR3 路径成为开销最低的最优路由。用户可以使用接口视图命令 ***ospf cost*** *cost* 来更改接口的 OSPF 开销，将其更改为 5，命令见例 3-19。

例 3-19　更改 AR1 G0/0/1 接口的 OSPF 开销为 5

```
[AR1]interface GigabitEthernet 0/0/1
[AR1-GigabitEthernet0/0/1]ospf cost 5
```

这时我们再查看 AR1 路由表中的 OSPF 路由条目，具体见例 3-20。读者可以将其与例 3-17 中的 OSPF 路由进行比较，会发现去往 10.10.0.3/32 和 172.16.30.0/24 的两条负载分担路由只剩下以 AR2 为下一跳的路由了。

步骤 2　验证路由表中的选路

例 3-20　查看变更后 AR1 路由表中的 OSPF 路由

```
[AR1]display ip routing-table protocol ospf
Route Flags: R - relay, D - download to fib
------------------------------------------------------------------------------
Public routing table : OSPF
         Destinations : 6        Routes : 6

OSPF routing table status : <Active>
         Destinations : 6        Routes : 6

Destination/Mask    Proto   Pre  Cost      Flags NextHop         Interface

    10.10.0.2/32    OSPF    10   1           D   10.10.12.2      GigabitEthernet0/0/2
    10.10.0.3/32    OSPF    10   2           D   10.10.12.2      GigabitEthernet0/0/2
    10.10.0.4/32    OSPF    10   3           D   10.10.12.2      GigabitEthernet0/0/2
    10.10.23.0/24   OSPF    10   2           D   10.10.12.2      GigabitEthernet0/0/2
    10.10.34.0/24   OSPF    10   3           D   10.10.12.2      GigabitEthernet0/0/2
   172.16.30.0/24   OSPF    10   3           D   10.10.12.2      GigabitEthernet0/0/2

OSPF routing table status : <Inactive>
         Destinations : 0        Routes : 0
```

除了去往 10.10.0.3/32 和 172.16.30.0/24 的两条路由发生变化外，读者通过比较还可以看出去往 AR3 与 AR4 之间 10.10.34.0/24 网段的路由也发生了变化。去往这个目的地的路由从下一跳接口 G0/0/1 变为了 G0/0/2，并且开销值从 2 变为了 3。

另外，我们可以从另一个角度来进一步检查上述操作得到的结果，读者可以查看 AR3 的路由表，会发现它依然会使用两条负载分担路由去往 PC1，具体见例 3-21。

例 3-21　查看 AR3 路由表中的 OSPF 路由

```
[AR3]display ip routing-table protocol ospf
Route Flags: R - relay, D - download to fib
------------------------------------------------------------------------------
Public routing table : OSPF
         Destinations : 6        Routes : 8

OSPF routing table status : <Active>
         Destinations : 6        Routes : 8

Destination/Mask    Proto   Pre  Cost      Flags NextHop         Interface

    10.10.0.1/32    OSPF    10   2           D   10.10.23.2      GigabitEthernet0/0/2
                    OSPF    10   2           D   10.10.34.4      GigabitEthernet0/0/1
    10.10.0.2/32    OSPF    10   1           D   10.10.23.2      GigabitEthernet0/0/2
    10.10.0.4/32    OSPF    10   1           D   10.10.34.4      GigabitEthernet0/0/1
    10.10.12.0/24   OSPF    10   2           D   10.10.23.2      GigabitEthernet0/0/2
    10.10.14.0/24   OSPF    10   2           D   10.10.34.4      GigabitEthernet0/0/1
   172.16.10.0/24   OSPF    10   3           D   10.10.23.2      GigabitEthernet0/0/2
                    OSPF    10   3           D   10.10.34.4      GigabitEthernet0/0/1

OSPF routing table status : <Inactive>
         Destinations : 0        Routes : 0
```

如此可见，图 3-2 中又出现了不对等路由，我们可以通过额外配置，令 AR3 也将 AR3-AR2-AR1 当作去往 PC1 的路径首选路由，将 AR3-AR4-AR1 作为备用路由。读者可以在自己的实验环境中尝试对不对等路由进行优化，此处不再赘述。

步骤 3 验证备用路由的效果

读者可以在 PC1 上使用命令 **ping 172.16.30.10 -t** 对 PC2 发起持续 ping 测试。在此期间关闭 AR1 的 G0/0/2 接口（使用接口视图命令 **shutdown**）。观察 ping 是否会中断。

例 3-22 在 AR1 上关闭了 G0/0/2 接口，与静态浮动路由的配置相同，这个操作会让 AR1 启用去往 PC2 的备选路由（通过 AR4 去往 PC2）。例 3-22 中展示了两部分内容，第一部分展示了关闭接口的配置，第二部分展示了 OSPF 邻居关系的变化。

例 3-22 关闭 AR1 的 G0/0/2 接口

```
[AR1]interface GigabitEthernet 0/0/2
[AR1-GigabitEthernet0/0/2]shutdown
[AR1-GigabitEthernet0/0/2]quit
[AR1]
Mar 18 2021 06:54:06-08:00 AR1 %%01IFPDT/4/IF_STATE(l)[62]:Interface GigabitEthe
rnet0/0/2 has turned into DOWN state.
[AR1]
Mar 18 2021 06:54:06-08:00 AR1 %%01IFNET/4/LINK_STATE(l)[63]:The line protocol I
P on the interface GigabitEthernet0/0/2 has entered the DOWN state.
[AR1]
Mar 18 2021 06:54:06-08:00 AR1 %%01OSPF/3/NBR_CHG_DOWN(l)[64]:Neighbor event:nei
ghbor state changed to Down. (ProcessId=256, NeighborAddress=2.0.10.10, Neighbor
Event=KillNbr, NeighborPreviousState=Full, NeighborCurrentState=Down)
[AR1]
Mar 18 2021 06:54:06-08:00 AR1 %%01OSPF/3/NBR_DOWN_REASON(l)[65]:Neighbor state
leaves full or changed to Down. (ProcessId=256, NeighborRouterId=2.0.10.10, Neig
hborAreaId=0, NeighborInterface=GigabitEthernet0/0/2,NeighborDownImmediate reaso
n=Neighbor Down Due to Kill Neighbor, NeighborDownPrimeReason=Physical Interface
 State Change, NeighborChangeTime=2021-03-18 06:54:06-08:00)
[AR1]
```

另外，我们还可以从这个过程中看到动态路由协议的"智能"：它可以根据网络链路的状态，在整个路由域中自动调整路由。与静态路由一样，AR1 可以感知到本地链路的断开，因此它会重新进行选路。例 3-23 通过查看 AR1 路由表中的 OSPF 路由条目，确认备用路由已被启用，现在 AR1 启用了以 AR4 为下一跳的路由。

例 3-23 查看链路断开后的 AR1 路由表

```
[AR1]display ip routing-table protocol ospf
Route Flags: R - relay, D - download to fib
------------------------------------------------------------------------------
Public routing table : OSPF
         Destinations : 5        Routes : 5

OSPF routing table status : <Active>
         Destinations : 5        Routes : 5

Destination/Mask    Proto   Pre  Cost      Flags NextHop         Interface

     10.10.0.2/32   OSPF    10   7           D  10.10.14.4      GigabitEthernet0/0/1
     10.10.0.3/32   OSPF    10   6           D  10.10.14.4      GigabitEthernet0/0/1
    10.10.23.0/24   OSPF    10   7           D  10.10.14.4      GigabitEthernet0/0/1
    10.10.34.0/24   OSPF    10   6           D  10.10.14.4      GigabitEthernet0/0/1
   172.16.30.0/24   OSPF    10   7           D  10.10.14.4      GigabitEthernet0/0/1

OSPF routing table status : <Inactive>
         Destinations : 0        Routes : 0
```

AR2 也会重新进行选路，因为它也能感知本地链路的状态，例 3-24 展示了 AR2 的路由表。从中我们可以看出，10.10.12.0/24 已经消失，它去往所有远端网段的路由都指向了 G0/0/2。

例 3-24　查看链路断开后的 AR2 路由表

```
[AR2]display ip routing-table
Route Flags: R - relay, D - download to fib
------------------------------------------------------------------------------
Routing Tables: Public
         Destinations : 14       Routes : 14

Destination/Mask    Proto   Pre   Cost      Flags  NextHop         Interface
      10.10.0.1/32  OSPF    10    3           D    10.10.23.3      GigabitEthernet0/0/2
      10.10.0.2/32  Direct  0     0           D    127.0.0.1       LoopBack0
      10.10.0.3/32  OSPF    10    1           D    10.10.23.3      GigabitEthernet0/0/2
     10.10.14.0/24  OSPF    10    3           D    10.10.23.3      GigabitEthernet0/0/2
     10.10.23.0/24  Direct  0     0           D    10.10.23.2      GigabitEthernet0/0/2
     10.10.23.2/32  Direct  0     0           D    127.0.0.1       GigabitEthernet0/0/2
   10.10.23.255/32  Direct  0     0           D    127.0.0.1       GigabitEthernet0/0/2
     10.10.34.0/24  OSPF    10    2           D    10.10.23.3      GigabitEthernet0/0/2
       127.0.0.0/8  Direct  0     0           D    127.0.0.1       InLoopBack0
       127.0.0.1/32 Direct  0     0           D    127.0.0.1       InLoopBack0
   127.255.255.255/32 Direct 0    0           D    127.0.0.1       InLoopBack0
     172.16.10.0/24 OSPF    10    4           D    10.10.23.3      GigabitEthernet0/0/2
     172.16.30.0/24 OSPF    10    2           D    10.10.23.3      GigabitEthernet0/0/2
   255.255.255.255/32 Direct 0    0           D    127.0.0.1       InLoopBack0
```

动态路由协议还能让与该链路非直连的设备也感知到链路的状态变化，从而对路由进行重新计算。例 3-25 展示了此时 AR3 的路由表，我们发现之前的负载分担路由只剩下了指向 AR4 的单条路由，这是因为通过 AR2 已不能去往 PC1，AR3 根据这个情况调整了自己的选路。

例 3-25　查看链路断开后的 AR3 路由表

```
[AR3]display ip routing-table
Route Flags: R - relay, D - download to fib
------------------------------------------------------------------------------
Routing Tables: Public
         Destinations : 18       Routes : 18

Destination/Mask    Proto   Pre   Cost      Flags  NextHop         Interface
      10.10.0.1/32  OSPF    10    2           D    10.10.34.4      GigabitEthernet0/0/1
      10.10.0.2/32  OSPF    10    1           D    10.10.23.2      GigabitEthernet0/0/2
      10.10.0.3/32  Direct  0     0           D    127.0.0.1       LoopBack0
     10.10.14.0/24  OSPF    10    2           D    10.10.34.4      GigabitEthernet0/0/1
     10.10.23.0/24  Direct  0     0           D    10.10.23.3      GigabitEthernet0/0/2
     10.10.23.3/32  Direct  0     0           D    127.0.0.1       GigabitEthernet0/0/2
   10.10.23.255/32  Direct  0     0           D    127.0.0.1       GigabitEthernet0/0/2
     10.10.34.0/24  Direct  0     0           D    10.10.34.3      GigabitEthernet0/0/1
     10.10.34.3/32  Direct  0     0           D    127.0.0.1       GigabitEthernet0/0/1
   10.10.34.255/32  Direct  0     0           D    127.0.0.1       GigabitEthernet0/0/1
       127.0.0.0/8  Direct  0     0           D    127.0.0.1       InLoopBack0
       127.0.0.1/32 Direct  0     0           D    127.0.0.1       InLoopBack0
   127.255.255.255/32 Direct 0    0           D    127.0.0.1       InLoopBack0
     172.16.10.0/24 OSPF    10    3           D    10.10.34.4      GigabitEthernet0/0/1
     172.16.30.0/24 Direct  0     0           D    172.16.30.3     GigabitEthernet0/0/0
     172.16.30.3/32 Direct  0     0           D    127.0.0.1       GigabitEthernet0/0/0
   172.16.30.255/32 Direct  0     0           D    127.0.0.1       GigabitEthernet0/0/0
   255.255.255.255/32 Direct 0    0           D    127.0.0.1       InLoopBack0
```

静态路由的配置中，在同样的拓扑中发生同样的链路问题后，AR3 上通过 AR2 去往 PC1 的静态路由不会发生任何变化，除非管理员手动进行修改。在大规模网络中，一旦出现这种牵一发而动全身的情况，仅凭人力是无法进行及时逐一修复的。因此在实际工作中，静态路由主要用于默认路由的设置，或者企业网关连接 ISP（Internet Service Provider，互联网服务提供商）的设置。其他工作交给如 OSPF 等动态路由协议，既减轻了管理员的工作负担，也让网络更加安全和高效。

例 3-26 中展示了步骤 3 中 PC1 向 PC2 发起 ping 测试的结果。用户首先在 PC1

上发出命令 **ping 172.14.30.10 -t**，接着在 AR1 上使用命令 **shutdown** 关闭 G0/0/2 接口，然后观察 ping 测试是否会中断，从 ping 结果看来 PC1 与 PC2 之间的路由并没有中断。

例 3-26　PC1 上的持续 ping 测试

```
PC>ping 172.16.30.10 -t

Ping 172.16.30.10: 32 data bytes, Press Ctrl_C to break
From 172.16.30.10: bytes=32 seq=1 ttl=125 time=47 ms
From 172.16.30.10: bytes=32 seq=2 ttl=125 time=46 ms
From 172.16.30.10: bytes=32 seq=3 ttl=125 time=46 ms
From 172.16.30.10: bytes=32 seq=4 ttl=125 time=32 ms
From 172.16.30.10: bytes=32 seq=5 ttl=125 time=62 ms
From 172.16.30.10: bytes=32 seq=6 ttl=125 time=62 ms
From 172.16.30.10: bytes=32 seq=7 ttl=125 time=46 ms
From 172.16.30.10: bytes=32 seq=8 ttl=125 time=46 ms
From 172.16.30.10: bytes=32 seq=9 ttl=125 time=47 ms
From 172.16.30.10: bytes=32 seq=10 ttl=125 time=31 ms
From 172.16.30.10: bytes=32 seq=11 ttl=125 time=31 ms
From 172.16.30.10: bytes=32 seq=12 ttl=125 time=31 ms

--- 172.16.30.10 ping statistics ---
  12 packet(s) transmitted
  12 packet(s) received
  0.00% packet loss
  round-trip min/avg/max = 31/44/62 ms
```

测试完成后，读者可以将 AR1 上的 G0/0/2 接口再次启用。

3.2.3　通过 OSPF 发布默认路由

假设在本实验所示的网络中，AR1 是整个网络的出口，用户则需要让 AR1 向其他 OSPF 路由器发布默认路由，让其他 OSPF 路由器在路由表中没有更精确匹配的条目时，将数据包发送到 AR1。

步骤 1　发布默认路由

例 3-27 在 AR1 上配置了发布默认路由的命令。命令语法如下。

default-route-advertise [always]：默认情况下，**default-route-advertise** 命令会将路由器中的默认路由通告在 OSPF 中。由于当前 AR1 中没有默认路由，因此我们使用关键字 **always**，使路由表中无论有无活跃的默认路由，都向 OSPF 通告默认路由。

例 3-27　使 AR1 向 OSPF 发布默认路由

```
[AR1]ospf 1
[AR1-ospf-1]default-route-advertise always
```

在验证 OSPF 默认路由之前，我们先在 AR1 上创建一个环回接口，以此模拟 AR1 身后未加入 OSPF 的其他网段。例 3-28 中展示了创建新环回接口的命令。

例 3-28　在 AR1 上创建新的环回接口

```
[AR1]interface LoopBack 1
[AR1-LoopBack1]ip address 10.10.1.1 24
```

步骤 2　在其他 OSPF 路由器中查看默认路由

我们以 AR3 为例，查看其 IP 路由表，首先验证 AR3 上通过 OSPF 学到了默认路由，其次验证 AR3 上没有去往 AR1 Lo1 接口的明细路由，具体见例 3-29。

例 3-29 查看 AR3 的 IP 路由表

```
[AR3]display ip routing-table
Route Flags: R - relay, D - download to fib
------------------------------------------------------------------------------
Routing Tables: Public
         Destinations : 20     Routes : 23

Destination/Mask      Proto   Pre   Cost      Flags   NextHop         Interface

        0.0.0.0/0    O_ASE    150    1          D    10.10.23.2      GigabitEthernet0/0/2
                     O_ASE    150    1          D    10.10.34.4      GigabitEthernet0/0/1
      10.10.0.1/32   OSPF     10     2          D    10.10.34.4      GigabitEthernet0/0/1
                     OSPF     10     2          D    10.10.23.2      GigabitEthernet0/0/2
      10.10.0.2/32   OSPF     10     1          D    10.10.23.2      GigabitEthernet0/0/2
      10.10.0.3/32   Direct   0      0          D    127.0.0.1       LoopBack0
      10.10.12.0/24  OSPF     10     2          D    10.10.23.2      GigabitEthernet0/0/2
      10.10.14.0/24  OSPF     10     2          D    10.10.34.4      GigabitEthernet0/0/1

      10.10.34.0/24  Direct   0      0          D    10.10.34.3      GigabitEthernet0/0/1
      10.10.34.3/32  Direct   0      0          D    127.0.0.1       GigabitEthernet0/0/1
    10.10.34.255/32  Direct   0      0          D    127.0.0.1       GigabitEthernet0/0/1
      127.0.0.0/8    Direct   0      0          D    127.0.0.1       InLoopBack0
      127.0.0.1/32   Direct   0      0          D    127.0.0.1       InLoopBack0
  127.255.255.255/32 Direct   0      0          D    127.0.0.1       InLoopBack0
     172.16.10.0/24  OSPF     10     3          D    10.10.34.4      GigabitEthernet0/0/1
                     OSPF     10     3          D    10.10.23.2      GigabitEthernet0/0/2
     172.16.30.0/24  Direct   0      0          D    172.16.30.3     GigabitEthernet0/0/0
     172.16.30.3/32  Direct   0      0          D    127.0.0.1       GigabitEthernet0/0/0
   172.16.30.255/32  Direct   0      0          D    127.0.0.1       GigabitEthernet0/0/0
  255.255.255.255/32 Direct   0      0          D    127.0.0.1       InLoopBack0
```

通过查看路由表中的前两条路由，首先我们可以看出默认路由也是可以实现负载分担的。其次，OSPF 默认路由的路由优先级默认为 150，用户可以使用命令 **default-route-advertise cost** *cost* 对其进行手动修改，感兴趣的读者可以自行尝试，本章不进行演示。

另外，从 AR3 的路由表中我们也可以确定这里没有去往 10.10.1.1 的明细路由，接下来读者可以尝试从 PC2 向 10.10.1.1 发起 ping 测试，来验证默认路由是否生效。

步骤 3 通过 ping 测试验证默认路由

例 3-30 展示了在 PC2 上向 10.10.1.1 发起评测的结果。如果配置正确，读者会在这一步测试时得到相同的结果。

例 3-30 验证默认路由的效果

```
PC2>ping 10.10.1.1

Ping 10.10.1.1: 32 data bytes, Press Ctrl_C to break
From 10.10.1.1: bytes=32 seq=1 ttl=253 time=62 ms
From 10.10.1.1: bytes=32 seq=2 ttl=253 time=47 ms
From 10.10.1.1: bytes=32 seq=3 ttl=253 time=62 ms
From 10.10.1.1: bytes=32 seq=4 ttl=253 time=62 ms
From 10.10.1.1: bytes=32 seq=5 ttl=253 time=46 ms

--- 10.10.1.1 ping statistics ---
  5 packet(s) transmitted
  5 packet(s) received
  0.00% packet loss
  round-trip min/avg/max = 46/55/62 ms
```

3.2.4 配置 OSPF 认证

为了加强 OSPF 网络的安全性，使恶意设备无法随意加入企业的 OSPF 域，管理员可以在 OSPF 域内配置邻居之间的认证。只有通过了认证的路由器之间才会建立 OSPF 邻居关系，否则将无法建立邻居关系。用户可以使用区域认证和接口认证两种方法来设置 OSPF 认证，并且接口认证优先于区域认证。

本例将在 AR1 与 AR2 之间配置 OSPF 接口认证，并且在区域 0 中配置 OSPF 区域认证。

步骤 1　配置 OSPF 区域认证

要想配置 OSPF 区域认证，用户需要使用 **authentication-mode** OSPF 区域命令，具体命令语法如下。

authentication-mode {**md5**} [*key-id* {[**cipher**] *cipher-text*}]：在本例中，用户使用了 md5 关键字将 MD5 作为密文认证方法，并且使用了 **cipher** 关键字，在路由器配置中以密文的方式保存密码。

例 3-31 中展示了 AR1 上 OSPF 区域的认证配置并且展示了配置文件中的密码 Area-0-Auth，密码以密文形式保存，这样提升了路由器安全性。

例 3-31　在 AR1 上配置 OSPF 区域认证

```
[AR1-ospf-1]area 0
[AR1-ospf-1-area-0.0.0.0]authentication-mode md5 1 cipher Area-0-Auth
[AR1-ospf-1-area-0.0.0.0]display this
[V200R003C00]
#
 area 0.0.0.0
  authentication-mode md5 1 cipher %$%$)|gWIsaF7XA5(-1Ttq2~e/o'%$%$
  network 10.10.0.1 0.0.0.0
  network 10.10.12.0 0.0.0.255
  network 10.10.14.0 0.0.0.255
  network 172.16.10.0 0.0.0.255
#
return
```

配置完成后，AR1 与 AR2 和 AR4 之间的 OSPF 邻居关系会因为认证失败而无法继续交互 OSPF 消息，并最终导致邻居关系断开。从例 3-32 中我们可以看出，AR1 与邻居之间由于 Dead 计时器超时而中断：NeighborDownImmediate reason=Neighbor Down Due to Inactivity。

例 3-32　AR1 与 AR2 和 AR4 之间的 OSPF 邻居关系断开

```
[AR1]
Mar 18 2021 10:02:15-08:00 AR1 %%01OSPF/3/NBR_CHG_DOWN(l)[114]:Neighbor event:ne
ighbor state changed to Down. (ProcessId=256, NeighborAddress=4.0.10.10, Neighbo
rEvent=InactivityTimer, NeighborPreviousState=Full, NeighborCurrentState=Down)
[AR1]
Mar 18 2021 10:02:15-08:00 AR1 %%01OSPF/3/NBR_DOWN_REASON(l)[115]:Neighbor state
 leaves full or changed to Down. (ProcessId=256, NeighborRouterId=4.0.10.10, Nei
ghborAreaId=0, NeighborInterface=GigabitEthernet0/0/1,NeighborDownImmediate reas
on=Neighbor Down Due to Inactivity, NeighborDownPrimeReason=Hello Not Seen, Neig
hborChangeTime=2021-03-18 10:02:15-08:00)
[AR1]
Mar 18 2021 10:02:24-08:00 AR1 %%01OSPF/3/NBR_CHG_DOWN(l)[116]:Neighbor event:ne
ighbor state changed to Down. (ProcessId=256, NeighborAddress=2.0.10.10, Neighbo
rEvent=InactivityTimer, NeighborPreviousState=Full, NeighborCurrentState=Down)
[AR1]
Mar 18 2021 10:02:24-08:00 AR1 %%01OSPF/3/NBR_DOWN_REASON(l)[117]:Neighbor state
 leaves full or changed to Down. (ProcessId=256, NeighborRouterId=2.0.10.10, Nei
ghborAreaId=0, NeighborInterface=GigabitEthernet0/0/2,NeighborDownImmediate reas
on=Neighbor Down Due to Inactivity, NeighborDownPrimeReason=Hello Not Seen, Neig
hborChangeTime=2021-03-18 10:02:24-08:00)
```

接着我们在 AR2、AR3 和 AR4 上分别配置了 OSPF 区域认证，具体命令见例 3-33～例 3-35。

例 3-33　在 AR2 上配置 OSPF 区域认证

```
[AR2]ospf 2
[AR2-ospf-2]area 0
[AR2-ospf-2-area-0.0.0.0]authentication-mode md5 1 cipher Area-0-Auth
```

例 3-34　在 AR3 上配置 OSPF 区域认证

```
[AR3]ospf 3
[AR3-ospf-3]area 0
[AR3-ospf-3-area-0.0.0.0]authentication-mode md5 1 cipher Area-0-Auth
```

例 3-35　在 AR4 上配置 OSPF 区域认证

```
[AR4]ospf 4
[AR4-ospf-4]area 0
[AR4-ospf-4-area-0.0.0.0]authentication-mode md5 1 cipher Area-0-Auth
```

如果用户进行及时配置，这些路由器之间的 OSPF 邻居关系不会断开，而是以认证的方式进行交互并保持邻居关系的连续性。以 AR2 为例，展示当前 AR2 的 OSPF 邻居，从例 3-36 中我们可以看出，AR2 与 AR3 之间的邻居关系并没有因为添加了认证信息而断开。在 AR2 上，与 AR1 之间的邻居关系已建立 4 分 30 秒，而与 AR3 之间的邻居关系已建立 1 小时 49 分 56 秒。

例 3-36　AR2 的邻居关系建立时长

```
[AR2]display ospf peer
         OSPF Process 2 with Router ID 10.10.0.2
              Neighbors

Area 0.0.0.0 interface 10.10.12.2(GigabitEthernet0/0/1)'s neighbors
Router ID: 10.10.0.1       Address: 10.10.12.1
  State: Full  Mode:Nbr is  Slave  Priority: 1
  DR: 10.10.12.2  BDR: 10.10.12.1  MTU: 0
  Dead timer due in 37  sec
  Retrans timer interval: 5
  Neighbor is up for 00:04:30
  Authentication Sequence: [ 36937]

              Neighbors

Area 0.0.0.0 interface 10.10.23.2(GigabitEthernet0/0/2)'s neighbors
Router ID: 10.10.0.3       Address: 10.10.23.3
  State: Full  Mode:Nbr is  Master  Priority: 1
  DR: 10.10.23.2  BDR: 10.10.23.3  MTU: 0
  Dead timer due in 39  sec
  Retrans timer interval: 5
  Neighbor is up for 01:49:56
  Authentication Sequence: [ 40552]
```

读者在做实验的过程中需要注意，当 AR1 与 AR2 和 AR4 之间的邻居关系断开后，会在 AR2 和 AR3 上看到类似例 3-37 中的输出内容，该内容表示路由发生了变化。

例 3-37　AR2 上的默认路由发生了变化

```
Mar 18 2021 09:41:02-08:00 AR2 %%01RM/4/IPV4_DEFT_RT_CHG(l)[98]:IPV4 default Rou
te is changed. (ChangeType=Delete, InstanceId=0, Protocol=OSPF, ExitIf=GigabitEt
hernet0/0/1, Nexthop=10.10.12.1, Neighbour=0.0.0.0, Preference=2516582400, Label
=NULL, Metric=16777216)
```

步骤 2　配置 OSPF 接口认证

要想配置 OSPF 接口认证，用户需要使用 **authentication-mode** 接口命令，具体命令语法如下。

authentication-mode {**md5**} [*key-id* {**plain** *plain-text* | [**cipher**] *cipher-text*}]：在本例中，用户使用了 MD5 作为密文认证方法，并且使用 **plain** 关键字，在路由器配置中以明

文的方式保存密码。

如例 3-38 所示，在 AR1 的 G0/0/2 接口上配置了接口认证后，AR1 与 AR2 之间的 OSPF 邻居关系便断开了，因此印证接口认证配置优先于区域认证设置。同时，从路由器配置中我们可以看出，本例中的密码是以明文形式存在于配置中的，这种做法不安全，因此仅作命令展示，建议在生产环境中始终在配置中以加密形式保存密码。

例 3-38 在 AR1 的 G0/0/2 接口上配置接口认证

```
[AR1]interface GigabitEthernet 0/0/2
[AR1-GigabitEthernet0/0/2]ospf authentication-mode md5 1 plain AR1-AR2-Auth
[AR1-GigabitEthernet0/0/2]display this
[V200R003C00]
#
interface GigabitEthernet0/0/2
 ip address 10.10.12.1 255.255.255.0
 ospf authentication-mode md5 1 plain AR1-AR2-Auth
#
return
[AR1-GigabitEthernet0/0/2]quit
[AR1]
Mar 18 2021 10:36:31-08:00 AR1 %%01OSPF/3/NBR_CHG_DOWN(l)[128]:Neighbor event:ne
ighbor state changed to Down. (ProcessId=256, NeighborAddress=2.0.10.10, Neighbo
rEvent=InactivityTimer, NeighborPreviousState=Full, NeighborCurrentState=Down)
[AR1]
Mar 18 2021 10:36:31-08:00 AR1 %%01OSPF/3/NBR_DOWN_REASON(l)[129]:Neighbor state
 leaves full or changed to Down. (ProcessId=256, NeighborRouterId=2.0.10.10, Nei
ghborAreaId=0, NeighborInterface=GigabitEthernet0/0/2,NeighborDownImmediate reas
on=Neighbor Down Due to Inactivity, NeighborDownPrimeReason=Hello Not Seen, Neig
hborChangeTime=2021-03-18 10:36:31-08:00)
[AR1]
```

在配置中无论以明文形式保存密码还是密文形式保存密码，只要密码匹配，配置中保存密码的形式与认证是否成功无关。例 3-39 在 AR2 上以密文形式配置了 OSPF 接口认证。

例 3-39 在 AR2 的 G0/0/1 接口上配置接口认证

```
[AR2]interface GigabitEthernet 0/0/1
[AR2-GigabitEthernet0/0/1]ospf authentication-mode md5 1 cipher AR1-AR2-Auth
[AR2-GigabitEthernet0/0/1]display this
[V200R003C00]
#
interface GigabitEthernet0/0/1
 ip address 10.10.12.2 255.255.255.0
 ospf authentication-mode md5 1 cipher %$%$<]_9=.>&y.2f}+E-CC1OeCQ)%$%$
 ospf enable 2 area 0.0.0.0
#
return
```

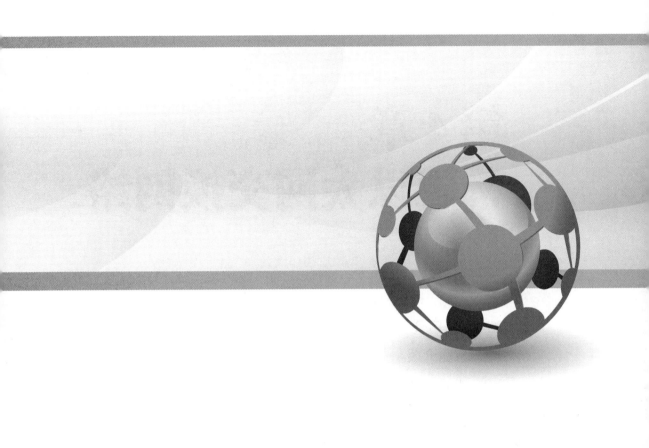

第4章
构建以太网交换网络

本章主要内容

4.1　实验介绍

4.2　实验配置任务

在以太网交换环境中，我们会使用 VLAN（Virtual Local Area Network，虚拟局域网）将一个大型广播域分隔为多个较小的广播域，这样既减轻了交换机的广播流量压力，也保障了安全性和可管理性。每个 VLAN 内部是一个广播域，VLAN 内部的所有主机之间的通信就像在一个局域网（LAN）中一样，交换机无须支持三层路由功能。不同 VLAN 之间不能直接通信，交换机也不会把一个 VLAN 中的广播数据包转发到另一个 VLAN（即不会转发给属于另一个 VLAN 的端口）中。

本质上，VLAN 是与端口相关联的，交换机需要知道哪个端口属于哪个 VLAN，或者允许这个端口传输哪些 VLAN 的流量，这样才可以正确地执行转发。让交换机明确地知道哪个端口属于哪个 VLAN 的过程称为 VLAN 划分，管理员可以根据实际情况，使用多种方法来划分 VLAN，其中最简单的划分方法是直接在端口上输入命令指定该端口所属的 VLAN，但这种方法过于静态，不适用于多种环境。多种 VLAN 划分的方法见表 4-1。

表 4-1　多种 VLAN 划分的方法

VLAN 划分方法	描述	适用场景
基于端口	直接指定该端口所属的 VLAN	适用于设备位置固定的安全环境
基于 MAC 地址	根据数据帧头部的 MAC 地址来划分 VLAN	适用于设备位置不固定，但设备 MAC 地址固定的小型网络
基于子网	根据数据包头部的源 IP 地址来划分 VLAN	适用于对移动性和简化管理有较高要求的安全环境
基于网络层协议	根据数据帧的协议类型和封装格式来划分 VLAN	适用于同时运行了多种协议的杂合网络
基于匹配策略	根据配置的策略来划分 VLAN，策略中可以组合多种参数，如端口、MAC 地址、IP 地址等	适用于需求较为复杂，无法通过单一方法划分 VLAN 的环境

本实验会介绍基于端口、基于 MAC 地址的划分方法，如果读者对其他方法的配置命令感兴趣，可以查询华为相关设备的配置指南。

4.1　实验介绍

4.1.1　关于本实验

在一家微型企业中，企业的办公区域分为两个房间，一个小房间为老板办公室，一个大房间为开放办公室，财务部和销售部的员工共同使用这个开放办公空间。

我们需要通过 VLAN 的划分，使老板 PC、财务部 PC 和销售部 PC 之间无法进行通信，保障其安全性和便捷性。同时在这个办公环境中，财务部和销售部各自拥有本部门的服务器，财务部 PC 可以访问财务部服务器，销售部 PC 可以访问销售部服务器，但不可跨部门访问。

在本实验中，我们着重关注 VLAN 的概念和配置，实验所需的两台交换机只做二层交换机使用，不启用路由功能，且不在交换机上为 PC 配置默认网关。我们不考虑 PC 连接外网的需求，只考虑公司内部终端之间的通信。

4.1.2 实验目的

- 掌握创建 VLAN 的命令，允许 VLAN 内部通信，不允许 VLAN 之间通信。
- 掌握将交换机端口配置为 Access、Trunk、Hybrid 端口的命令。
- 掌握限制 Trunk 链路上允许传输的 VLAN 的命令。
- 掌握根据端口、MAC 地址划分 VLAN 的方法。

4.1.3 实验组网介绍

构建以太网交换网络的实验拓扑如图 4-1 所示。

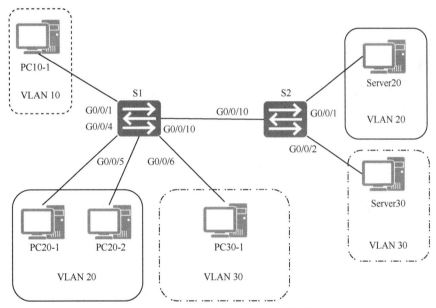

图 4-1 构建以太网交换网络的实验拓扑

设备连接说明如下。

① 两台交换机之间通过 G0/0/10 端口连接，允许传输 VLAN 20 和 VLAN 30 的流量，禁止传输其他 VLAN 的流量。

② 交换机 S1 的端口 G0/0/1～G0/0/3 用于老板办公室，当前只使用了一个端口（G0/0/1）连接 PC10-1，另外两个端口已连接线缆并预留到老板办公室内。

③ 交换机 S1 的端口 G0/0/4～G0/0/9 用于公司开放办公室内的其他员工，交换机需要自动区分销售部员工和财务部员工，并将他们的设备自动地分到相应的 VLAN 中（VLAN 20 和 VLAN 30）。

④ 交换机 S2 分别由 G0/0/1 连接销售部的服务器 Server20，由 G0/0/2 连接财务部的服务器 Server30。

⑤ 交换机需要根据某些设置（端口或 MAC 地址）来自动为所连终端分配 VLAN。

⑥ 每个 VLAN 中的主机只能与本 VLAN 中的其他主机进行通信，不允许跨 VLAN 通信。

本章使用的网络地址见表 4-2。

表 4-2　本章使用的网络地址

设备	接口	IP 地址	子网掩码	默认网关
PC10-1	E0/0/1	10.0.10.1	255.255.255.0	—
PC20-1	E0/0/1	10.0.20.1	255.255.255.0	—
PC20-2	E0/0/1	10.0.20.2	255.255.255.0	—
PC30-1	E0/0/1	10.0.30.1	255.255.255.0	—
Server20	E0/0/1	10.0.20.10	255.255.255.0	—
Server30	E0/0/1	10.0.30.10	255.255.255.0	—

4.1.4　实验任务列表

配置任务 1：创建 VLAN。
配置任务 2：基于端口划分 VLAN。
配置任务 3：基于 MAC 地址划分 VLAN。
配置任务 4：验证配置效果。

4.2　实验配置任务

4.2.1　创建 VLAN

我们先要在两台交换机 S1 和 S2 上分别创建公司需要的 VLAN。按照图 4-1 所示的部分规划，我们将公司的 VLAN、PC、交换机端口的规划总结如下，具体见表 4-3。

表 4-3　VLAN 规划

VLAN ID	交换机端口	PC 或服务器
10	S1：G0/0/1～G0/0/3	PC10-1
20	S1：G0/0/4～G0/0/9	PC20-1、PC20-2
	S2：G0/0/1	Server20
30	S1：G0/0/4～G0/0/9	PC30-1
	S2：G0/0/2	Server30

在本实验环境中，VLAN 10 是专用于老板办公室的，VLAN 20 是专用于销售部的，VLAN 30 是专用于财务部的。从表 4-3 中我们可以看出，交换机 S1 和交换机 S2 上分别需要配置以下 VLAN。

① 交换机 S1：VLAN 10、VLAN 20、VLAN 30。
② 交换机 S2：VLAN 20、VLAN 30。

要想在交换机上配置 VLAN，我们可以使用两种方法。第一种方法是逐个创建 VLAN，在输入创建一个 VLAN 的命令后，用户会进入 VLAN 配置视图。第二种方法是通过一条命令批量创建多个 VLAN，这条命令不会使用户进入某个 VLAN 配置视图。我们会通过交换机 S1 和 S2 分别展示这两种配置方法，具体见例 4-1 和例 4-2。

vlan *vlan-id*：创建 VLAN 的第一种方法是逐个创建 VLAN，*vlan-id* 的取值范围是 2~4094。若重复创建了已存在的 VLAN，则会保留原 VLAN 的相关设置。

例 4-1　在交换机 S1 上创建 VLAN

```
[S1]vlan 10
[S1-vlan10]quit
[S1]vlan 20
[S1-vlan20]quit
[S1]vlan 30
```

vlan batch [*vlan-id*1 *vlan-id*2 ... *vlan-idn*]：创建 VLAN 的第二种方法是批量创建多个 VLAN。如需创建连续的 VLAN，如 VLAN 2、VLAN 3、VLAN 4、VLAN5，我们可以在要创建的第一个 VLAN 和最后一个 VLAN 之间使用关键字 **to**。本例未使用连续编号的 VLAN，因此不展示这个关键字的用法，感兴趣的读者可以自行尝试。

例 4-2　在交换机 S2 上创建 VLAN

```
[S2]vlan batch 20 30
Info: This operation may take a few seconds. Please wait for a moment...done.
[S2]
```

创建完 VLAN 后，我们可以使用 **display** 命令来查看 VLAN 是否创建成功，首先通过 **display vlan summary** 查看已存在的 VLAN 汇总情况，具体见例 4-3。在这条命令中，我们可以看到交换机 S1 上共有 4 个静态（static）VLAN，其中 VLAN 1 是缺省 VLAN，默认存在且所有端口默认都属于 VLAN 1，VLAN 10、VLAN 20 和 VLAN 30 是我们刚创建的 VLAN，说明 VLAN 已创建完成。

例 4-3　查看 VLAN 汇总信息

```
[S1]display vlan summary
static vlan:
Total 4 static vlan.
  1 10 20 30

dynamic vlan:
Total 0 dynamic vlan.

reserved vlan:
Total 0 reserved vlan.
```

为了查看端口与 VLAN 的对应关系，我们需要使用一条能够显示更详细信息的命令：**display vlan**，具体见例 4-4。从命令的输出中我们可以看出，目前所有端口都属于 VLAN 1，刚创建的 3 个 VLAN 中没有端口。

例 4-4　查看 VLAN 与端口的对应关系

```
[S1]display vlan
The total number of vlans is : 4
--------------------------------------------------------------------------------
U: Up;         D: Down;         TG: Tagged;         UT: Untagged;
MP: Vlan-mapping;               ST: Vlan-stacking;
#: ProtocolTransparent-vlan;    *: Management-vlan;
--------------------------------------------------------------------------------

VID  Type   Ports
--------------------------------------------------------------------------------
```

```
1    common  UT:GE0/0/1(U)    GE0/0/2(D)     GE0/0/3(D)     GE0/0/4(U)
                GE0/0/5(U)    GE0/0/6(U)     GE0/0/7(D)     GE0/0/8(D)
                GE0/0/9(D)    GE0/0/10(U)    GE0/0/11(D)    GE0/0/12(D)
                GE0/0/13(D)   GE0/0/14(D)    GE0/0/15(D)    GE0/0/16(D)
                GE0/0/17(D)   GE0/0/18(D)    GE0/0/19(D)    GE0/0/20(D)
                GE0/0/21(D)   GE0/0/22(D)    GE0/0/23(D)    GE0/0/24(D)
10   common
20   common
30   common

VID  Status  Property      MAC-LRN  Statistics  Description
--------------------------------------------------------------------------
1    enable  default       enable   disable     VLAN 0001
10   enable  default       enable   disable     VLAN 0010
20   enable  default       enable   disable     VLAN 0020
30   enable  default       enable   disable     VLAN 0030
```

至此公司环境中需要的 VLAN 就创建完成了，读者可以自行在交换机 S2 上使用上述 **display** 命令进行检查。

4.2.2 基于端口划分 VLAN

基于端口划分 VLAN 的方法适用于设备位置固定的安全环境，因为这种方法会在交换机端口上静态指定该端口所属的 VLAN，即无论终端设备是否属于该 VLAN，只要它连接在这个端口上，就可以访问这个 VLAN 中的资源。

在本实验环境中，老板办公室是一个相对安全的环境，因此我们对其进行基于端口的划分。本环境只在 VLAN 10 中连接了一台 PC，但需要为老板多预留出两个端口，以便连接多台设备，因此我们将交换机 S1 的 G0/0/1～G0/0/3 端口静态地划分到 VLAN 10。另外，连接两台交换机的端口，以及连接服务器的端口一般是比较固定的，不会轻易改动，因此这些端口也适用于基于端口划分。本小节会展示表 4-4 所示端口的配置。

表 4-4 基于端口划分 VLAN

交换机	端口	类型	VLAN
S1	G0/0/1 – G0/0/3	Access	10
	G0/0/10	Trunk	20 和 30
S2	G0/0/1	Access	20
	G0/0/2	Access	30
	G0/0/10	Trunk	20 和 30

步骤 1 设置 Access 端口并绑定 VLAN

在交换机 S1 上，我们需要将 G0/0/1～G0/0/3 端口设置为 Access 端口，并静态关联到 VLAN 10。读者可以按照以下步骤进行配置。

① 使用命令 **interface** *interface-type interface-number* 进入相应端口的接口配置视图。

② 使用命令 **port link-type access** 将端口设置为 Access 模式。

③ 使用命令 **port default vlan** *vlan-id* 将端口添加到某个 VLAN 中。

例 4-5 中展示了交换机 S1 上的配置。

例 4-5 在 S1 上配置 G0/0/1～G0/0/3 端口

```
[S1]interface GigabitEthernet 0/0/1
[S1-GigabitEthernet0/0/1]port link-type access
[S1-GigabitEthernet0/0/1]port default vlan 10
[S1-GigabitEthernet0/0/1]quit
[S1]interface GigabitEthernet 0/0/2
[S1-GigabitEthernet0/0/2]port link-type access
[S1-GigabitEthernet0/0/2]port default vlan 10
[S1-GigabitEthernet0/0/2]quit
[S1]interface GigabitEthernet 0/0/3
[S1-GigabitEthernet0/0/3]port link-type access
[S1-GigabitEthernet0/0/3]port default vlan 10
```

配置完成后，我们可以通过 **display vlan** 命令来验证配置结果，具体见例 4-6。

例 4-6 查看 G0/0/1～G0/0/3 的配置结果

```
[S1]display vlan
The total number of vlans is : 4
--------------------------------------------------------------------------------
U: Up;         D: Down;         TG: Tagged;         UT: Untagged;
MP: Vlan-mapping;                ST: Vlan-stacking;
#: ProtocolTransparent-vlan;     *: Management-vlan;
--------------------------------------------------------------------------------

VID  Type    Ports
--------------------------------------------------------------------------------
1    common  UT:GE0/0/4(U)    GE0/0/5(U)     GE0/0/6(U)     GE0/0/7(D)
               GE0/0/8(D)     GE0/0/9(D)     GE0/0/10(U)    GE0/0/11(D)
               GE0/0/12(D)    GE0/0/13(D)    GE0/0/14(D)    GE0/0/15(D)
               GE0/0/16(D)    GE0/0/17(D)    GE0/0/18(D)    GE0/0/19(D)
               GE0/0/20(D)    GE0/0/21(D)    GE0/0/22(D)    GE0/0/23(D)
               GE0/0/24(D)

10   common  UT:GE0/0/1(U)    GE0/0/2(D)     GE0/0/3(D)

20   common
30   common

VID  Status  Property    MAC-LRN Statistics Description
--------------------------------------------------------------------------------
1    enable  default     enable  disable    VLAN 0001
10   enable  default     enable  disable    VLAN 0010
20   enable  default     enable  disable    VLAN 0020
30   enable  default     enable  disable    VLAN 0030
```

交换机 S2 上有两个端口分别连接了一台服务器，我们将它们设置为 Access 端口并静态配置 VLAN。例 4-7 中展示了交换机 S2 上的配置命令。

例 4-7 在 S2 上配置 G0/0/1、G0/0/2 端口

```
[S2]interface GigabitEthernet 0/0/1
[S2-GigabitEthernet0/0/1]port link-type access
[S2-GigabitEthernet0/0/1]port default vlan 20
[S2-GigabitEthernet0/0/1]quit
[S2]interface GigabitEthernet 0/0/2
[S2-GigabitEthernet0/0/2]port link-type access
[S2-GigabitEthernet0/0/2]port default vlan 30
```

配置完成后，我们可以使用 **display vlan** 命令来验证配置结果，具体见例 4-8。

例 4-8 查看 G0/0/1 和 G0/0/2 的配置结果

```
[S2]displayvlan
The total number of vlansis : 3
--------------------------------------------------------------------------------
U: Up;         D: Down;         TG: Tagged;         UT: Untagged;
MP: Vlan-mapping;                ST: Vlan-stacking;
#: ProtocolTransparent-vlan;     *: Management-vlan;
--------------------------------------------------------------------------------

VID  Type    Ports
--------------------------------------------------------------------------------
1    common  UT:GE0/0/3(D)    GE0/0/4(D)     GE0/0/5(D)     GE0/0/6(D)
               GE0/0/7(D)     GE0/0/8(D)     GE0/0/9(D)     GE0/0/10(U)
```

```
                        GE0/0/11(D)     GE0/0/12(D)     GE0/0/13(D)     GE0/0/14(D)
                        GE0/0/15(D)     GE0/0/16(D)     GE0/0/17(D)     GE0/0/18(D)
                        GE0/0/19(D)     GE0/0/20(D)     GE0/0/21(D)     GE0/0/22(D)
                        GE0/0/23(D)     GE0/0/24(D)

20      common  UT:GE0/0/1(U)

30      common  UT:GE0/0/2(U)

VID  Status  Property      MAC-LRN Statistics Description
--------------------------------------------------------------------------------
1    enable  default       enable  disable    VLAN 0001
20   enable  default       enable  disable    VLAN 0020
30   enable  default       enable  disable    VLAN 0030
```

步骤 2 设置 Trunk 端口并允许 VLAN

交换机之间的互联链路上通常需要承载多个 VLAN 的流量，我们会把这条链路设置为 Trunk 模式，并且限制它只能传输 VLAN 20 和 VLAN 30 的流量。读者可以按照以下步骤进行配置。

① 使用命令 **interface** *interface-type interface-number* 进入相应端口的接口配置视图。

② 使用命令 **port link-type trunk** 将端口设置为 Trunk 模式。

③ 使用命令 **port trunk allow-pass vlan** {{*vlan-id*1 [**to** *vlan-id*2]} | **all**} 设置允许 Trunk 传输的 VLAN。如需允许编号连续的 VLAN，我们可以在允许的第一个 VLAN 与最后一个 VLAN 之间使用关键字 **to**；如需允许所有 VLAN，可以使用关键字 **all**。

例 4-9 中展示了交换机 S1 上的配置。

例 4-9 配置 S1 上的 Trunk 端口

```
[S1]interface GigabitEthernet 0/0/10
[S1-GigabitEthernet0/0/10]port link-type trunk
[S1-GigabitEthernet0/0/10]port trunk allow-pass vlan 20 30
```

我们通过例 4-10 确认当前的配置，从命令输出中可以发现，G0/0/10 端口除了可以传输 VLAN 20 和 VLAN 30 的流量外，还可以传输 VLAN 1 的流量。这是因为 VLAN 1 是缺省 VLAN，在实际工作中，出于安全性的考虑，我们会根据需要在 Trunk 链路上移除 VLAN 1。

例 4-10 查看 Trunk 端口支持的 VLAN

```
[S1]display vlan
The total number of vlans is : 4
--------------------------------------------------------------------------------
U: Up;          D: Down;        TG: Tagged;          UT: Untagged;
MP: Vlan-mapping;               ST: Vlan-stacking;
#: ProtocolTransparent-vlan;    *: Management-vlan;
--------------------------------------------------------------------------------

VID  Type    Ports
--------------------------------------------------------------------------------
1    common  UT:GE0/0/4(U)   GE0/0/5(U)      GE0/0/6(U)      GE0/0/7(U)
                GE0/0/8(D)      GE0/0/9(D)      GE0/0/10(U)     GE0/0/11(D)
                GE0/0/12(D)     GE0/0/13(D)     GE0/0/14(D)     GE0/0/15(D)
                GE0/0/16(D)     GE0/0/17(D)     GE0/0/18(D)     GE0/0/19(D)
                GE0/0/20(D)     GE0/0/21(D)     GE0/0/22(D)     GE0/0/23(D)
                GE0/0/24(D)

10   common  UT:GE0/0/1(U)   GE0/0/2(D)      GE0/0/3(D)

20   common  TG:GE0/0/10(U)
30   common  TG:GE0/0/10(U)

VID  Status  Property      MAC-LRN Statistics Description
--------------------------------------------------------------------------------
```

```
1      enable   default           enable   disable    VLAN 0001
10     enable   default           enable   disable    VLAN 0010
20     enable   default           enable   disable    VLAN 0020
30     enable   default           enable   disable    VLAN 0030
```

例 4-11 中展示了从 Trunk 链路上移除 VLAN 1 的命令和结果。从命令输出中我们可以看出现在 VLAN 1 中已经没有了端口 G0/0/10。

例 4-11　从 Trunk 上移除 VLAN 1

```
[S1]interface GigabitEthernet 0/0/10
[S1-GigabitEthernet0/0/10]undo port trunk allow-pass vlan 1
[S1-GigabitEthernet0/0/10]quit
[S1]display vlan
The total number of vlans is : 4
--------------------------------------------------------------------------------
U: Up;         D: Down;         TG: Tagged;         UT: Untagged;
MP: Vlan-mapping;               ST: Vlan-stacking;
#: ProtocolTransparent-vlan;    *: Management-vlan;
--------------------------------------------------------------------------------

VID  Type    Ports
--------------------------------------------------------------------------------
1    common  UT:GE0/0/4(U)     GE0/0/5(U)      GE0/0/6(D)      GE0/0/7(D)
             GE0/0/8(D)        GE0/0/9(D)      GE0/0/11(D)     GE0/0/12(D)
             GE0/0/13(D)       GE0/0/14(D)     GE0/0/15(D)     GE0/0/16(D)
             GE0/0/17(D)       GE0/0/18(D)     GE0/0/19(D)     GE0/0/20(D)
             GE0/0/21(D)       GE0/0/22(D)     GE0/0/23(D)     GE0/0/24(D)

10   common  UT:GE0/0/1(U)     GE0/0/2(D)      GE0/0/3(D)

20   common  TG:GE0/0/10(U)

30   common  TG:GE0/0/10(U)


VID  Status  Property      MAC-LRN Statistics Description
--------------------------------------------------------------------------------
1      enable   default           enable   disable    VLAN 0001
10     enable   default           enable   disable    VLAN 0010
20     enable   default           enable   disable    VLAN 0020
30     enable   default           enable   disable    VLAN 0030
```

读者可以按照交换机 S1 的 G0/0/10 端口的配置命令来配置交换机 S2 的 G0/0/10 端口。例 4-12 中展示了 S2 上相应的配置命令。

例 4-12　配置 S2 上的 Trunk 端口

```
[S2]interface GigabitEthernet 0/0/10
[S2-GigabitEthernet0/0/10]port link-type trunk
[S2-GigabitEthernet0/0/10]port trunk allow-pass vlan 20 30
[S2-GigabitEthernet0/0/10]undo port trunk allow-pass vlan 1
```

4.2.3　基于 MAC 地址划分 VLAN

本实验环境所模拟的开放办公室中，基于 MAC 地址划分 VLAN 的做法可以提高终端接入的灵活性和安全性，无须关注物理端口（或者说工位）的位置。当交换机端口接收到终端 PC 发来的数据帧时，它会以数据帧的源 MAC 地址为依据在 MAC-VLAN 表中查找匹配项，并根据匹配的 VLAN ID 和优先级进行转发。

根据实验要求，交换机 S1 的 G0/0/4～G0/0/9 端口需要根据 PC20-1、PC20-2 和 PC30-1 的 MAC 地址为其分配特定的 VLAN。读者可以按照以下步骤进行配置。

① 使用命令 **vlan** *vlan-id* 创建 VLAN 进入 VLAN 视图。由于在前文步骤中我们已经创建好 VLAN，因此此时会进入 VLAN 视图。

使用命令 **mac-vlan mac-address** *mac-address* 将 MAC 地址与该 VLAN 进行关联。命令中 MAC 地址的格式为 H-H-H。

② 使用命令 **interface** *interface-type interface-number* 进入要配置基于 MAC 地址的 VLAN 划分的端口。

a. 使用命令 **port link-type hybrid** 将端口设置为 Hybrid 模式。

b. 使用命令 **port hybrid untagged vlan** {{*vlan-id*1 [**to** *vlan-id*2]} | **all**}设置允许这个 Hybrid 端口传输的 VLAN。如需允许编号连续的 VLAN，我们可以在允许的第一个 VLAN 与最后一个 VLAN 之间使用关键字 **to**；如需允许所有 VLAN，可以使用关键字 **all**。

c. 使用命令 **mac-vlan enable** 在端口上启用基于 MAC 地址的 VLAN 划分。

在本节中，我们需要在交换机 S1 的 G0/0/4~G0/0/9 端口上进行相同的配置。我们会先按照上述步骤展示如何根据 MAC 地址划分 VLAN，以及如何在端口 G0/0/4 上启用根据 MAC 地址划分 VLAN 的功能。接着我们会展示一种同时配置多个端口的简便配置方法，在 G0/0/5~G0/0/9 上进行相同的配置。

步骤 1　将 MAC 地址与 VLAN 进行关联

读者可以从 PC 的配置界面查看 PC 的 MAC 地址，或者更改 MAC 地址为易于识别的值，实验按照以下规则设置了 PC 的 MAC 地址：前 6 位十六进制为分配给华为使用的 OUI（Organizationally Unique Identifier，组织唯一标识符）00-E0-FC，后 6 位与 PC 的点分十进制 IP 地址后 2 位的数字相对应。PC20-1、PC20-2 和 PC30-1 的 MAC 地址和所属 VLAN 如下。

① PC20-1：00-E0-FC-00-20-01，VLAN 20。
② PC20-2：00-E0-FC-00-20-02，VLAN 20。
③ PC30-1：00-E0-FC-00-30-01，VLAN 30。

例 4-13 中展示了在交换机 S1 上将 MAC 地址与 VLAN 进行关联的配置。

例 4-13　在 S1 上关联 MAC 地址和 VLAN

```
[S1]vlan 20
[S1-vlan20]mac-vlan mac-address 00e0-fc00-2001
[S1-vlan20]mac-vlan mac-address 00e0-fc00-2002
[S1-vlan20]quit
[S1]vlan 30
[S1-vlan30]mac-vlan mac-address 00e0-fc00-3001
```

配置完成后，读者可以使用命令 **display mac-vlan mac-address all** 来查看 MAC-VLAN 表来确认配置，具体见例 4-14。

例 4-14　验证 MAC 地址与 VLAN 的关联

```
[S1]display mac-vlan mac-address all
---------------------------------------------
MAC Address      MASK            VLAN   Priority
---------------------------------------------
00e0-fc00-2001   ffff-ffff-ffff  20     0
00e0-fc00-2002   ffff-ffff-ffff  20     0
00e0-fc00-3001   ffff-ffff-ffff  30     0

Total MAC VLAN address count: 3
```

我们接下来会在步骤 2 中完成端口上的配置，然后通过步骤 3 来展示批量配置端口的方式。

步骤 2　配置 G0/0/4 端口

第 4 章 构建以太网交换网络

在端口 G0/0/4 上,我们需要允许它传输 VLAN 20 和 VLAN 30 的流量,并且启用根据 MAC 地址划分 VLAN 的功能,具体见例 4-15。

例 4-15 在 G0/0/4 上完成配置

```
[S1]interface GigabitEthernet 0/0/4
[S1-GigabitEthernet0/0/4]port link-type hybrid
[S1-GigabitEthernet0/0/4]port hybrid untagged vlan 20 30
[S1-GigabitEthernet0/0/4]mac-vlan enable
Info: This operation may take a few seconds. Please wait for a moment...done.
[S1-GigabitEthernet0/0/4]
```

读者可以使用 **display current-configuration** 来查看端口配置,从例 4-16 的命令输出中我们可以看出,端口下没有我们配置的第一条命令 **port link-type hybrid**,这是因为端口的默认模式为 Hybrid。

例 4-16 查看 G0/0/4 的配置

```
[S1]display current-configuration interface GigabitEthernet 0/0/4
#
interface GigabitEthernet0/0/4
 port hybrid untagged vlan 20 30
 mac-vlan enable
#
return
```

步骤 3 批量配置 G0/0/5~G0/0/9 端口

要想批量配置多个端口,我们需要创建一个端口组,在其中指定要配置的端口范围,以及具体的配置命令。读者可以按照以下命令进行配置。

① 使用命令 **port-group** *port-group-name* 创建并进入端口组。

② 使用命令 **group-member** *interface-type interface-number* **to** *interface-type interface-number* 添加端口范围,本实验中添加的是端口 G0/0/5~G0/0/9。

③ 按需添加其他端口命令。本实验中添加了命令 **port hybrid untagged vlan 20 30** 和 **mac-vlan enable**。

例 4-17 中展示了本步骤的配置。

例 4-17 批量配置 G0/0/5~G0/0/9 端口

```
[S1]port-group port5-9
[S1-port-group-port5-9]group-member GigabitEthernet 0/0/5 to GigabitEthernet 0/0/9
[S1-port-group-port5-9]port hybrid untagged vlan 20 30
[S1-GigabitEthernet0/0/5]port hybrid untagged vlan 20 30
[S1-GigabitEthernet0/0/6]port hybrid untagged vlan 20 30
[S1-GigabitEthernet0/0/7]port hybrid untagged vlan 20 30
[S1-GigabitEthernet0/0/8]port hybrid untagged vlan 20 30
[S1-GigabitEthernet0/0/9]port hybrid untagged vlan 20 30
[S1-port-group-port5-9]mac-vlan enable
[S1-GigabitEthernet0/0/5]mac-vlan enable
Info: This operation may take a few seconds. Please wait for a moment...done.
[S1-GigabitEthernet0/0/6]mac-vlan enable
Info: This operation may take a few seconds. Please wait for a moment...done.
[S1-GigabitEthernet0/0/7]mac-vlan enable
Info: This operation may take a few seconds. Please wait for a moment...done.
[S1-GigabitEthernet0/0/8]mac-vlan enable
Info: This operation may take a few seconds. Please wait for a moment...done.
[S1-GigabitEthernet0/0/9]mac-vlan enable
Info: This operation may take a few seconds. Please wait for a moment...done.
[S1-port-group-port5-9]
```

4.2.4 验证配置效果

至此本实验的全部配置均已完成,我们可以通过在 PC 上发出 ping 命令来验证配置效果是否与需求相同。公司的通信需求如下。

① 相同 VLAN 中的 PC 之间可以相互访问。

② 不同 VLAN 中的 PC 之间不可以相互访问。

我们在 PC20-1 上分别对 PC20-2、PC20-10 和 PC30-1 发起 ping 测试，测试结果应该为前两个 ping 成功，最后一个 ping 失败。例 4-18 中展示了测试结果，与通信需求相同。

例 4-18　测试结果

```
PC20-1>ping 10.0.20.2

Ping 10.0.20.2: 32 data bytes, Press Ctrl_C to break
From 10.0.20.2: bytes=32 seq=1 ttl=128 time=31 ms
From 10.0.20.2: bytes=32 seq=2 ttl=128 time=31 ms
From 10.0.20.2: bytes=32 seq=3 ttl=128 time=31 ms
From 10.0.20.2: bytes=32 seq=4 ttl=128 time=32 ms
From 10.0.20.2: bytes=32 seq=5 ttl=128 time=47 ms

--- 10.0.20.2 ping statistics ---
  5 packet(s) transmitted
  5 packet(s) received
  0.00% packet loss
  round-trip min/avg/max = 31/34/47 ms

PC20-1>ping 10.0.20.10

Ping 10.0.20.10: 32 data bytes, Press Ctrl_C to break
From 10.0.20.10: bytes=32 seq=1 ttl=128 time=62 ms
From 10.0.20.10: bytes=32 seq=2 ttl=128 time=78 ms
From 10.0.20.10: bytes=32 seq=3 ttl=128 time=62 ms
From 10.0.20.10: bytes=32 seq=4 ttl=128 time=47 ms
From 10.0.20.10: bytes=32 seq=5 ttl=128 time=31 ms

--- 10.0.20.10 ping statistics ---
  5 packet(s) transmitted
  5 packet(s) received
  0.00% packet loss
  round-trip min/avg/max = 31/56/78 ms

PC20-1>ping 10.0.30.1

Ping 10.0.30.1: 32 data bytes, Press Ctrl_C to break
From 10.0.20.1: Destination host unreachable
From 10.0.20.1: Destination host unreachable
From 10.0.20.1: Destination host unreachable
From 10.0.20.1: Destination host unreachable
From 10.0.20.1: Destination host unreachable

--- 10.0.20.254 ping statistics ---
  5 packet(s) transmitted
  0 packet(s) received
  100.00% packet loss
```

看到这里可能有读者会质疑：不同 IP 网段中的主机需要借助路由器才能够实现相互之间的通信，上述实验虽然能够验证相同 VLAN 中的主机之间能够相互通信，但不足以说明不同 VLAN 之间的主机无法进行通信，因为 PC20-1 和 Server30 本就不处于同一个 IP 子网中。考虑到在真实网络环境中，我们一般会将 VLAN 与 IP 网段进行关联和区分，比如一个 VLAN 对应一个 IP 网段，因此本实验的 VLAN 设计与 IP 地址规划遵循了这种一般性原则。为了验证不同 VLAN 中的主机无法进行通信，读者可以在自己的实验环境中，将 PC20-2 更改到 VLAN 10 中，再从 PC20-1 上执行命令 ping 10.0.20.2，此时 ping 测试结果应为失败。

回到实验中，当 PC 向其直连的交换机端口发送数据帧后，交换机会记录从该端口接收到的 MAC 地址，并且对于 PC20-1 所连接的交换机 S1 G0/0/4 来说，该端口不仅会记录 PC20-1 的 MAC 地址，还会根据这个 MAC 地址为 PC20-1 分配 VLAN。

为了确认这一点，读者可以使用命令 **display mac-address dynamic GigabitEthernet 0/0/4** 看到交换机 S1 从该端口学习到了 PC20-1 的 MAC 地址，并且将其分配到 VLAN 20，具体见例 4-19。

例 4-19　查看 S1 在 G0/0/4 上学到的 MAC 地址

```
[S1]display mac-address dynamic GigabitEthernet 0/0/4
MAC address table of slot 0:
-------------------------------------------------------------------------------
MAC Address    VLAN/        PEVLAN CEVLAN Port              Type      LSP/LSR-ID
               VSI/SI                                                 MAC-Tunnel
-------------------------------------------------------------------------------
00e0-fc00-2001 20           -      -      GE0/0/4           dynamic   0/-
-------------------------------------------------------------------------------
Total matching items on slot 0 displayed = 1
```

另外，读者还可以使用命令 **display mac-address dynamic vlan 20** 来查看交换机 S1 通过上述连通性测试学习到的所有 MAC 地址（仅限 VLAN 20）。从命令的输出内容中我们可以看出，交换机 S1 不但学习到 PC20-1、PC20-2 的 MAC 地址，而且将其正确分配到 VLAN 20。

例 4-20　查看 S1 在 VLAN 20 中学习到的 MAC 地址

```
[S1]display mac-address dynamic vlan 20
MAC address table of slot 0:
-------------------------------------------------------------------------------
MAC Address    VLAN/        PEVLAN CEVLAN Port              Type      LSP/LSR-ID
               VSI/SI                                                 MAC-Tunnel
-------------------------------------------------------------------------------
00e0-fc00-2001 20           -      -      GE0/0/4           dynamic   0/-
00e0-fc00-2010 20           -      -      GE0/0/10          dynamic   0/-
00e0-fc00-2002 20           -      -      GE0/0/5           dynamic   0/-
-------------------------------------------------------------------------------
Total matching items on slot 0 displayed = 3
```

第 5 章
生成树基础实验

本章主要内容

5.1　实验介绍

5.2　实验配置任务

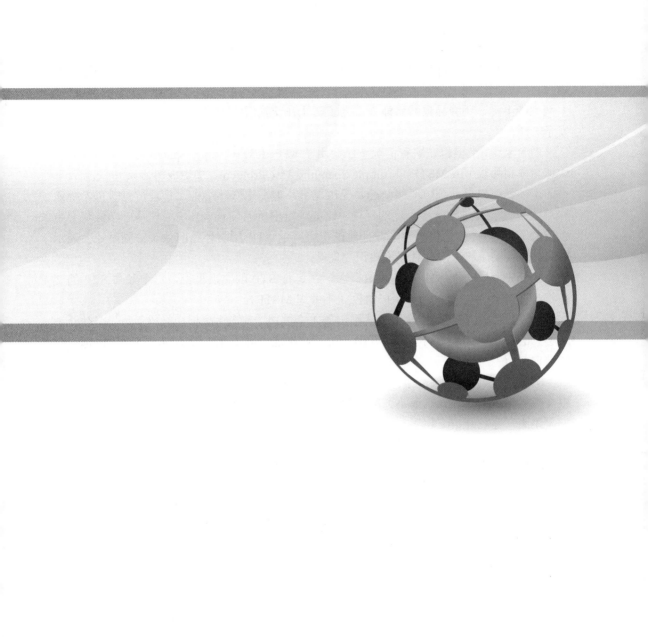

在网络环境中，我们通常会通过各种冗余部署来避免单点故障，包括设备冗余、链路冗余等。这个冗余部署需要避免单点故障，提供可恢复性，同时还不能为网络带来问题。在以太网环境中，冗余部署最容易产生的问题是交换环路，交换环路不仅会导致以太网性能下降，还会导致网络瘫痪。为了防止在交换网络中产生交换环路问题，人们开发了 STP。

通常来说，交换机在缺省情况下已启用了 STP、RSTP（Rapid Spanning Tree Protocol，快速生成树协议）或 MSTP（Multiple Spanning Tree Protocol，多生成树协议），无须管理员的干涉就可以自动预防环路问题并提供冗余。但在生产环境中，管理员有时仍希望能够对 STP 的操作进行一定程度的优化，比如确定根桥的位置等。生成树基础实验将会带领读者了解 STP 的优化配置，并且通过实验展示 STP 预防环路的效果，同时还会展示 MSTP 的效果，加深读者对理论知识的理解。

我们先回顾一下 STP 中的重要概念，在运行 STP 的交换机上，每个参与 STP 的交换机端口都会拥有一个 STP 端口角色（不考虑 MSTP）。

① 根端口：在非根交换机上选举出的 1 个端口，首选距离根交换机最近（路径开销最小）的端口，如果距离（路径开销）相等，次选对端桥 ID 最小的端口，否则选择对端端口 ID 最小的端口。

② 指定端口：在同一个网段上选举出的 1 个端口（根端口不参与选举），首选距离根交换机最近（路径开销最小）的端口（由此，根交换机上的所有端口都是指定端口），如果距离（路径开销）相等，次选桥 ID 最小的端口；否则选择端口 ID 最小的端口。

③ 预备端口：非根端口且非指定端口的端口，会侦听网段中传输的 BPDU（Bridge Protocol Data Unit，网桥协议数据单元），但不会转发数据。

表 5-1 对比了这 3 种端口角色。

表 5-1　STP 端口角色

端口角色	发送 BPDU	接收 BPDU	发送数据	接收数据
根端口	是	是	是	是
指定端口	是	是	是	是
预备端口	否	是	否	否

接下来本实验会带领读者直观地通过配置和 **display** 命令来理解 STP 的工作原理。

5.1　实验介绍

5.1.1　关于本实验

在本实验展示的微型企业环境中，公司网络分为两层结构：核心层和接入层。在多数情况下，根交换机位于核心层，以优化交换效率。实际上，在交换网络中，STP 默认可以正常工作，它会自动打破环路，并在网络出现故障时提供一定程度的冗余性和可恢复性。然而，如果没有人为干预，STP 在运行时有可能会产生次优的交换路径或较长的等待时间,因此管理员希望使用一些参数对 STP 的运行进行调整。本实验将会逐步带领读者了解这些参数对 STP 运行的影响,以及如何调整这些参数。

5.1.2 实验目的

- 了解如何切换 STP 模式:STP、RSTP、MSTP。
- 指定根交换机和备份根交换机。
- 调整交换机设备优先级。
- 调整端口路径开销。
- 调整端口优先级。
- 配置根保护。
- 配置 BPDU 保护。
- 配置边缘端口。
- 配置 MSTP。

5.1.3 实验组网介绍

生成树基础实验拓扑如图 5-1 所示。

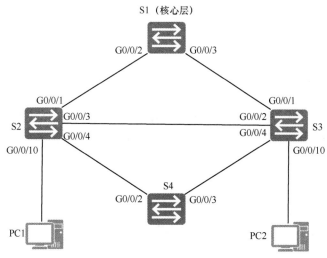

图 5-1 生成树基础实验拓扑

5.1.4 实验任务列表

配置任务 1:搭建拓扑,确认当前环境。
配置任务 2:启用 STP,并将 S1 设置为根交换机。
配置任务 3:启用交换机 S4,并观察 STP 端口状态机。
配置任务 4:调整 STP 参数,影响选举和收敛时间。
配置任务 5:配置 STP 保护参数。
配置任务 6:配置 MSTP。

5.2 实验配置任务

5.2.1 搭建拓扑，确认当前环境

读者可以按照图 5-1 搭建本实验拓扑。在本配置任务中，读者需要先开启交换机 S1、S2 和 S3，暂时不开启交换机 S4、PC1 和 PC2，并且将核心层交换机 S1 设置为根交换机，相应地让 STP 将交换机 S2 和 S3 之间的链路阻塞，以防止环路问题产生。

在本配置任务中，我们需要查看在交换机 S1、S2 和 S3 组成的交换网络中，STP 的默认运行结果，即 STP 端口角色。例 5-1～例 5-3 分别展示了交换机 S1、S2 和 S3 上 **display stp brief** 命令的输出内容。在读者自己搭建的实验环境中，初始状态可能与本例所示有所区别。

例 5-1 查看交换机 S1 上的 STP 端口角色

```
[S1]display stp brief
 MSTID  Port                        Role  STP State    Protection
   0    GigabitEthernet0/0/2        DESI  FORWARDING   NONE
   0    GigabitEthernet0/0/3        ROOT  FORWARDING   NONE
```

例 5-2 查看交换机 S2 上的 STP 端口角色

```
[S2]display stp brief
 MSTID  Port                        Role  STP State    Protection
   0    GigabitEthernet0/0/1        ALTE  DISCARDING   NONE
   0    GigabitEthernet0/0/3        ROOT  FORWARDING   NONE
```

例 5-3 查看交换机 S3 上的 STP 端口角色

```
[S3]display stp brief
 MSTID  Port                        Role  STP State    Protection
   0    GigabitEthernet0/0/1        DESI  FORWARDING   NONE
   0    GigabitEthernet0/0/2        DESI  FORWARDING   NONE
```

通过上述 3 个示例中的命令输出信息，我们可以看出，交换机 S3 是这个 STP 域中的根交换机，交换机 S2 的 G0/0/1 端口为阻塞状态，以防止环路，由此我们可以得出图 5-2 所示的初始的 STP 状态。

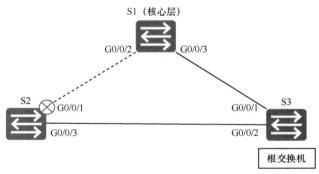

图 5-2 初始的 STP 状态

5.2.2 启用 STP，并将 S1 设置为根交换机

在这个实验网络中，交换机 S1 充当核心层交换机，它可能负责连接网络的其他部分或者连接外部网络（图中未展示），因此使 S1 成为根交换机是最优的做法。我们现在

手动干预网络中的 STP 运行。华为交换机默认运行的 STP 模式为 MSTP，本实验最后会展示 MSTP 的配置和效果，目前我们会将 3 台交换机的 STP 模式更改为 STP，同时将交换机 S1 设置为根交换机。例 5-4～例 5-6 分别展示了交换机 S1、S2 和 S3 上相应的配置命令。

例 5-4 在 S1 上配置 STP 并将其设置为根交换机

```
<S1>system-view
Enter system view, return user view with Ctrl+Z.
[S1]stp enable
[S1]stp mode stp
Info: This operation may take a few seconds. Please wait for a moment...done.
[S1]stp root primary
```

例 5-5 在 S2 上配置 STP

```
<S2>system-view
Enter system view, return user view with Ctrl+Z.
[S2]stp mode stp
Info: This operation may take a few seconds. Please wait for a moment...done.
[S2]
```

例 5-6 在 S3 上配置 STP

```
<S3>system-view
Enter system view, return user view with Ctrl+Z.
[S3]stp mode stp
Info: This operation may take a few seconds. Please wait for a moment...done.
[S3]
```

在例 5-4 所示的 S1 配置中，管理员分别配置了以下命令。

① **stp enable**：全局启用 STP 的命令。默认情况下，华为交换机已启用 STP，因此本例中该命令仅作为展示，并不会改变任何配置。用户可以在接口配置视图中使用相同的命令，在接口上启用 STP。**undo stp enable** 命令为禁用 STP，用户可以在系统视图中使用该命令全局禁用 STP，也可以在接口配置视图中使用该命令在该接口上禁用 STP。

② **stp mode stp**：改变 STP 的模式，可选关键字包括 **stp**、**rstp** 和 **mstp**，默认配置为 mstp。

③ **stp root primary**：将交换机设置为 STP 根交换机。实际上，它的效果是将设备优先级的值更改为 0，优先级的值默认为 32768，值越低，优先级越高，因此这台交换机会成为 STP 根交换机。若管理员在多台交换机上都配置了该命令，则系统 MAC 地址最小的交换机会成为根交换机。以下两条命令与系统优先级相关。

a. **stp root secondary**：将交换机设置为 STP 备用根交换机，即将设备优先级更改为 4096。

b. **stp priority** *priority*：可以配置交换机在 STP 域中的设备优先级，需要以 4096 的倍数进行设置。在通过上述两条命令设置为根交换机或备用根交换机的设备上，无法使用该命令更改设备优先级，也就是说若用户已经使用 **stp root primary** 或 **stp root secondary** 命令进行了配置，则在使用 **stp priority** *priority* 命令更改优先级之前，需要先执行 **undo stp root** 命令。

交换机 S2 和 S3 的配置与 S1 类似，仅更改了 STP 模式，保留默认的系统优先级。在这个设置的基础上，我们可以再次查看 3 台交换机上的 STP 端口角色，具体见例 5-7～例 5-9。

例 5-7 再次查看 S1 上的 STP 端口角色

```
[S1]display stp brief
 MSTID  Port                     Role  STP State    Protection
   0       GigabitEthernet0/0/2   DESI  FORWARDING   NONE
   0       GigabitEthernet0/0/3   DESI  FORWARDING   NONE
```

例 5-8 再次查看 S2 上的 STP 端口角色

```
[S2]display stp brief
 MSTID  Port                     Role  STP State    Protection
   0       GigabitEthernet0/0/1   ROOT  FORWARDING   NONE
   0       GigabitEthernet0/0/3   ALTE  DISCARDING   NONE
```

例 5-9 再次查看 S3 上的 STP 端口角色

```
[S3]display stp brief
 MSTID  Port                     Role  STP State    Protection
   0       GigabitEthernet0/0/1   ROOT  FORWARDING   NONE
   0       GigabitEthernet0/0/2   DESI  FORWARDING   NONE
```

现在我们成功将交换机 S1 设置为 STP 根交换机，S2 和 S3 上与 S1 相连的端口成为本地交换机上的根端口，S2 上的 G0/0/3 端口被阻塞以打破环路，如图 5-3 所示。

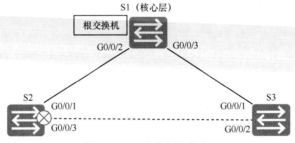

图 5-3 S1 成为根交换机

读者还可以使用 **display stp** 来查看更多有关 STP 的详细信息，具体见例 5-10。

例 5-10 在 S2 上使用 display stp 命令查看更多详情

```
[S2]display stp
-------[CIST Global Info][Mode STP]-------
CIST Bridge          :32768.4c1f-cc99-1add
Config Times         :Hello 2s MaxAge 20s FwDly 15s MaxHop 20
Active Times         :Hello 2s MaxAge 20s FwDly 15s MaxHop 20
CIST Root/ERPC       :0    .4c1f-cc24-61e2 / 20000
CIST RegRoot/IRPC    :32768.4c1f-cc99-1add / 0
CIST RootPortId      :128.1
BPDU-Protection      :Disabled
TC or TCN received   :81
TC count per hello   :0
STP Converge Mode    :Normal
Time since last TC   :0 days 2h:50m:35s
Number of TC         :19
Last TC occurred     :GigabitEthernet0/0/1
----[Port1(GigabitEthernet0/0/1)][FORWARDING]----
 Port Protocol         :Enabled
 Port Role             :Root Port
 Port Priority         :128
 Port Cost(Dot1T )     :Config=auto / Active=20000
 Designated Bridge/Port :0.4c1f-cc24-61e2 / 128.2
 Port Edged            :Config=default / Active=disabled
 Point-to-point        :Config=auto / Active=true
 Transit Limit         :147 packets/hello-time
 Protection Type       :None
 Port STP Mode         :STP
 Port Protocol Type    :Config=auto / Active=dot1s
 BPDU Encapsulation    :Config=stp / Active=stp
 PortTimes             :Hello 2s MaxAge 20s FwDly 15s RemHop 20
 TC or TCN send        :1
 TC or TCN received    :18
```

```
 BPDU Sent              :3
        TCN: 1, Config: 2, RST: 0, MST: 0
 BPDU Received          :3787
        TCN: 0, Config: 3787, RST: 0, MST: 0
（省略部分输出）
----[Port3(GigabitEthernet0/0/3)][DISCARDING]----
 Port Protocol          :Enabled
 Port Role              :Alternate Port
 Port Priority          :128
 Port Cost(Dot1T )      :Config=auto / Active=20000
 Designated Bridge/Port :32768.4c1f-cc04-583f / 128.2
 Port Edged             :Config=default / Active=disabled
 Point-to-point         :Config=auto / Active=true
 Transit Limit          :147 packets/hello-time
 Protection Type        :None
 Port STP Mode          :STP
 Port Protocol Type     :Config=auto / Active=dot1s
 BPDU Encapsulation     :Config=stp / Active=stp
 PortTimes              :Hello 2s MaxAge 20s FwDly 15s RemHop 20
 TC or TCN send         :1
 TC or TCN received     :23
 BPDU Sent              :2
        TCN: 1, Config: 1, RST: 0, MST: 0
 BPDU Received          :3774
        TCN: 0, Config: 3774, RST: 0, MST: 0
（省略部分输出）
```

命令输出中显示了三部分信息：全局信息、G0/0/1 和 G0/0/3 端口信息，示例中使用阴影标注了每部分的标题，并省略了其他信息来突出重点。display stp 命令中的全局信息见表 5-2。

表 5-2　display stp 命令中的全局信息

项目	本例中的值	描述
[CIST Global Info] [Mode STP]	Mode STP	STP 模式为 STP，默认为 MSTP
CIST Bridge	32768.4c1f-cc99-1add	交换机 BID（Bridge ID，网桥 ID）由优先级和交换机 MAC 地址组成
Config Times	Hello 2s MaxAge 20s FwDly 15s MaxHop 20	交换机本地配置的时间值如下。 • Hello：BPDU 发送周期，默认为 2s。 • MaxAge：BPDU 最大生存时间，默认为 20s。 • FwDly：端口状态转换时延，默认为 15s。 • MaxHop：最大支持跳数，默认为 20 跳
Active Times	Hello 2s MaxAge 20s FwDly 15s MaxHop 20	交换机实际使用的时间值如下。 • Hello：BPDU 发送周期，默认为 2s。 • MaxAge：BPDU 最大生存时间，默认为 20s。 • FwDly：端口状态转换时延，默认为 15s。 • MaxHop：最大支持跳数，默认为 20 跳
CIST Root/ERPC	0.4c1f-cc24-61e2 / 20000	CIST 总根交换机的 BID 及相关开销
CIST RegRoot/IRPC	32768.4c1f-cc99-1add / 0	CIST 域根交换机的 BID 及相关开销
CIST RootPortId	128.1	CIST 根端口 ID，在根交换机上，该值为 "0.0"，表示没有根端口

表 5-2　display stp 命令中的全局信息（续）

项目	本例中的值	描述
BPDU-Protection	Disabled	BPDU 保护功能如下。 • Disabled：未启用 BPDU 保护。 • Enabled：已启用 BPDU 保护
TC or TCN received	81	接收到的 TC（Topology Change，拓扑变更）或 TCN（Topology Change Notification，拓扑变更通知）报文数量
TC count per hello	0	每个 Hello 时间接收的 TC 报文总数
STP Converge Mode	Normal	STP 的收敛方式如下。 • Normal：正常。 • Fast：快速
Time since last TC	0 days 2h:50m:35s	自上次拓扑变更后，到现在所经过的时间
Number of TC	19	拓扑变更的次数
Last TC occurred	GigabitEthernet0/0/1	最后一次引起拓扑变更的端口

接下来交换机会按顺序显示各个端口的信息，每部分的格式都是相同的，display stp 命令中的端口信息见表 5-3。

表 5-3　display stp 命令中的端口信息

项目	本例中的值	描述
Portx	GigabitEthernet0/0/1 和 GigabitEthernet0/0/3	端口号
端口状态	FORWARDING 和 DISCARDING	端口状态包括以下内容。 • DOWN：端口的物理状态为 Down。 • LEARNING：学习状态的端口，接收且发送 BPDU，但不转发数据。 • FORWARDING：转发状态的端口，接收且发送 BPDU，接收且转发数据。 • DISCARDING：被阻塞的端口，接收 BPDU，但不发送 BPDU，也不转发数据
Port Protocol	Enabled	端口 STP 状态如下。 • Enabled：已启用。 • Disabled：未启用
Port Role	Root Port 和 Alternate Port	端口 STP 角色包括以下内容。 • Root Port：根端口。 • Designated Port：指定端口。 • Alternate Port：替换端口。 • Backup Port：备份端口
Port Priority	128	端口优先级，默认值为 128

表 5-3 display stp 命令中的端口信息（续）

项目	本例中的值	描述
Port Cost(Dot1T)	Config=auto/Active=20000	端口路径开销（采用 Dot1T 计算方法），Config 是管理员手动配置的路径开销，Active 是实际使用的路径开销
Designated Bridge/Port	0.4c1f-cc24-61e2 / 128.2 和 32768.4c1f-cc04-583f / 128.2	指定交换机 BID 和端口
Port Edged	Config=default/Active=disabled	边缘端口特性，默认未启用
Point-to-point	Config=auto / Active=true	端口的链路类型。STP 会自动检测链路类型是否为点对点链路
Transit Limit	147 packets/hello-time	端口在每 Hello 时间内发送 BPDU 的最大数目
Protection Type	None	保护类型包括以下内容。 • None：未启用。 • Root：根保护，仅适用于指定端口。 • Loop：环路保护，仅适用于根端口或替换端口。
Port STP Mode	STP	端口的 STP 模式
Port Protocol Type	Config=auto/Active=dot1s	端口接收和发送报文的格式，包括： • auto（默认值）； • dot1s； • legacy
PortTimes	Hello 2s MaxAge 20s FwDly 15s RemHop 0	端口使用的 STP 时间如下。 • Hello：BPDU 发送周期，默认为 2s。 • MaxAge：BPDU 最大生存时间，默认为 20s。 • FwDly：端口状态转换时延，默认为 15s。 • RemHop：最大支持跳数，默认为 20 跳
TC or TCN send	1	端口发送的 TC 或 TCN 报文的数量
TC or TCN received	18 和 23	端口接收的 TC 或 TCN 报文的数量
BPDU Sent	TCN: 1, Config: 2, RST: 0, MST: 0 和 TCN: 1, Config: 1, RST: 0, MST: 0	发送的 BPDU 信息如下。 • TCN：拓扑变化通知报文。 • Config：STP 报文。 • RST：RSTP 报文。 • MST：MSTP 报文
BPDU Received	TCN: 0, Config: 3787, RST: 0, MST: 0 和 TCN: 0, Config: 3774, RST: 0, MST: 0	接收的 BPDU 信息如下。 • TCN：拓扑变化通知报文。 • Config：STP 报文。 • RST：RSTP 报文。 • MST：MSTP 报文

在这个由 3 台交换机构成的简单 STP 域中，我们按照选举根交换机、根端口、指定

端口以及阻塞剩余端口的顺序，分别为读者描述每一步的选举是基于哪些参数进行的。

① STP 域中根交换机的选举方式如下。

STP 域中的交换机通过 BPDU 对比彼此的 BID，BID 是由设备优先级和系统 MAC 地址构成的组合值。我们在交换机 S1 上使用命令 **stp root primary** 将 S1 的设备优先级更改为 0，因此它拥有最高优先级（值越小，优先级越高），从而成为根交换机。

② 每台交换机上根端口的选举方式如下。

根端口是通过对比本地交换机每个端口到达根交换机的路径开销来进行选举的，开销越小，优先级越高。交换机会对比它从每个端口接收的 BPDU 中的根路径开销（Root Path Cost，RPC）。交换机 S1 作为根交换机，它向 S2 和 S3 发送的 BPDU 中的 RPC 为 0，因此 S2 和 S3 上连接 S1 的端口会成为本地交换机上的根端口（本实验拓扑中交换机 S4 上的根端口无法只通过这一个步骤选举出来，因为它从 S2 和 S3 接收的 BPDU 中的 RPC 是相同的，需要进行第二步对比才能选出自己的根端口，详见下一个实验配置任务）。

③ 每个网段上指定端口的选举方式如下。

通过对比这个网段上所有 BPDU 中的 RPC，RPC 最小的端口成为指定端口。STP 域中有三个网段：S1 与 S2 之间的网段、S1 与 S3 之间的网段、S2 与 S3 之间的网段。其中前两个网段都可以通过这一步选出指定端口，即 S1 上的端口分别为这两个网段中的指定端口，因为从这两个端口发出的 BPDU 中的 RPC 为 0，开销值越小，优先级越高。

对于 S2 与 S3 之间的网段，两个端口发出的 BPDU 中的 RPC 都为 20000，因此它们需要进行下一步对比，即对比 BID。因为实验中没有更改过 S2 与 S3 的设备优先级，所以需要对比两者的系统 MAC 地址。从例 5-10 的输出内容中我们可以看出，S2 的 MAC 地址是 4c1f-cc99-1add，S3 的 MAC 地址是 4c1f-cc04-583f，值较小的端口成为指定端口，则 S3 的端口获选。

④ 阻塞剩余的端口。

最后一步是将非根端口、非指定端口的端口阻塞，此时只有交换机 S2 上用来连接 S3 的端口没有赢得任何选举，因此它会被阻塞。

5.2.3　启用交换机 S4，并观察 STP 端口状态机

当一台新交换机加入 STP 域时，相关端口会经历 STP 端口状态机，并最终达到稳定状态。在这个实验中我们要启用交换机 S4，并将其加入网络中。读者可以先不连接 S4 上的任何链路，启动并进行初始化配置，然后再按照拓扑所示进行连接。例 5-11 展示了 S4 上的初始配置。

例 5-11　交换机 S4 的初始化配置

```
<Huawei>system-view
Enter system view, return user view with Ctrl+Z.
[Huawei]sysname S4
[S4]stp mode stp
Info: This operation may take a few seconds. Please wait for a moment...done.
[S4]
```

配置完成后，读者可以按照图 5-1 连接线缆。在交换机 S4 分别与 S2 和 S3 相连后，有两个网段加入了 STP 域。以交换机 S2 为例介绍 STP 的选举过程。

① 根交换机：仍为 S1。S2 在确定谁是根交换机的过程中，会对比 3 个接口（G0/0/1、

G0/0/3 和 G0/0/4）接收到的 BPDU 中的"根交换机 BID"。也就是 S1、S3 和 S4 发出的 BPDU 中的根交换机信息（0.4c1f-cc24-61e2、0.4c1f-cc24-61e2 和 32768.4c1f-cc04-3e1b，S4 会以自身为根），通过优先级可以判断出，0.4c1f-cc24-61e2（S1）优先级更高，因此 S1 为根交换机。

② 根端口：仍为 G0/0/1。S2 在确定谁是根端口的过程中，会对比 3 个接口（G0/0/1、G0/0/3 和 G0/0/4）接收到的 BPDU 中的"RPC"。也就是 S1、S3 和 S4 发出的 BPDU 中的根路径开销（0、20000 和 40000），通过对比发现，S1 的 RPC 更小，因此连接 S1 的端口 G0/0/1 为根端口。

③ 指定端口：G0/0/4。S2 在确定谁是指定端口的过程中，会通过它与 S4 之间的网段上的 BPDU 进行判断，此时对比的参数为"RPC"。S2 BPDU 中的 RPC 为 20000，S4 BPDU 中的 RPC 为 40000，因此 S2 的 G0/0/4 端口成为这个网段中的指定端口。图 5-4 展示了这个网段抓包的截图，其中显示出的两个 BPDU 分别是 S2 和 S4 发送的。从中我们可以看出，S2 的 Cost 为 20000，S4 的 Cost 为 40000（以阴影显示）。

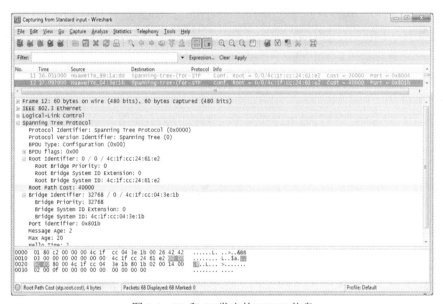

图 5-4　S2 和 S4 发出的 BPDU 信息

④ 至此 S2 上的端口角色都已确定，G0/0/4 端口作为指定端口，需要进入转发状态。但这个端口的 STP 状态会在进入 FORWARDING 状态之前，先经历 DISCARDING、LEARNING 状态，这是为了避免在网络收敛完成之前将错误的端口置于转发状态，而引发交换环路问题。例 5-12 展示了在交换机 S2 上使用 **display stp brief** 观察端口状态变化的过程（华为交换机在缺省情况下运行 MSTP 模式，当交换机从 MSTP 模式切换到 STP 模式时，运行 STP 的交换机端口所支持的端口状态仍然保持和 MSTP 支持的端口状态一致，即支持的状态仅包括 FORWARDING、LEARNING 和 DISCARDING。华为交换机将 LISTENING 和 LEARNING 状态合并显示为 LEARNING，因此读者在实验过程中，无法通过 **display stp brief** 命令看到 LISTENING 状态，但 LISTENING 状态的 15s 时延仍然存在）。

例 5-12　通过交换机 S2 观察端口 STP 状态变化

```
[S2]display stp brief
MSTID   Port                       Role   STP State    Protection
  0     GigabitEthernet0/0/1       ROOT   FORWARDING   NONE
  0     GigabitEthernet0/0/3       ALTE   DISCARDING   NONE
  0     GigabitEthernet0/0/4       DESI   DISCARDING   NONE
[S2]display stp brief
MSTID   Port                       Role   STP State    Protection
  0     GigabitEthernet0/0/1       ROOT   FORWARDING   NONE
  0     GigabitEthernet0/0/3       ALTE   DISCARDING   NONE
  0     GigabitEthernet0/0/4       DESI   LEARNING     NONE
[S2]display stp brief
MSTID   Port                       Role   STP State    Protection
  0     GigabitEthernet0/0/1       ROOT   FORWARDING   NONE
  0     GigabitEthernet0/0/3       ALTE   DISCARDING   NONE
  0     GigabitEthernet0/0/4       DESI   FORWARDING   NONE
```

交换机 S3 上的 STP 收敛过程与 S2 相同，新加入 STP 域的端口 G0/0/4 会成为指定端口，并最终进入转发状态，在此不详细描述。以下是交换机 S4 发生的选举过程。

① 根交换机：S1。选举过程不做赘述。

② 根端口：S4 无法通过第 1 步的 RPC 选举出根端口，因为 S2 和 S3 BPDU 中的 RPC 都为 20000。S4 需要进行第 2 步选举，即对比 S2 和 S3 的 BID：32768.4c1f-cc99-1add 和 32768.4c1f-cc04-583f，BID 较小的 S3 获选，因此与 S3 相连的 G0/0/3 端口成为 S4 上的根端口，并最终进入转发状态。

③ 指定端口：在 G0/0/2 所连网段的指定端口选举中，G0/0/2 端口落选，从而成为被阻塞的端口。

最终，在 S4 加入网络并且 STP 收敛后，网络的状态如图 5-5 所示。

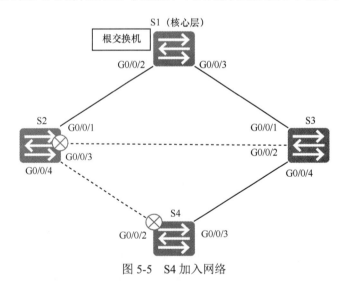

图 5-5　S4 加入网络

5.2.4　调整 STP 参数，影响选举和收敛时间

在 5.2.2 配置任务中，我们展示了如何通过更改设备优先级来指定 STP 域中的根交换机。本部分将首先展示通过 STP 参数将 S4 的根端口变更为 G0/0/2，然后展示如何统一更改 STP 域中的 STP 计时器。

根端口的选举规则为：首选与根交换机路径开销最小的端口；如果路径开销相等，

次选对端桥 ID 最小的端口,否则选择对端端口 ID 最小的端口。S4 是根据第 2 步选出的根端口,在本实验中,我们首先通过更改设备优先级,使 S4 认为 G0/0/2 为根端口,再通过更改端口路径开销,将 G0/0/3 再次选举为根端口。

步骤 1 更改设备优先级

首先,更改端口优先级需要在系统视图下使用命令 **stp priority** *priority*。其次,在根端口的选举过程中,交换机 S4 会对从本地所有 STP 端口(G0/0/2 和 G0/0/3)接收的 BPDU 中的对端交换机 BID 进行对比。我们可以降低 S2 的设备优先级,使 S4 将根端口更改为 G0/0/2,例 5-13 展示了 S2 的配置。

例 5-13 更改 S2 的设备优先级

```
[S2]stp priority 8192
```

在更改设备优先级后,我们在 S2 的 G0/0/4 端口上开启抓包,可以看到 S2 发送的 BPDU 的设备 BID 的优先级值不再是 32768,而是 8192,如图 5-6 中的阴影部分所示。在图中,读者还可以发现编号为 21 的数据包是由 S4 发出的 STP TCN 报文,S4 上的拓扑发生变化时,它会通过根端口向上游交换机发送 TCN 报文,告知下游的拓扑发生了变化。

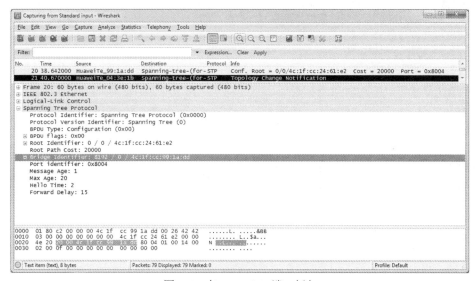

图 5-6 在 S2 G0/0/4 端口抓包

当交换机 S4 重新选举出新的根端口后,新的根端口也同样需要经历端口状态转换时延,并最终进入转发状态,见例 5-14。

例 5-14 观察 S4 的根端口变化

```
[S4]display stp brief
 MSTID  Port                        Role  STP State     Protection
   0    GigabitEthernet0/0/2        ALTE  DISCARDING    NONE
   0    GigabitEthernet0/0/3        ROOT  FORWARDING    NONE
[S4]display stp brief
 MSTID  Port                        Role  STP State     Protection
   0    GigabitEthernet0/0/2        ROOT  DISCARDING    NONE
   0    GigabitEthernet0/0/3        ALTE  DISCARDING    NONE
[S4]display stp brief
 MSTID  Port                        Role  STP State     Protection
   0    GigabitEthernet0/0/2        ROOT  LEARNING      NONE
   0    GigabitEthernet0/0/3        ALTE  DISCARDING    NONE
[S4]display stp brief
```

```
MSTID    Port                          Role  STP State   Protection
  0      GigabitEthernet0/0/2          ROOT  FORWARDING  NONE
  0      GigabitEthernet0/0/3          ALTE  DISCARDING  NONE
```

步骤 2 更改端口路径开销

根端口选举的第 1 步是对比根路径开销，从拓扑中可以看出，S4 去往根交换机的两条路径拥有相同的根路径开销。我们要通过配置减小 G0/0/3 的端口路径开销，让 S4 恢复使用 G0/0/3 端口作为根端口。

首先我们通过命令 **display stp interface GigabitEthernet 0/0/3** 查看 S4 G0/0/3 端口当前（默认）的路径开销和根路径开销，具体见例 5-15。第 1 个阴影部分中的 40000 是当前端口的根路径开销，它是 S4 的 G0/0/3 端口路径开销与 S3 的 G0/0/1 端口路径开销之和；第 2 个阴影部分显示出当前的配置是默认值（Config=auto），且开销值为 20000（Active=20000）。

例 5-15 查看默认开销

```
[S4]display stp interface GigabitEthernet 0/0/3
-------[CIST Global Info][Mode STP]-------
CIST Bridge         :32768.4c1f-cc04-3e1b
Config Times        :Hello 2s MaxAge 20s FwDly 15s MaxHop 20
Active Times        :Hello 2s MaxAge 20s FwDly 15s MaxHop 20
CIST Root/ERPC      :0    .4c1f-cc24-61e2 / 40000
CIST RegRoot/IRPC   :32768.4c1f-cc04-3e1b / 0
CIST RootPortId     :128.27
BPDU-Protection     :Disabled
TC or TCN received  :315
TC count per hello  :0
STP Converge Mode   :Normal
Time since last TC  :0 days 3h:48m:20s
Number of TC        :14
Last TC occurred    :GigabitEthernet0/0/2
 ----[Port3(GigabitEthernet0/0/3)][DISCARDING]----
 Port Protocol       :Enabled
 Port Role           :Alternate Port
 Port Priority       :128
 Port Cost(Dot1T )   :Config=auto / Active=20000
 Designated Bridge/Port   :32768.4c1f-cc04-583f / 128.4
 Port Edged          :Config=default / Active=disabled
 Point-to-point      :Config=auto / Active=true
 Transit Limit       :147 packets/hello-time
 Protection Type     :None
 Port STP Mode       :STP
 Port Protocol Type  :Config=auto / Active=dot1s
 BPDU Encapsulation  :Config=stp / Active=stp
 PortTimes           :Hello 2s MaxAge 20s FwDly 15s RemHop 0
 TC or TCN send      :1
 TC or TCN received  :122
 BPDU Sent           :3
         TCN: 1, Config: 2, RST: 0, MST: 0
 BPDU Received       :10911
         TCN: 0, Config: 10911, RST: 0, MST: 0
```

为了调整端口的路径开销，我们需要使用接口配置视图的命令 **stp cost** *cost*，具体见例 5-16。

例 5-16 更改 S4 G0/0/3 端口开销

```
[S4]interface GigabitEthernet 0/0/3
[S4-GigabitEthernet0/0/3]stp cost 10000
```

更改后，S4 G0/0/3 端口的 STP 开销为 10000，根路径开销变为 30000，从而它会重新被选举为 S4 的根端口。例 5-17 中以阴影显示了根路径开销和端口开销，另外，读者还可以看到这个端口的角色已经从替换端口（Port Role :Alternate Port）变为了根端口（Port Role :Root Port）。

例 5-17 查看更改后的开销

```
[S4]display stp interface GigabitEthernet 0/0/3
-------[CIST Global Info][Mode STP]-------
CIST Bridge            :32768.4c1f-cc04-3e1b
Config Times           :Hello 2s MaxAge 20s FwDly 15s MaxHop 20
Active Times           :Hello 2s MaxAge 20s FwDly 15s MaxHop 20
CIST Root/ERPC         :0       .4c1f-cc24-61e2 / 30000
CIST RegRoot/IRPC      :32768.4c1f-cc04-3e1b / 0
CIST RootPortId        :128.3
BPDU-Protection        :Disabled
TC or TCN received     :342
TC count per hello     :0
STP Converge Mode      :Normal
Time since last TC     :0 days 3h:49m:25s
Number of TC           :14
Last TC occurred       :GigabitEthernet0/0/2
----[Port3(GigabitEthernet0/0/3)][LEARNING]----
 Port Protocol          :Enabled
 Port Role              :Root Port
 Port Priority          :128
 Port Cost(Dot1T )      :Config=10000 / Active=10000
 Designated Bridge/Port :32768.4c1f-cc04-583f / 128.4
 Port Edged             :Config=default / Active=disabled
 Point-to-point         :Config=auto / Active=true
 Transit Limit          :147 packets/hello-time
 Protection Type        :None
 Port STP Mode          :STP
 Port Protocol Type     :Config=auto / Active=dot1s
 BPDU Encapsulation     :Config=stp / Active=stp
 PortTimes              :Hello 2s MaxAge 20s FwDly 15s RemHop 0
 TC or TCN send         :2
 TC or TCN received     :137
 BPDU Sent              :4
         TCN: 2, Config: 2, RST: 0, MST: 0
 BPDU Received          :10942
         TCN: 0, Config: 10942, RST: 0, MST: 0
```

步骤 3 更改 STP 计时器值

STP 的收敛速度较慢，但仍有一些参数能够影响它的收敛时间，这些参数包括 Hello、MaxAge、FwDly、网络直径等。默认情况下，使用 STP 的默认计时器设置就可以让 STP 顺利地工作，因此除非管理员熟知计时器的设置规则及其对网络的影响，否则不建议更改默认值。本例仅作为参考，为读者展示如何更改计时器值。例 5-18 展示了默认的 STP 计时器值。

例 5-18 默认的 STP 计时器值

```
[S1]display stp
-------[CIST Global Info][Mode STP]-------
CIST Bridge            :0       .4c1f-cc24-61e2
Config Times           :Hello 2s MaxAge 20s FwDly 15s MaxHop 20
Active Times           :Hello 2s MaxAge 20s FwDly 15s MaxHop 20
CIST Root/ERPC         :0       .4c1f-cc24-61e2 / 0
CIST RegRoot/IRPC      :0       .4c1f-cc24-61e2 / 0
CIST RootPortId        :0.0
BPDU-Protection        :Disabled
CIST Root Type         :Primary root
TC or TCN received     :81
TC count per hello     :0
STP Converge Mode      :Normal
Time since last TC     :0 days 2h:45m:7s
Number of TC           :32
Last TC occurred       :GigabitEthernet0/0/3
(省略部分输出)
```

如需更改 STP 计时器值，就要确保在 STP 域中参与 STP 运行的所有交换机使用统一的值，因此，我们只需要在 STP 根交换机上进行配置，STP 根交换机可以自动将更改后的计时器值同步给 STP 域中的其他非根交换机。

在更改根交换机上的计时器值时，我们需要遵守以下规则，否则网络会频繁震荡。

- 2 × (FwDly – 1.0 秒)≥MaxAge
- MaxAge≥2 × (Hello + 1.0 秒)

我们可以使用以下命令来更改计时器值。

- **stp timer forward-delay** *forward-delay*：配置设备的转发时延，以厘秒为单位，默认值为 1500 厘秒，即 15 秒。
- **stp timer hello** *hello-time*：配置设备的 Hello 时间，以厘秒为单位，默认值为 200，即 2 秒。
- **stp timer max-age** *max-age*：设置设备的 MaxAge 时间，以厘秒为单位，默认值为 2000 厘秒，即 20 秒。

我们在 S1 上将 STP 计时器值更改为 Hello 4 秒、MaxAge 10 秒、FwDly 6 秒，具体见例 5-19。

例 5-19　在根交换机上更改 STP 计时器值

```
[S1]stp timer hello 400
[S1]stp timer max-age 1000
[S1]stp timer forward-delay 600
```

更改后，我们可以通过命令 **display stp** 查看当前的计时器值，具体见例 5-20，配置的计时器值和实际使用的计时器值相同。

例 5-20　验证 S1 上的 STP 计时器值

```
[S1]display stp
-------[CIST Global Info][Mode STP]-------
CIST Bridge         :0    .4c1f-cc24-61e2
Config Times        :Hello 4s MaxAge 10s FwDly 6s MaxHop 20
Active Times        :Hello 4s MaxAge 10s FwDly 6s MaxHop 20
CIST Root/ERPC      :0    .4c1f-cc24-61e2 / 0
CIST RegRoot/IRPC   :0    .4c1f-cc24-61e2 / 0
CIST RootPortId     :0.0
BPDU-Protection     :Disabled
CIST Root Type      :Primary root
TC or TCN received  :81
TC count per hello  :0
STP Converge Mode   :Normal
Time since last TC  :0 days 3h:2m:20s
Number of TC        :32
Last TC occurred    :GigabitEthernet0/0/3
（省略部分输出）
```

为了验证根交换机会自动将新的计时器值同步给 STP 域中的所有交换机，我们可以使用 **display stp** 命令查看 S4 上的计时器值，具体见例 5-21。从命令输出内容上，我们可以看出，S4 上实际使用的值已经发生了变化，并与 S1 进行了同步。

例 5-21　验证 S4 上的 STP 计时器值

```
[S4]display stp
-------[CIST Global Info][Mode STP]-------
CIST Bridge         :32768.4c1f-cc04-3e1b
Config Times        :Hello 2s MaxAge 20s FwDly 15s MaxHop 20
Active Times        :Hello 4s MaxAge 10s FwDly 6s MaxHop 20
CIST Root/ERPC      :0    .4c1f-cc24-61e2 / 30000
CIST RegRoot/IRPC   :32768.4c1f-cc04-3e1b / 0
CIST RootPortId     :128.3
BPDU-Protection     :Disabled
TC or TCN received  :369
TC count per hello  :0
STP Converge Mode   :Normal
Time since last TC  :0 days 3h:3m:20s
Number of TC        :16
Last TC occurred    :GigabitEthernet0/0/3
（省略部分输出）
```

步骤 4 更改 STP 网络直径

STP 网络直径是指 STP 网络中任意两台终端设备之间间隔的最大设备数量，网络规模越大，网络直径就越大。在小型网络中，为了加快收敛速度，我们可以调整 STP 网络直径，使交换机自动优化上述计时器值。更改 STP 网络直径的命令如下。

stp bridge-diameter *diameter*：配置 STP 网络直径，默认值为 7。

例 5-22 展示了如何在根交换机 S1 上将网络直径配置为 3，以及查看更改后的计时器。通过命令的输出内容，我们可以看出，交换机根据 STP 网络直径为 3，自动计算出最优的计时器设置：Hello 2 秒、MaxAge 12 秒、FwDly 9 秒。

例 5-22 在根交换机上配置 STP 网络直径

```
[S1]stp bridge-diameter 3
[S1]display stp
-------[CIST Global Info][Mode STP]-------
CIST Bridge            :0      .4c1f-cc24-61e2
Config Times           :Hello 2s MaxAge 12s FwDly 9s MaxHop 20
Active Times           :Hello 2s MaxAge 12s FwDly 9s MaxHop 20
CIST Root/ERPC         :0      .4c1f-cc24-61e2 / 0
CIST RegRoot/IRPC      :0      .4c1f-cc24-61e2 / 0
CIST RootPortId        :0.0
BPDU-Protection        :Disabled
CIST Root Type         :Primary root
TC or TCN received     :81
TC count per hello     :0
STP Converge Mode      :Normal
Time since last TC     :0 days 3h:14m:2s
Number of TC           :32
Last TC occurred       :GigabitEthernet0/0/3
----[Port1(GigabitEthernet0/0/1)][DOWN]----
 Port Protocol         :Enabled
 Port Role             :Disabled Port
 Port Priority         :128
 Port Cost(Dot1T )     :Config=auto / Active=200000000
 Designated Bridge/Port   :0.4c1f-cc24-61e2 / 128.1
 Port Edged            :Config=default / Active=disabled
 Point-to-point        :Config=auto / Active=false
 Transit Limit         :147 packets/hello-time
（省略部分输出）
```

在根交换机上，更改 STP 网络直径导致计时器变更后，新的计时器设置会同步到 STP 域中的非根交换机。感兴趣的读者可以在交换机 S2、S3 或 S4 上自行查看。

5.2.5 配置 STP 保护参数

为了防止有人意外或恶意将具有较高优先级的交换机接入网络，导致 STP 运行出现问题，我们可以配置一些保护参数。

步骤 1 配置根保护特性

在一个 STP 网络中，根交换机的位置一般是固定的，管理员会通过 5.2.2 中演示的命令强制某台交换机成为根交换机。为了防止新加入的交换机剥夺合法根交换机的角色并引起错误的网络波动，我们可以在某些指定端口上配置根保护特性。

根保护特性只能配置在指定端口上，它会使这个端口只能成为指定端口。若配置了根保护的指定端口接收了更优的 BPDU，该端口就会进入 DISCARDING 状态，不再转发数据。如果在一段时间内（通常是 FwDly 的两倍），该端口没有再接收更优的 BPDU，它会自动恢复为 FORWARDING 状态，但在恢复为转发状态之前，仍需要经历转发时延。

我们选择在交换机 S3 的 G0/0/4 端口上启用根保护特性，例 5-23 显示了 S3 上的配置，配置命令如下。

stp root-protection：接口配置视图命令，在端口上启用 STP 根保护。

例 5-23 启用根保护

```
[S3]interface GigabitEthernet 0/0/4
[S3-GigabitEthernet0/0/4]stp root-protection
```

启用根保护后，我们可以通过执行 **display stp brief** 命令查看 G0/0/4 端口的根保护状态，具体见例 5-24，ROOT 表示该端口上配置了根保护。

例 5-24 查看根保护状态

```
[S3]display stp brief
 MSTID  Port                  Role  STP State    Protection
    0   GigabitEthernet0/0/1  ROOT  FORWARDING   NONE
    0   GigabitEthernet0/0/2  ALTE  DISCARDING   NONE
    0   GigabitEthernet0/0/4  DESI  FORWARDING   ROOT
```

对效果感兴趣的读者可以提高 S4 的优先级值，使它认为自己是优于 S1 的根，并向 S3 发送 BPDU。这种做法会使 S3 的 G0/0/4 端口进入 DISCARDING 状态。若在两倍的 FwDly 时间内，G0/0/4 没有再次收到高优先级的 BPDU，端口就会进入 LEARNING 状态，并最终恢复为 FORWARDING 状态。

步骤 2 配置环路保护参数

若由于链路阻塞、单向链路故障等问题导致非根交换机的根端口没有及时接收上游发来的 BPDU，非根交换机就会重新选举并启用新的根端口，此时很有可能会造成交换环路。为了避免这种情况发生，管理员可以在相应的根端口和替换端口上启用环路保护特性，启用后，交换机感知到链路问题时会做出以下反应：

① 将根端口转换为指定端口，并使其处于 DISCARDING 状态；
② 将替换端口转换为根端口，并使其处于 DISCARDING 状态；
③ 生成日志消息并对管理员进行提示。

我们选择在交换机 S3 的 G0/0/1 和 G0/0/2 端口上启用环路保护特性，例 5-25 显示了 S3 的配置，配置命令如下。

stp loop-protection：接口配置视图命令，在端口上启用 STP 环路保护。

例 5-25 启用环路保护

```
[S3]interface GigabitEthernet 0/0/1
[S3-GigabitEthernet0/0/1]stp loop-protection
[S3-GigabitEthernet0/0/1]quit
[S3]interface GigabitEthernet 0/0/2
[S3-GigabitEthernet0/0/2]stp loop-protection
```

启用环路保护后，我们可以通过执行 **display stp brief** 命令查看 G0/0/1 和 G0/0/2 两个端口的环路保护状态，具体见例 5-26，LOOP 表示该端口上配置了环路保护。

例 5-26 查看环路保护状态

```
[S3]display stp brief
 MSTID  Port                  Role  STP State    Protection
    0   GigabitEthernet0/0/1  ROOT  FORWARDING   LOOP
    0   GigabitEthernet0/0/2  ALTE  DISCARDING   LOOP
    0   GigabitEthernet0/0/4  DESI  FORWARDING   ROOT
```

为了验证环路保护的效果，我们暂时在 S1 的 G0/0/3 端口上禁用 STP，读者可以在该端口上使用命令 **stp disable**。MaxAge 时间过后，S3 会生成日志消息，并转换端口状态。例 5-27 首先展示了网络正常时 S3 的端口状态，还展示了在生成日志消息后，管理员再次查看 S3 的端口状态。

例 5-27 查看 S3 的端口状态

```
[S3]display stp brief
 MSTID  Port                         Role  STP State    Protection
   0    GigabitEthernet0/0/1         ROOT  FORWARDING   LOOP
   0    GigabitEthernet0/0/2         ALTE  DISCARDING   LOOP
   0    GigabitEthernet0/0/4         DESI  FORWARDING   ROOT
[S3]
Apr 25 2021 07:05:02-08:00 S3 %%01MSTP/4/LOOP_GUARD(l)[0]:MSTP process 0 Instanc
e0's LOOP-Protection port GigabitEthernet0/0/1 did not receive message in prescr
iptive time!
[S3]
[S3]display stp brief
 MSTID  Port                         Role  STP State    Protection
   0    GigabitEthernet0/0/1         DESI  DISCARDING   LOOP
   0    GigabitEthernet0/0/2         ROOT  DISCARDING   LOOP
   0    GigabitEthernet0/0/4         DESI  DISCARDING   ROOT
```

验证环路保护的效果后，读者需要在 S1 的 G0/0/3 端口上使用命令 **stp enable** 启用 STP，使网络恢复正常。

步骤 3 配置 BPDU 保护和边缘端口

在接入层交换机上，大多数端口负责把各种各样的终端设备连接到网络，这些设备不会（也不应该）发送 BPDU 及参与 STP 运行。通过配置 BPDU 保护和边缘端口特性，可以在本应该连接终端设备的端口接收 BPDU 时，将端口关闭并生成日志消息。

我们选择对 S3 及其 G0/0/10 端口进行配置，读者需要先启动 PC2。例 5-28 显示了 S3 上的配置，配置命令如下。

① **stp bpdu-protection**：系统视图命令，在交换机上启用 BPDU 保护。

② **stp edged-port enable**：接口配置视图命令，可将端口配置为边缘端口。

例 5-28 启用 BPDU 保护并指定边缘端口

```
[S3]stp bpdu-protection
[S3]interface GigabitEthernet 0/0/10
[S3-GigabitEthernet0/0/10]stp edged-port enable
```

配置完成后，我们可以通过执行 **display stp** 命令看到交换机上启用了 BPDU 保护，也可以通过执行 **display stp brief** 命令看到哪些端口被配置为边缘端口，具体见例 5-29。

例 5-29 查看 BPDU 保护状态和边缘端口

```
[S3]display stp interface GigabitEthernet 0/0/10
-------[CIST Global Info][Mode STP]-------
CIST Bridge          :32768.4c1f-cc04-583f
Config Times         :Hello 2s MaxAge 20s FwDly 15s MaxHop 20
Active Times         :Hello 2s MaxAge 12s FwDly 9s MaxHop 20
CIST Root/ERPC       :0       .4c1f-cc24-61e2 / 20000
CIST RegRoot/IRPC    :32768.4c1f-cc04-583f / 0
CIST RootPortId      :128.1
BPDU-Protection      :Enabled
TC or TCN received   :424
TC count per hello   :0
STP Converge Mode    :Normal
Time since last TC   :0 days 0h:40m:38s
Number of TC         :48
Last TC occurred     :GigabitEthernet0/0/4
----[Port10(GigabitEthernet0/0/10)][FORWARDING]----
 Port Protocol       :Enabled
 Port Role           :Designated Port
 Port Priority       :128
 Port Cost(Dot1T )   :Config=auto / Active=20000
 Designated Bridge/Port   :32768.4c1f-cc04-583f / 128.10
 Port Edged          :Config=enabled / Active=enabled
 BPDU-Protection     :Enabled
 Point-to-point      :Config=auto / Active=true
 Transit Limit       :147 packets/hello-time
 Protection Type     :None
 Port STP Mode       :STP
```

```
Port Protocol Type     :Config=auto / Active=dot1s
BPDU Encapsulation     :Config=stp / Active=stp
PortTimes              :Hello 2s MaxAge 12s FwDly 9s RemHop 20
TC or TCN send         :0
TC or TCN received     :0
BPDU Sent              :6
        TCN: 0, Config: 6, RST: 0, MST: 0
BPDU Received          :0
        TCN: 0, Config: 0, RST: 0, MST: 0

[S3]display stp brief
MSTID  Port                      Role  STP State    Protection
  0    GigabitEthernet0/0/1      ROOT  FORWARDING   LOOP
  0    GigabitEthernet0/0/2      ALTE  DISCARDING   LOOP
  0    GigabitEthernet0/0/4      DESI  FORWARDING   ROOT
  0    GigabitEthernet0/0/10     DESI  FORWARDING   BPDU
```

要想验证配置效果，读者可以将 S3 的 G0/0/10 端口与 S1 的 G0/0/1 端口相连，此时会看到例 5-30 中的日志消息，并且端口会被关闭。

例 5-30 验证 BPDU 保护和边缘端口的效果

```
[S3]
Apr 25 2021 07:40:56-08:00 S3 %%01MSTP/4/BPDU_PROTECTION(l)[5]:This edged-port G
igabitEthernet0/0/10 that enabled BPDU-Protection will be shutdown, because it r
eceived BPDU packet!
Apr 25 2021 07:40:57-08:00 S3 %%01PHY/1/PHY(l)[6]:    GigabitEthernet0/0/10: cha
nge status to down
[S3]
```

此时，我们查看 G0/0/10 端口的状态时，会发现它的状态为 Down，相当于管理员在接口配置视图下配置了 **shutdown** 命令，我们只能手动在 G0/0/10 下使用 **undo shutdown** 命令，才能将其启用，具体见例 5-31。

例 5-31 G0/0/10 端口进入 DOWN 状态

```
[S3]display interface GigabitEthernet 0/0/10
GigabitEthernet0/0/10 current state : Administratively DOWN
Line protocol current state : DOWN
Description:
Switch Port, PVID :    1, TPID : 8100(Hex), The Maximum Frame Length is 9216
IP Sending Frames' Format is PKTFMT_ETHNT_2, Hardware address is 4c1f-cc04-583f
Last physical up time   : 2021-04-25 07:40:38 UTC-08:00
Last physical down time : 2021-04-25 07:40:57 UTC-08:00
Current system time: 2021-04-25 07:45:01-08:00
Hardware address is 4c1f-cc04-583f
    Last 300 seconds input rate 0 bytes/sec, 0 packets/sec
    Last 300 seconds output rate 0 bytes/sec, 0 packets/sec
    Input: 120 bytes, 2 packets
    Output: 165960 bytes, 2766 packets
    Input:
      Unicast: 0 packets, Multicast: 2 packets
      Broadcast: 0 packets
    Output:
      Unicast: 0 packets, Multicast: 2766 packets
      Broadcast: 0 packets
    Input bandwidth utilization  :      0%
    Output bandwidth utilization :      0%
```

华为提供了自动恢复机制，我们可以通过配置，使启用了 BPDU 保护的边缘端口在收到 BPDU 后进入 ERROR DOWN 状态，并在一段时间后自动恢复为 UP 状态。例 5-32 显示了 S3 上的配置，配置命令如下。

error-down auto-recovery cause bpdu-protection interval *interval-value*：系统视图命令，可启用端口自动恢复功能，并设置自动恢复时延。

例 5-32 设置 BPDU 保护自动恢复

```
[S3]interface GigabitEthernet 0/0/10
[S3-GigabitEthernet0/0/10]undo shutdown
[S3-GigabitEthernet0/0/10]quit
[S3]error-down auto-recovery cause bpdu-protection interval 30
```

BPDU 自动恢复要在端口触发违规行为之前进行配置，当端口触发违规行为后 BPDU 才可以自动恢复，因此我们需要手动将 G0/0/10 端口开启。

再次将 S3 的 G0/0/10 端口与 S1 的 G0/0/1 端口相连，例 5-33 展示了此时的日志消息，除了有与例 5-30 相同的两条日志消息，还生成了两条与 ERRDOWN 相关的消息。

例 5-33 日志消息

```
[S3]
Apr 25 2021 07:53:21-08:00 S3 %%01MSTP/4/BPDU_PROTECTION(l)[10]:This edged-port
GigabitEthernet0/0/10 that enabled BPDU-Protection will be shutdown, because it
received BPDU packet!
Apr 25 2021 07:53:21-08:00 S3 %%01ERRDOWN/4/ERRDOWN_DOWNNOTIFY(l)[11]:Notify int
erface to change status to error-down. (InterfaceName=GigabitEthernet0/0/10, Cau
se=bpdu-protection)
Apr 25 2021 07:53:21-08:00 S3 ERRDOWN/4/ErrordownOccur:OID 1.3.6.1.4.1.2011.5.25
.257.2.1 Error-down occured. (Ifindex=15, Ifname=GigabitEthernet0/0/10, Cause=bp
du-protection)
Apr 25 2021 07:53:21-08:00 S3 %%01PHY/1/PHY(l)[12]:    GigabitEthernet0/0/10: ch
ange status to down
[S3]
```

此时，查看 G0/0/10 端口的状态，具体见例 5-34。与例 5-31 进行对比，端口的状态不是 DOWN，而是因违反 BPDU 保护而进入了 ERROR DOWN 状态。

例 5-34 G0/0/10 端口进入 ERROR DOWN 状态

```
[S3]display interface GigabitEthernet 0/0/10
GigabitEthernet0/0/10 current state : ERROR DOWN(bpdu-protection)
Line protocol current state : DOWN
Description:
Switch Port, PVID :     1, TPID : 8100(Hex), The Maximum Frame Length is 9216
IP Sending Frames' Format is PKTFMT_ETHNT_2, Hardware address is 4c1f-cc04-583f
Last physical up time   : 2021-04-25 07:52:42 UTC-08:00
Last physical down time : 2021-04-25 07:53:22 UTC-08:00
Current system time: 2021-04-25 07:53:46-08:00
Hardware address is 4c1f-cc04-583f
    Last 300 seconds input rate 0 bytes/sec, 0 packets/sec
    Last 300 seconds output rate 0 bytes/sec, 0 packets/sec
    Input: 180 bytes, 3 packets
    Output: 177120 bytes, 2952 packets
    Input:
      Unicast: 0 packets, Multicast: 3 packets
      Broadcast: 0 packets
    Output:
      Unicast: 0 packets, Multicast: 2952 packets
      Broadcast: 0 packets
    Input bandwidth utilization  :     0%
    Output bandwidth utilization :     0%
```

经过 30 秒后，端口会自动恢复为 UP 状态，具体见例 5-35，但若此时再次收到 BPDU，端口会再次进入 ERROR DOWN 状态。

例 5-35 30 秒后自动恢复

```
[S3]
Apr 25 2021 07:53:50-08:00 S3 %%01ERRDOWN/4/ERRDOWN_DOWNRECOVER(l)[13]:Notify in
terface to recover state from error-down. (InterfaceName=GigabitEthernet0/0/10)
Apr 25 2021 07:53:50-08:00 S3 ERRDOWN/4/ErrordownRecover:OID 1.3.6.1.4.1.2011.5.
25.257.2.2 Error-down recovered. (Ifindex=15, Ifname=GigabitEthernet0/0/10, Caus
e=bpdu-protection, RecoverType=auto recovery)
Apr 25 2021 07:53:51-08:00 S3 %%01PHY/1/PHY(l)[14]:    GigabitEthernet0/0/10: ch
ange status to up
[S3]
```

5.2.6 配置 MSTP

STP（或 RSTP）会在整个 STP 域中形成一颗无环路径树，只有当网络中出现问题时，被阻塞的链路才会作为备用链路被启用。MSTP 的优势是让管理员可以根据 VLAN 来构建无环路径树，MSTP 不仅可以提供备份路径，还可以实现负载分担。

在本实验中，为了突出重点，我们使用交换机 S1、S2 和 S3 来进行演示（清除前文的配置），其中 S1 为总根，S2 为 VLAN 2 的域根，S3 为 VLAN 3 的域根，以此展示 MSTP 的作用及其带来的优势。

步骤 1 基础配置

首先在 3 台交换机上创建 VLAN 2 和 VLAN 3，将交换机互联链路配置为 Trunk，并允许所有 VLAN 通过。例 5-36～例 5-38 分别展示了 S1、S2 和 S3 的基础配置。

例 5-36 S1 的基础配置

```
[S1]vlan batch 2 3
Info: This operation may take a few seconds. Please wait for a moment...done.
[S1]interface GigabitEthernet 0/0/2
[S1-GigabitEthernet0/0/2]port link-type trunk
[S1-GigabitEthernet0/0/2]port trunk allow vlan all
[S1-GigabitEthernet0/0/2]quit
[S1]interface GigabitEthernet 0/0/3
[S1-GigabitEthernet0/0/3]port link-type trunk
[S1-GigabitEthernet0/0/3]port trunk allow vlan all
```

例 5-37 S2 的基础配置

```
[S2]vlan batch 2 3
Info: This operation may take a few seconds. Please wait for a moment...done.
[S2]interface GigabitEthernet 0/0/1
[S2-GigabitEthernet0/0/1]port link-type trunk
[S2-GigabitEthernet0/0/1]port trunk allow vlan all
[S2-GigabitEthernet0/0/1]quit
[S2]interface GigabitEthernet 0/0/3
[S2-GigabitEthernet0/0/3]port link-type trunk
[S2-GigabitEthernet0/0/3]port trunk allow vlan all
```

例 5-38 S3 的基础配置

```
[S3]vlan batch 2 3
Info: This operation may take a few seconds. Please wait for a moment...done.
[S3]interface GigabitEthernet 0/0/1
[S3-GigabitEthernet0/0/1]port link-type trunk
[S3-GigabitEthernet0/0/1]port trunk allow vlan all
[S3-GigabitEthernet0/0/1]quit
[S3]interface GigabitEthernet 0/0/2
[S3-GigabitEthernet0/0/2]port link-type trunk
[S3-GigabitEthernet0/0/2]port trunk allow vlan all
```

步骤 2 实施 MSTP 基础配置

如果读者新建拓扑进行实验，无须更改 STP 运行模式，华为交换机默认的 STP 运行模式就是 MSTP。如果是在前面实验步骤的基础上继续配置，需要先将 STP 模式更改为 MSTP，可以选择以下两条命令之一进行更改。

① **stp mode mstp**：将 STP 模式更改为 MSTP。

② **undo stp mode**：将 STP 模式恢复为默认的 MSTP。

MSTP 中加入了域的概念，因此要想使多台交换机属于同一个 MST 域，则需要使它们具有一些相同的配置。

① MST 域的域名：默认情况下，MST 域名是设备管理口的 MAC 地址。

② MST 实例与 VLAN 的映射关系：默认情况下，MST 域内的所有 VLAN 都映射到 MST0。

③ MST 域的修订级别：默认情况下，MST 域的修订级别为 0。此为可选配置，若 MST 域中所有设备的修订级别都为 0，则无须修改。

配置 MSTP 的步骤如下：

① 使用系统视图命令 **stp region-configuration** 进入 MST 域视图；
② 使用 MST 域视图命令 **region-name** *name* 设置 MST 域的域名，所有设备保持一致；
③ 使用 MST 域视图命令 **instance** *instance-id* **vlan** {*vlan-id*1 [**to** *vlan-id*2]}配置 MST 实例与 VLAN 的映射关系，所有设备保持一致；
④ （可选）使用 MST 域视图命令 **revision-level** *level* 配置 MST 域的修订级别，所有设备保持一致；
⑤ 使用 MST 域视图命令 **active region-configuration** 激活 MST 域的配置，使上述配置生效。

下面我们在交换机上对 MSTP 进行配置，将 MST 域名设置为 HCIA，实例 2 与 VLAN 2 映射，实例 3 与 VLAN 3 映射，更改修订级别为 1。例 5-39～例 5-41 分别展示了 S1、S2 和 S3 的相关配置。

例 5-39　在 S1 上完成 MSTP 基础配置

```
[S1]stp region-configuration
[S1-mst-region]region-name HCIA
[S1-mst-region]instance 2 vlan 2
[S1-mst-region]instance 3 vlan 3
[S1-mst-region]revision-level 1
[S1-mst-region]active region-configuration
Info: This operation may take a few seconds. Please wait for a moment...done.
[S1-mst-region]
```

例 5-40　在 S2 上完成 MSTP 基础配置

```
[S2]stp region-configuration
[S2-mst-region]region-name HCIA
[S2-mst-region]instance 2 vlan 2
[S2-mst-region]instance 3 vlan 3
[S2-mst-region]revision-level 1
[S2-mst-region]active region-configuration
Info: This operation may take a few seconds. Please wait for a moment...done.
[S2-mst-region]
```

例 5-41　在 S3 上完成 MSTP 基础配置

```
[S3]stp region-configuration
[S3-mst-region]region-name HCIA
[S3-mst-region]instance 2 vlan 2
[S3-mst-region]instance 3 vlan 3
[S3-mst-region]revision-level 1
[S3-mst-region]active region-configuration
Info: This operation may take a few seconds. Please wait for a moment...done.
[S3-mst-region]
```

配置完成后，读者可以使用命令 **display stp region-configuration** 来查看实例与 VLAN 的映射。下面以 S1 为例展示这条命令的输出内容，具体见例 5-42。

例 5-42　查看 MSTP 参数

```
[S1]display stp region-configuration
 Oper configuration
  Format selector      :0
  Region name          :HCIA
  Revision level       :1

  Instance   VLANs Mapped
     0       1, 4 to 4094
     2       2
     3       3
```

启用 MSTP 后，读者可以先通过命令确认当前环境中的根交换机。查找根交换机的命令有多种，在这里我们使用的是 **display stp brief** 命令，该命令不仅可以判断出根交换机的位置，还可以在后续配置中观察端口角色的转变。例 5-43～例 5-45 分别展示了 S1、S2 和 S3 的端口状态。

例 5-43 查看 S1 的端口状态

```
[S1]display stp brief
 MSTID  Port                       Role  STP State    Protection
   0    GigabitEthernet0/0/2       DESI  FORWARDING   NONE
   0    GigabitEthernet0/0/3       ROOT  FORWARDING   NONE
   2    GigabitEthernet0/0/2       DESI  FORWARDING   NONE
   2    GigabitEthernet0/0/3       ROOT  FORWARDING   NONE
   3    GigabitEthernet0/0/2       DESI  FORWARDING   NONE
   3    GigabitEthernet0/0/3       ROOT  FORWARDING   NONE
```

例 5-44 查看 S2 的端口状态

```
[S2]display stp brief
 MSTID  Port                       Role  STP State    Protection
   0    GigabitEthernet0/0/1       ALTE  DISCARDING   NONE
   0    GigabitEthernet0/0/3       ROOT  FORWARDING   NONE
   2    GigabitEthernet0/0/1       ALTE  DISCARDING   NONE
   2    GigabitEthernet0/0/3       ROOT  FORWARDING   NONE
   3    GigabitEthernet0/0/1       ALTE  DISCARDING   NONE
   3    GigabitEthernet0/0/3       ROOT  FORWARDING   NONE
```

例 5-45 查看 S3 的端口状态

```
[S3]display stp brief
 MSTID  Port                       Role  STP State    Protection
   0    GigabitEthernet0/0/1       DESI  FORWARDING   NONE
   0    GigabitEthernet0/0/2       DESI  FORWARDING   NONE
   2    GigabitEthernet0/0/1       DESI  FORWARDING   NONE
   2    GigabitEthernet0/0/2       DESI  FORWARDING   NONE
   3    GigabitEthernet0/0/1       DESI  FORWARDING   NONE
   3    GigabitEthernet0/0/2       DESI  FORWARDING   NONE
```

通过 3 台交换机的命令输出，我们可以看出，在所有实例中，根交换机都是 S3。我们通过命令，将 S1 设置为总根交换机，S2 设置为实例 2 的根交换机，S3 设置为实例 3 的根交换机。

步骤 3 根据实例设置根交换机

要想根据实例来设置根交换机，需要在命令中添加 **instance** 关键字和参数，具体如下。

① **stp [instance** *instance-id*] **root primary**：系统视图命令，将当前设备设置为根交换机，如果不指定 **instance**，则会将设备配置为实例 0 的根交换机。

② **stp [instance** *instance-id*] **root secondary**：系统视图命令，将当前设备设置为备用根交换机，如果不指定 **instance**，则会将设备配置为实例 0 的备用根交换机。

例 5-46～例 5-48 分别展示了 S1、S2 和 S3 的相应配置。

例 5-46 将 S1 设置为总根交换机

```
[S1]stp root primary
```

例 5-47 将 S2 设置为实例 2 的根交换机

```
[S2]stp instance 2 root primary
```

例 5-48 将 S3 设置为实例 3 的根交换机

```
[S3]stp instance 3 root primary
```

配置完成后，再次查看每台交换机上 STP 端口的角色，可以发现每个实例的根交换机已经与设计相同，例 5-49～例 5-51 分别展示了 S1、S2 和 S3 的端口角色。

例 5-49 查看 S1 的端口角色

```
[S1]display stp brief
MSTID  Port                  Role  STP State    Protection
  0    GigabitEthernet0/0/2  DESI  FORWARDING   NONE
  0    GigabitEthernet0/0/3  DESI  FORWARDING   NONE
  2    GigabitEthernet0/0/2  ROOT  FORWARDING   NONE
  2    GigabitEthernet0/0/3  ALTE  DISCARDING   NONE
  3    GigabitEthernet0/0/2  ALTE  DISCARDING   NONE
  3    GigabitEthernet0/0/3  ROOT  FORWARDING   NONE
```

例 5-50 查看 S2 的端口角色

```
[S2]display stp brief
MSTID  Port                  Role  STP State    Protection
  0    GigabitEthernet0/0/1  ROOT  FORWARDING   NONE
  0    GigabitEthernet0/0/3  ALTE  DISCARDING   NONE
  2    GigabitEthernet0/0/1  DESI  FORWARDING   NONE
  2    GigabitEthernet0/0/3  DESI  FORWARDING   NONE
  3    GigabitEthernet0/0/1  ALTE  DISCARDING   NONE
  3    GigabitEthernet0/0/3  ROOT  FORWARDING   NONE
```

例 5-51 查看 S3 的端口角色

```
[S3]display stp brief
MSTID  Port                  Role  STP State    Protection
  0    GigabitEthernet0/0/1  ROOT  FORWARDING   NONE
  0    GigabitEthernet0/0/2  DESI  FORWARDING   NONE
  2    GigabitEthernet0/0/1  DESI  FORWARDING   NONE
  2    GigabitEthernet0/0/2  ROOT  FORWARDING   NONE
  3    GigabitEthernet0/0/1  DESI  FORWARDING   NONE
  3    GigabitEthernet0/0/2  DESI  FORWARDING   NONE
```

根据实例设置，读者可以得到根交换机的结果，如图 5-7 所示。在实际实验环境中，被阻塞的端口可能与本实验有所不同，但被阻塞的链路是一样的，这就实现了 3 条链路同时启用，进行负载分担，同时又提供备用链路来应对故障的目标。

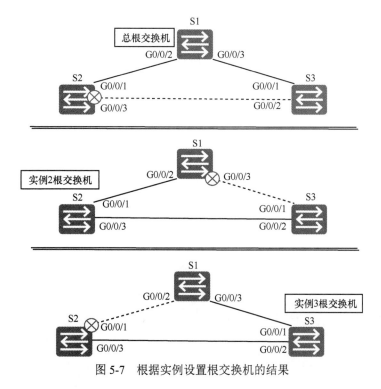

图 5-7 根据实例设置根交换机的结果

读者在运行 MSTP 的环境中还可以调整影响选举和收敛时间的参数，以及配置保护功能。这些参数和特性的运行原理与 STP 相同，只是所有计算都是以实例为单位，而不是以整个 STP 域为单位。感兴趣的读者可以在华为设备配置指南中查询具体命令。

第 6 章
以太网链路聚合实验

本章主要内容

6.1　实验介绍

6.2　实验配置任务

上一个实验所介绍的 STP 可以防止交换网络产生环路。想要在两台交换机之间通过增加互联链路来增加带宽，实现负载均衡，那么通过 STP 的计算后只能有一条链路传输数据。以太网聚合设置可以让 STP 将多条链路捆绑为一条聚合链路，并将其当作单条路径进行计算，从而不会阻塞其中的任何物理端口。以太网聚合带来了以下好处。

① 增加带宽：链路聚合接口的最大带宽是其成员接口带宽之和。

② 提高可靠性：当一条活跃的物理链路发生故障时，相关流量会自动切换到其他可用的成员链路上。

③ 负载分担：一个虚拟链路组合中的多条活跃的物理链路可以实现负载分担。

在进入实验之前，我们可以了解一下链路聚合中的基本概念。

① 链路聚合组：多条以太网链路捆绑在一起形成的逻辑链路。

② 链路聚合接口（Eth-Trunk 接口）：一个链路聚合组在一台设备上对应一个链路聚合接口，它是由多个物理接口捆绑在一起形成的逻辑接口。

③ 成员接口和成员链路：Eth-Trunk 接口中的每个物理接口被称为成员接口，成员接口所连接的链路被称为成员链路。

④ 活动接口和非活动接口，活动链路和非活动链路：链路聚合组中的成员分为活动和非活动两种状态，活动接口负责转发数据，非活动接口不转发数据；活动接口对应的链路被称为活动链路，非活动接口对应的链路被称为非活动链路。

⑤ 活动接口上限阈值：一个链路聚合组中活动接口的数量由管理员根据实际情况进行调整。当 3G 带宽的链路聚合组中捆绑了 5 个 1G 物理接口时，管理员可以将活动接口的最大数量设置为 3，此时会有 3 条链路进入活动状态，另外 2 条链路为非活跃链路并提供备份。手动模式不支持该设置。

⑥ 活动接口下限阈值：为保障最小带宽，管理员可以设置活动接口下限阈值，当活动链路的数量小于该值时，Eth-Trunk 接口会进入 Down 状态。

⑦ 链路聚合工作模式：手动模式和 LACP 模式。

6.1 实验介绍

6.1.1 关于本实验

本实验旨在为读者介绍以太网链路聚合的配置和验证,其中会展示手动模式的配置、LACP 模式的配置，以及以太网链路聚合在 LACP 配置环境中的一些特性。

6.1.2 实验目的

- 掌握如何配置链路聚合的手动模式。
- 掌握如何配置链路聚合的 LACP 模式。
- 掌握如何在 LACP 模式中控制活动链路。
- 掌握 LACP 的特性。

6.1.3 实验组网介绍

设备连接说明：两台交换机 S1 和 S2 之间通过 G0/0/10、G0/0/11 和 G0/0/12 端口互联，并且这三个端口构成以太网聚合链路，如图 6-1 所示。

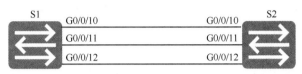

图 6-1 设备连接说明

6.1.4 实验任务列表

配置任务 1：配置手动模式。
配置任务 2：配置 LACP 模式。
配置任务 3：设置 LACP 参数。

6.2 实验配置任务

6.2.1 配置手动模式

读者需要先构建图 6-1 所示的拓扑，构建完成后，可以通过前一章学习的 STP 命令确认这 3 条链路中只有一条链路能够转发数据，STP 阻塞了两个端口，具体见例 6-1。

例 6-1 STP 阻塞了两个端口

```
[S2]display stp brief
 MSTID  Port                        Role  STP State    Protection
    0   GigabitEthernet0/0/10       ROOT  FORWARDING   NONE
    0   GigabitEthernet0/0/11       ALTE  DISCARDING   NONE
    0   GigabitEthernet0/0/12       ALTE  DISCARDING   NONE
```

链路聚合配置可以将这 3 个物理端口捆绑为一个逻辑接口，从而避免 STP 将物理端口阻塞。图 6-2 展示了配置链路聚合前后的对比。

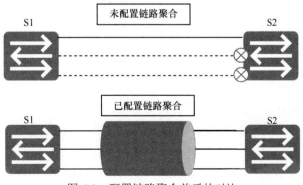

图 6-2 配置链路聚合前后的对比

以太网链路聚合手动配置需要以下两个步骤。

① 使用系统视图命令 **interface eth-trunk** *trunk-id* 创建 Eth-Trunk 接口并进入接口视图。

② 使用以下两种方式之一添加成员接口：

- 使用 Eth-Trunk 接口视图命令 **trunkport** *interface-type* {*interface-number1* [**to** *interface-number2*]} 添加成员接口，在添加多个接口时，如果某个接口添加失败，则它后面的接口也不会被加入 Eth-Trunk；
- 进入某成员接口视图，使用命令 **eth-trunk** *trunk-id* 将当前接口加入相应的 Eth-Trunk。

例 6-2 和例 6-3 分别展示了 S1 和 S2 的手动配置命令，S1 和 S2 分别创建 Eth-Trunk 1，并将 G0/0/10、G0/0/11 和 G0/0/12 端口加入 Eth-Trunk 1。S1 采用上述第一种端口添加方式，S2 采用上述第二种端口添加方式。

例 6-2 在 S1 上手动配置 Eth-Trunk

```
[S1]interface Eth-Trunk 1
[S1-Eth-Trunk1]trunkport GigabitEthernet 0/0/10 to 0/0/12
Info: This operation may take a few seconds. Please wait for a moment...done.
[S1-Eth-Trunk1]
Apr 27 2021 17:57:02-08:00 S1 %%01IFNET/4/IF_STATE(l)[0]:Interface Eth-Trunk1 ha
s turned into UP state.
```

例 6-3 在 S2 上手动配置 Eth-Trunk

```
[S2]interface Eth-Trunk 1
[S2-Eth-Trunk1]quit
[S2]interface GigabitEthernet 0/0/10
[S2-GigabitEthernet0/0/10]eth-trunk 1
Info: This operation may take a few seconds. Please wait for a moment...done.
[S2-GigabitEthernet0/0/10]quit
[S2]
Apr 27 2021 18:00:18-08:00 S2 %%01IFNET/4/IF_STATE(l)[0]:Interface Eth-Trunk1 ha
s turned into UP state.
[S2]interface GigabitEthernet 0/0/11
[S2-GigabitEthernet0/0/11]eth-trunk 1
Info: This operation may take a few seconds. Please wait for a moment...done.
[S2-GigabitEthernet0/0/11]quit
[S2]interface GigabitEthernet 0/0/12
[S2-GigabitEthernet0/0/12]eth-trunk 1
Info: This operation may take a few seconds. Please wait for a moment...done.
```

在将交换机端口加入 Eth-Trunk 时，读者需要注意以下事项。

① 每个 Eth-Trunk 接口中最多可以包含 8 个成员接口。

② Eth-Trunk 接口的成员接口不能是其他 Eth-Trunk 接口。

③ 一个以太网端口只能加入一个 Eth-Trunk 接口，若一个以太网端口已经是某个 Eth-Trunk 接口成员，那么其加入其他 Eth-Trunk 接口之前，需要先退出当前的 Eth-Trunk 接口。

④ 当一个以太网端口被配置为 Eth-Trunk 接口成员后，与它直连的对端以太网端口也必须被配置为 Eth-Trunk 接口成员，以便两端能正常通信。

⑤ Eth-Trunk 链路两端的参数必须保持一致，这些参数包括物理端口数量、速率、双工方式等。

读者可以使用 **display eth-trunk** *trunk-id* 来查看配置的 Eth-Trunk，例 6-4 和例 6-5 分别展示了 S1 和 S2 上的 Eth-Trunk 1，命令中显示的工作模式（NORMAL）表示这个 Eth-Trunk 是手动配置的。

例 6-4　在 S1 上查看 Eth-Trunk 1

```
[S1]display eth-trunk 1
Eth-Trunk1's state information is:
WorkingMode: NORMAL         Hash arithmetic: According to SIP-XOR-DIP
Least Active-linknumber: 1  Max Bandwidth-affected-linknumber: 8
Operate status: up          Number Of Up Port In Trunk: 3
--------------------------------------------------------------------------
PortName                    Status        Weight
GigabitEthernet0/0/10       Up            1
GigabitEthernet0/0/11       Up            1
GigabitEthernet0/0/12       Up            1
```

例 6-5　在 S2 上查看 Eth-Trunk 1

```
[S2]display eth-trunk 1
Eth-Trunk1's state information is:
WorkingMode: NORMAL         Hash arithmetic: According to SIP-XOR-DIP
Least Active-linknumber: 1  Max Bandwidth-affected-linknumber: 8
Operate status: up          Number Of Up Port In Trunk: 3
--------------------------------------------------------------------------
PortName                    Status        Weight
GigabitEthernet0/0/10       Up            1
GigabitEthernet0/0/11       Up            1
GigabitEthernet0/0/12       Up            1
```

这时我们再查看 STP 状态，会发现 STP 不再基于每个物理端口进行计算，而是基于 Eth-Trunk 进行计算，具体见例 6-6。

例 6-6　基于 Eth-Trunk 计算 STP

```
[S2]display stp brief
MSTID   Port                   Role   STP State     Protection
  0     Eth-Trunk1             ROOT   FORWARDING    NONE
```

在以太网链路聚合接口中，每个成员接口都是活动接口，因此没有备用接口/链路的概念。当一条活跃链路出现故障时，Eth-Trunk 的带宽就会相应地减少。如果有需要，管理员可以使用 LACP 模式。

6.2.2　配置 LACP 模式

在更改 Eth-Trunk 接口的工作模式时，我们需要先删除其中的成员接口。有些交换机类型在从手动模式向 LACP 模式转换时无须删除已有成员接口，读者需要查看相应的设备配置手册来确认具体操作。本实验采用稳妥的方式，先删除已有的成员接口，再改变工作模式。例 6-7 和例 6-8 分别展示了交换机 S1 和 S2 的配置。更改 Eth-Trunk 接口运行模式的命令如下。

mode lacp-static：Eth-Trunk 接口视图命令，将本接口配置为 LACP 模式，默认的工作模式为手动模式（**manual**）。有些版本的命令语法为 **mode lacp**，读者需要根据实际的设备型号和 VRP 版本，查询华为设备配置指南。

例 6-7　在 S1 上配置 LACP

```
[S1]interface Eth-Trunk 1
[S1-Eth-Trunk1]undo trunkport GigabitEthernet 0/0/10 to 0/0/12
Info: This operation may take a few seconds. Please wait for a moment...done.
[S1-Eth-Trunk1]
Apr 27 2021 22:16:02-08:00 S1 %%01IFNET/4/IF_STATE(l)[3]:Interface Eth-Trunk1 ha
s turned into DOWN state.
[S1-Eth-Trunk1]mode lacp-static
[S1-Eth-Trunk1]trunkport GigabitEthernet 0/0/10 to 0/0/12
Info: This operation may take a few seconds. Please wait for a moment...done.
[S1-Eth-Trunk1]
Apr 27 2021 22:17:12-08:00 S1 %%01IFNET/4/IF_STATE(l)[6]:Interface Eth-Trunk1 ha
s turned into UP state.
```

例 6-8 在 S2 上配置 LACP

```
[S2]interface Eth-Trunk 1
[S2-Eth-Trunk1]undo trunkport GigabitEthernet 0/0/10 to 0/0/12
Info: This operation may take a few seconds. Please wait for a moment...done.
[S2-Eth-Trunk1]
Apr 27 2021 22:16:56-08:00 S2 %%01IFNET/4/IF_STATE(l)[0]:Interface Eth-Trunk1 ha
s turned into DOWN state.
[S2-Eth-Trunk1]mode lacp-static
[S2-Eth-Trunk1]trunkport GigabitEthernet 0/0/10 to 0/0/12
Info: This operation may take a few seconds. Please wait for a moment...done.
[S2-Eth-Trunk1]
Apr 27 2021 22:17:12-08:00 S2 %%01IFNET/4/IF_STATE(l)[3]:Interface Eth-Trunk1 ha
s turned into UP state.
```

通过对比 LACP 模式的配置和手动模式的配置，我们发现：在配置手动模式时，只要 Eth-Trunk 接口中加入了 UP 状态的成员接口，Eth-Trunk 接口就会进入 UP 状态；在配置 LACP 模式时，只有当链路两端都配置完成，并且两端接口参数相同，LACP 才会将 Eth-Trunk 接口置为 UP 状态。

读者此时使用 **display eth-trunk 1** 命令查看 Eth-Trunk 信息时，会发现命令输出中的内容比手动配置丰富，例 6-9 和例 6-10 分别展示了 S1 和 S2 的 Eth-Trunk 状态。

例 6-9 查看 S1 上的 Eth-Trunk 状态

```
[S1]display eth-trunk 1
Eth-Trunk1's state information is:
Local:
LAG ID: 1                    WorkingMode: STATIC
Preempt Delay: Disabled      Hash arithmetic: According to SIP-XOR-DIP
System Priority: 32768       System ID: 4c1f-cc29-4724
Least Active-linknumber: 1   Max Active-linknumber: 8
Operate status: up           Number Of Up Port In Trunk: 3
--------------------------------------------------------------------------------
ActorPortName          Status   PortType PortPri PortNo PortKey PortState Weight
GigabitEthernet0/0/10  Selected 1GE      32768   11     305     10111100  1
GigabitEthernet0/0/11  Selected 1GE      32768   12     305     10111100  1
GigabitEthernet0/0/12  Selected 1GE      32768   13     305     10111100  1

Partner:
--------------------------------------------------------------------------------
ActorPortName          SysPri  SystemID        PortPri PortNo PortKey PortState
GigabitEthernet0/0/10  32768   4c1f-cc6c-03de  32768   11     305     10111100
GigabitEthernet0/0/11  32768   4c1f-cc6c-03de  32768   12     305     10111100
GigabitEthernet0/0/12  32768   4c1f-cc6c-03de  32768   13     305     10111100
```

例 6-10 查看 S2 上的 Eth-Trunk 状态

```
[S2]display eth-trunk 1
Eth-Trunk1's state information is:
Local:
LAG ID: 1                    WorkingMode: STATIC
Preempt Delay: Disabled      Hash arithmetic: According to SIP-XOR-DIP
System Priority: 32768       System ID: 4c1f-cc6c-03de
Least Active-linknumber: 1   Max Active-linknumber: 8
Operate status: up           Number Of Up Port In Trunk: 3
--------------------------------------------------------------------------------
ActorPortName          Status   PortType PortPri PortNo PortKey PortState Weight
GigabitEthernet0/0/10  Selected 1GE      32768   11     305     10111100  1
GigabitEthernet0/0/11  Selected 1GE      32768   12     305     10111100  1
GigabitEthernet0/0/12  Selected 1GE      32768   13     305     10111100  1

Partner:
--------------------------------------------------------------------------------
ActorPortName          SysPri  SystemID        PortPri PortNo PortKey PortState
GigabitEthernet0/0/10  32768   4c1f-cc29-4724  32768   11     305     10111100
GigabitEthernet0/0/11  32768   4c1f-cc29-4724  32768   12     305     10111100
GigabitEthernet0/0/12  32768   4c1f-cc29-4724  32768   13     305     10111100
```

阴影标注的工作模式 STATIC 表示当前使用的是 LACP 模式（较新 VRP 版本中此处会显示为 LACP），其他重要的参数会在后续的实验中陆续介绍。

6.2.3 设置 LACP 参数

步骤 1 设置最小和最大活动链路数量

前文提到，LACP 模式的优点之一就是它可以启用备份链路功能。正常情况下我们只需要在 Eth-Trunk 链路中启用 2 个物理端口就可以满足带宽需求，并且当活动链路数量小于 2 时，可关闭整个 Eth-Trunk 链路。例 6-11 展示了在 S1 上设置最小和最大活动链路数量，命令语法如下。

① **least active-linknumber** *link-number*：Eth-Trunk 接口视图命令，设置链路聚合活动接口数量的下限阈值，默认情况下这个阈值为 1，当链路两端的下限阈值设置不同时，以数值较大的为准。

② **max active-linknumber** *link-number*：Eth-Trunk 接口视图命令，设置链路聚合活动接口数量的上限阈值，默认情况下这个阈值为 8，当链路两端的下限阈值设置不同时，以数值较小的为准。

例 6-11 在 S1 上设置最小和最大活动链路数量

```
[S1]interface Eth-Trunk 1
[S1-Eth-Trunk1]least active-linknumber 2
[S1-Eth-Trunk1]max active-linknumber 2
```

最大活动链路数量会以较小的一端为准，因此 Eth-Trunk 1 中的活动端口变为两个，具体见例 6-12，G0/0/12 成为非活动端口。

例 6-12 查看当前的活动端口

```
[S1]display eth-trunk 1
Eth-Trunk1's state information is:
Local:
LAG ID: 1                     WorkingMode: STATIC
Preempt Delay: Disabled       Hash arithmetic: According to SIP-XOR-DIP
System Priority: 32768        System ID: 4c1f-cc29-4724
Least Active-linknumber: 2    Max Active-linknumber: 2
Operate status: up            Number Of Up Port In Trunk: 2
--------------------------------------------------------------------------
ActorPortName            Status   PortType PortPri PortNo PortKey PortState Weight
GigabitEthernet0/0/10    Selected 1GE      32768   11     305     10111100  1
GigabitEthernet0/0/11    Selected 1GE      32768   12     305     10111100  1
GigabitEthernet0/0/12    Unselect 1GE      32768   13     305     10100000  1

Partner:
--------------------------------------------------------------------------
ActorPortName            SysPri   SystemID        PortPri PortNo PortKey PortState
GigabitEthernet0/0/10    32768    4c1f-cc6c-03de  32768   11     305     10111100
GigabitEthernet0/0/11    32768    4c1f-cc6c-03de  32768   12     305     10111100
GigabitEthernet0/0/12    32768    4c1f-cc6c-03de  32768   13     305     10110000
```

例 6-13 展示了 S2 上的 Eth-Trunk 1 信息，尽管 S2 上的参数没有发生改变，但它的 G0/0/12 端口成为了非活动接口，这个 Eth-Trunk 中只有两个活动接口。

例 6-13 查看 S2 上的活动端口

```
[S2]display eth-trunk 1
Eth-Trunk1's state information is:
Local:
LAG ID: 1                     WorkingMode: STATIC
Preempt Delay: Disabled       Hash arithmetic: According to SIP-XOR-DIP
System Priority: 32768        System ID: 4c1f-cc6c-03de
Least Active-linknumber: 1    Max Active-linknumber: 8
Operate status: up            Number Of Up Port In Trunk: 2
--------------------------------------------------------------------------
ActorPortName            Status   PortType PortPri PortNo PortKey PortState Weight
GigabitEthernet0/0/10    Selected 1GE      32768   11     305     10111100  1
GigabitEthernet0/0/11    Selected 1GE      32768   12     305     10111100  1
GigabitEthernet0/0/12    Unselect 1GE      32768   13     305     10110000  1
```

```
Partner:
--------------------------------------------------------------------------------
ActorPortName           SysPri     SystemID            PortPri  PortNo  PortKey  PortState
GigabitEthernet0/0/10   32768      4c1f-cc29-4724      32768    11      305      10111100
GigabitEthernet0/0/11   32768      4c1f-cc29-4724      32768    12      305      10111100
GigabitEthernet0/0/12   32768      4c1f-cc29-4724      32768    13      305      10100000
```

为了使两端的设置保持一致，我们在 S2 上进行了相同的设置，具体见例 6-14。

例 6-14 在 S2 上设置最小和最大活动链路数量

```
[S2]interface Eth-Trunk 1
[S2-Eth-Trunk1]least active-linknumber 2
[S2-Eth-Trunk1]max active-linknumber 2
```

在满足上述需求之后，我们需要对启用了 LACP 的设备配置一些通用设置，使 LACP 的运行处于管理员的设计中。

步骤 2 设置 LACP 系统优先级

在 LACP 模式中，链路两端的设备会根据管理员的配置或默认规则，选择 Eth-Trunk 中的活动接口，只有当两端选择的活动接口一致时，才能够正常工作。LACP 系统优先级的作用是设定以链路哪端的设备为主动端，并由它来决定活动接口的选择。我们要将交换机 S1 设置为主动端，例 6-15 展示了 S1 的设置命令。

lacp priority *priority*：系统视图命令，设置当前设备的 LACP 系统优先级，默认值为 32768，值越低，优先级越高。如果 LACP 系统优先级相同，则 MAC 地址较小的一端会成为主动端。

例 6-15 将 S1 设置为 LACP 主动端

```
[S1]lacp priority 30000
```

使用命令 **display eth-trunk 1** 查看 S1 的设置结果，具体见例 6-16 中的阴影部分。

例 6-16 验证 S1 中 LACP 系统优先级的设置

```
[S1]display eth-trunk 1
Eth-Trunk1's state information is:
Local:
LAG ID: 1                        WorkingMode: STATIC
Preempt Delay: Disabled          Hash arithmetic: According to SIP-XOR-DIP
System Priority: 30000           System ID: 4c1f-cc29-4724
Least Active-linknumber: 2       Max Active-linknumber: 2
Operate status: up               Number Of Up Port In Trunk: 2
--------------------------------------------------------------------------------
ActorPortName           Status    PortType  PortPri  PortNo  PortKey  PortState  Weight
GigabitEthernet0/0/10   Selected  1GE       32768    11      305      10111100   1
GigabitEthernet0/0/11   Selected  1GE       32768    12      305      10111100   1
GigabitEthernet0/0/12   Unselect  1GE       32768    13      305      10100000   1

Partner:
--------------------------------------------------------------------------------
ActorPortName           SysPri    SystemID            PortPri  PortNo  PortKey  PortState
GigabitEthernet0/0/10   32768     4c1f-cc6c-03de      32768    11      305      10111100
GigabitEthernet0/0/11   32768     4c1f-cc6c-03de      32768    12      305      10111100
GigabitEthernet0/0/12   32768     4c1f-cc6c-03de      32768    13      305      10100000
```

步骤 3 设置 LACP 接口优先级

LACP 接口优先级是选择活动接口的标准。由于当前 G0/0/12 为非活动接口，为了体现区别，本实验将 G0/0/11 和 G0/0/12 作为正常情况下的活动接口，将 G0/0/10 作为备用端口（非活动接口）。例 6-17 展示了 S1 的设置，命令语法如下。

lacp priority *priority*：接口视图命令，设置当前接口的 LACP 优先级，默认值为 32768，值越低，优先级越高。如果优先级相同，接口 ID 较小的优先级较高（通常情况

下,建议将多个具有相同速率的接口捆绑为聚合链路;如果成员接口的速率不同,可能会选择低速率的接口成为活动接口,比如 G0/0/1 为百兆端口,G0/2/1 为千兆端口,如果在 LACP 优先级相同的情况下根据接口 ID 进行选择,百兆端口 G0/0/1 就会成为活动接口。为了应对这种特殊情况,我们可以使用命令 **lacp selected {priority | speed}**,令 LACP 根据接口速率来选择活动接口,本实验不进行展示,感兴趣的读者可以自行进行实验)。

例 6-17 在 S1 上设置 G0/0/10 端口的 LACP 优先级

```
[S1]interface GigabitEthernet 0/0/10
[S1-GigabitEthernet0/0/10]lacp priority 40000
```

G0/0/10 端口的 LACP 优先级值增大后,它的优先级会低于默认优先级,从而成为非活动接口,但是从例 6-18 中的命令输出来看,非活动接口仍然是 G0/0/12。

例 6-18 G0/0/10 没有成为非活动接口

```
[S1]display eth-trunk 1
Eth-Trunk1's state information is:
Local:
LAG ID: 1                    WorkingMode: STATIC
Preempt Delay: Disabled      Hash arithmetic: According to SIP-XOR-DIP
System Priority: 30000       System ID: 4c1f-cc29-4724
Least Active-linknumber: 2   Max Active-linknumber: 2
Operate status: up           Number Of Up Port In Trunk: 2
--------------------------------------------------------------------------------
ActorPortName         Status    PortType PortPri PortNo PortKey PortState Weight
GigabitEthernet0/0/10 Selected  1GE      40000   11     305     10111100  1
GigabitEthernet0/0/11 Selected  1GE      32768   12     305     10111100  1
GigabitEthernet0/0/12 Unselect  1GE      32768   13     305     10100000  1

Partner:
--------------------------------------------------------------------------------
ActorPortName         SysPri  SystemID        PortPri PortNo PortKey PortState
GigabitEthernet0/0/10 32768   4c1f-cc6c-03de  32768   11     305     10111100
GigabitEthernet0/0/11 32768   4c1f-cc6c-03de  32768   12     305     10111100
GigabitEthernet0/0/12 32768   4c1f-cc6c-03de  32768   13     305     10100000
```

步骤 4 配置 LACP 抢占

步骤 3 中的命令之所以没有效果,是因为设备上默认没有开启 LACP 抢占功能。LACP 抢占功能可以使高优先级的接口总是能够成为活动接口。当高优先级接口由于故障切换到非活动状态后再次恢复时,抢占功能可以使高优先级接口抢占低优先级接口的角色,成为活动接口。

我们还可以在配置抢占功能的同时,为其配置一个时延,也就是非活动接口切换为活动接口之前需要等待的时间,时延的作用是避免链路切换而导致 Eth-Trunk 中的数据传输不稳定。

例 6-19 展示了 S1 中的配置,命令语法如下。

① **lacp preempt enable**:Eth-Trunk 接口视图命令,为当前 Eth-Trunk 接口启用 LACP 抢占功能。

② **lacp preempt delay** *delay-time*:Eth-Trunk 接口视图命令,为当前 Eth-Trunk 接口配置 LACP 抢占时延,默认的 LACP 抢占时延为 30s。

例 6-19 在 S1 上配置 LACP 抢占

```
[S1]interface Eth-Trunk 1
[S1-Eth-Trunk1]lacp preempt enable
[S1-Eth-Trunk1]lacp preempt delay 10
```

至此，Eth-Trunk 1 的活动接口仍然不会进行切换，我们需要在交换机 S2 上完成相同的配置，具体见例 6-20。

例 6-20　在 S2 上完成接口优先级和 LACP 抢占配置

```
[S2]interface GigabitEthernet 0/0/10
[S2-GigabitEthernet0/0/10]lacp priority 40000
[S2-GigabitEthernet0/0/10]quit
[S2]interface Eth-Trunk 1
[S2-Eth-Trunk1]lacp preempt enable
[S2-Eth-Trunk1]lacp preempt delay 10
```

在 LACP 抢占时延 10s 后，Eth-Trunk 1 中的活动接口就会进行切换，具体见例 6-21。

例 6-21　活动接口进行切换

```
[S2]display eth-trunk 1
Eth-Trunk1's state information is:
Local:
LAG ID: 1                    WorkingMode: STATIC
Preempt Delay Time: 10       Hash arithmetic: According to SIP-XOR-DIP
System Priority: 32768       System ID: 4c1f-cc6c-03de
Least Active-linknumber: 2   Max Active-linknumber: 2
Operate status: up           Number Of Up Port In Trunk: 2
--------------------------------------------------------------------------------
ActorPortName            Status   PortType PortPri PortNo PortKey PortState Weight
GigabitEthernet0/0/10    Unselect 1GE      40000   11     305     10100000  1
GigabitEthernet0/0/11    Selected 1GE      32768   12     305     10111100  1
GigabitEthernet0/0/12    Selected 1GE      32768   13     305     10111100  1

Partner:
--------------------------------------------------------------------------------
ActorPortName            SysPri   SystemID         PortPri PortNo PortKey PortState
GigabitEthernet0/0/10    30000    4c1f-cc29-4724   40000   11     305     10100000
GigabitEthernet0/0/11    30000    4c1f-cc29-4724   32768   12     305     10111100
GigabitEthernet0/0/12    30000    4c1f-cc29-4724   32768   13     305     10111100
```

步骤 5　验证备用接口的效果

要想测试 LACP 备用接口的效果，我们可以通过关闭交换机 S1 上的 G0/0/11 接口，模拟链路故障，观察 Eth-Trunk 中活动接口的切换。例 6-22 展示了测试过程：在 **display eth-trunk 1** 命令的输出内容中可以看到，当前 G0/0/10 为非活动接口；管理员使用 **shutdown** 命令关闭 G0/0/11 端口来模拟链路故障，从系统日志的输出中可以看出，在物理端口变为 DOWN 后，Eth-Trunk 1 端口的状态瞬间进行了切换；从最后 **display eth-trunk 1** 命令的输出内容中可以看到，当前的非活动接口已经变为 G0/0/11，同时 G0/0/10 已经切换到活动状态。此时，G0/0/10 从非活动状态切换到活动状态无须等待 10s，只有当 G0/0/11 恢复后，由于 G0/0/11 优先级高于 G0/0/10 而要抢占活动接口，LACP 抢占才会时延 10s。

例 6-22　测试 LACP 备用接口

```
[S1]display eth-trunk 1
Eth-Trunk1's state information is:
Local:
LAG ID: 1                    WorkingMode: STATIC
Preempt Delay Time: 10       Hash arithmetic: According to SIP-XOR-DIP
System Priority: 30000       System ID: 4c1f-cc29-4724
Least Active-linknumber: 2   Max Active-linknumber: 2
Operate status: up           Number Of Up Port In Trunk: 2
--------------------------------------------------------------------------------
ActorPortName            Status   PortType PortPri PortNo PortKey PortState Weight
GigabitEthernet0/0/10    Unselect 1GE      40000   11     305     10100000  1
GigabitEthernet0/0/11    Selected 1GE      32768   12     305     10111100  1
GigabitEthernet0/0/12    Selected 1GE      32768   13     305     10111100  1

Partner:
--------------------------------------------------------------------------------
```

```
ActorPortName            SysPri    SystemID              PortPri  PortNo  PortKey  PortState
GigabitEthernet0/0/10    32768     4c1f-cc6c-03de        40000    11      305      10100000
GigabitEthernet0/0/11    32768     4c1f-cc6c-03de        32768    12      305      10111100
GigabitEthernet0/0/12    32768     4c1f-cc6c-03de        32768    13      305      10111100

[S1]interface GigabitEthernet 0/0/11
[S1-GigabitEthernet0/0/11]shutdown
[S1-GigabitEthernet0/0/11]
Apr 28 2021 04:30:19-08:00 S1 %%01PHY/1/PHY(l)[11]:    GigabitEthernet0/0/11: ch
ange status to down
Apr 28 2021 04:30:19-08:00 S1 %%01IFNET/4/IF_STATE(l)[13]:Interface Eth-Trunk1 h
as turned into DOWN state.
Apr 28 2021 04:30:20-08:00 S1 %%01IFNET/4/IF_STATE(l)[15]:Interface Eth-Trunk1 h
as turned into UP state.
[S1-GigabitEthernet0/0/11]quit
[S1]display eth-trunk 1
Eth-Trunk1's state information is:
Local:
LAG ID: 1                       WorkingMode: STATIC
Preempt Delay Time: 10          Hash arithmetic: According to SIP-XOR-DIP
System Priority: 30000          System ID: 4c1f-cc29-4724
Least Active-linknumber: 2      Max Active-linknumber: 2
Operate status: up              Number Of Up Port In Trunk: 2
--------------------------------------------------------------------------------
ActorPortName            Status    PortType  PortPri  PortNo  PortKey  PortState  Weight
GigabitEthernet0/0/10    Selected  1GE       40000    11      305      10111100   1
GigabitEthernet0/0/11    Unselect  1GE       32768    12      305      10100010   1
GigabitEthernet0/0/12    Selected  1GE       32768    13      305      10111100   1

Partner:
--------------------------------------------------------------------------------
ActorPortName            SysPri    SystemID              PortPri  PortNo  PortKey  PortState
GigabitEthernet0/0/10    32768     4c1f-cc6c-03de        40000    11      305      10111100
GigabitEthernet0/0/11    0         0000-0000-0000        0        0       0        10100011
GigabitEthernet0/0/12    32768     4c1f-cc6c-03de        32768    13      305      10111100
```

步骤 6 测试活动链路下限阈值

在前面步骤中，我们将 Eth-Trunk 1 的活动链路下限阈值设置为 2，若在当前配置的基础上再关闭一个端口，会导致整个 Eth-Trunk 1 变为 DOWN 状态。本步骤通过关闭 S1 上的 G0/0/12 端口来查看效果，具体见例 6-23。在本例中，管理员关闭了 G0/0/12 端口后，Eth-Trunk 1 变为 DOWN 状态，没有再恢复 UP 状态。通过 **display eth-trunk 1** 命令查看状态时，发现 Eth-Trunk 接口的运行状态已经是 down，所有成员接口都变为非活动接口。

例 6-23 测试活动链路下限阈值

```
[S1]interface GigabitEthernet 0/0/12
[S1-GigabitEthernet0/0/12]shutdown
[S1-GigabitEthernet0/0/12]
Apr 28 2021 04:43:36-08:00 S1 %%01PHY/1/PHY(l)[0]:     GigabitEthernet0/0/12: cha
nge status to down
Apr 28 2021 04:43:36-08:00 S1 %%01IFNET/4/IF_STATE(l)[2]:Interface Eth-Trunk1 ha
s turned into DOWN state.
[S1-GigabitEthernet0/0/12]quit
[S1]display eth-trunk 1
Eth-Trunk1's state information is:
Local:
LAG ID: 1                       WorkingMode: STATIC
Preempt Delay Time: 10          Hash arithmetic: According to SIP-XOR-DIP
System Priority: 30000          System ID: 4c1f-cc29-4724
Least Active-linknumber: 2      Max Active-linknumber: 2
Operate status: down            Number Of Up Port In Trunk: 0
--------------------------------------------------------------------------------
ActorPortName            Status    PortType  PortPri  PortNo  PortKey  PortState  Weight
GigabitEthernet0/0/10    Unselect  1GE       40000    11      305      10100000   1
GigabitEthernet0/0/11    Unselect  1GE       32768    12      305      10100010   1
GigabitEthernet0/0/12    Unselect  1GE       32768    13      305      10100010   1
Partner:
--------------------------------------------------------------------------------
ActorPortName            SysPri    SystemID              PortPri  PortNo  PortKey  PortState
GigabitEthernet0/0/10    32768     4c1f-cc6c-03de        40000    11      305      10100000
GigabitEthernet0/0/11    0         0000-0000-0000        0        0       0        10100011
GigabitEthernet0/0/12    0         0000-0000-0000        0        0       0        10100011
```

我们还可以使用 **display trunkmembership eth-trunk 1** 来查看 Eth-Trunk 接口的状态及其成员接口的状态，具体见例 6-24。

例 6-24 快速查看 Eth-Trunk 接口及其成员接口状态

```
[S1]display trunkmembership eth-trunk 1
Trunk ID: 1
Used status: VALID
TYPE: ethernet
Working Mode : Static
Number Of Ports in Trunk = 3
Number Of Up Ports in Trunk = 0
Operate status: down

Interface GigabitEthernet0/0/10, valid, operate down, weight=1
Interface GigabitEthernet0/0/11, valid, operate down, weight=1
Interface GigabitEthernet0/0/12, valid, operate down, weight=1
```

我们再次启用 S1 的 G0/0/11 和 G0/0/12 端口，观察 Eth-Trunk 1 的状态，具体见例 6-25 和例 6-26。

例 6-25 启用 S1 的 G0/0/11 和 G0/0/12 端口

```
[S1]interface GigabitEthernet 0/0/11
[S1-GigabitEthernet0/0/11]undo shutdown
[S1-GigabitEthernet0/0/11]quit
[S1]interface GigabitEthernet 0/0/12
[S1-GigabitEthernet0/0/12]undo shutdown
```

例 6-26 Eth-Trunk 1 的运行状态恢复正常

```
[S2]display eth-trunk 1
Eth-Trunk1's state information is:
Local:
LAG ID: 1                    WorkingMode: STATIC
Preempt Delay Time: 10       Hash arithmetic: According to SIP-XOR-DIP
System Priority: 32768       System ID: 4c1f-cc6c-03de
Least Active-linknumber: 2   Max Active-linknumber: 2
Operate status: up           Number Of Up Port In Trunk: 2
--------------------------------------------------------------------------------
ActorPortName           Status     PortType  PortPri  PortNo  PortKey  PortState  Weight
GigabitEthernet0/0/10   Unselect   1GE       40000    11      305      10100000   1
GigabitEthernet0/0/11   Selected   1GE       32768    12      305      10111100   1
GigabitEthernet0/0/12   Selected   1GE       32768    13      305      10111100   1

Partner:
--------------------------------------------------------------------------------
ActorPortName           SysPri    SystemID            PortPri  PortNo  PortKey  PortState
GigabitEthernet0/0/10   30000     4c1f-cc29-4724      40000    11      305      10100000
GigabitEthernet0/0/11   30000     4c1f-cc29-4724      32768    12      305      10111100
GigabitEthernet0/0/12   30000     4c1f-cc29-4724      32768    13      305      10111100
```

步骤 7　配置 LACP 负载分担方式

无论 Eth-Trunk 中有几条活动链路，管理员都可以配置 LACP 负载分担方式，将不同的数据流分布在不同的活动链路上。为了避免发生数据包乱序的情况，应设置通过同一条活动链路发送同一个数据流。这种逐流进行负载分担的方式会根据数据帧中的某些参数，使用散列算法生成数据流的散列值，交换机根据散列值在 Eth-Trunk 中选择对应的出接口。每条数据流计算出的散列值不同，因此交换机选择的出接口不同，这样既保证了通过同一条活动链路发送同一个数据流的数据帧，也能够让多条活动链路负载分担不同数据流的数据帧。

不同型号交换机默认采用的负载分担方式不同，管理员可以令交换机根据以下参数的组合来计算散列值：

① 目的 IP 地址；

② 目的 MAC 地址;
③ 源 IP 地址;
④ 源 MAC 地址;
⑤ 源 IP 地址异或目的 IP 地址;
⑥ 源 MAC 地址异或目的 MAC 地址。

Eth-Trunk 链路两端设备上的负载分担方式可以不一致,因为负载分担只对本地设备出方向的流量有效,所以两端的配置互不影响。有些交换机型号只支持同时运行一种负载分担方式,也就是说,在一个 Eth-Trunk 接口下,配置的负载分担方式会在所有 Eth-Trunk 接口上生效。当创建新的 Eth-Trunk 接口时,交换机会将已有 Eth-Trunk 的负载分担方式更改为默认值。

我们以交换机 S1 为例展示 LACP 负载分担的配置命令,将负载分担方式更改为基于源 MAC 地址计算散列值,具体见例 6-27,命令语法如下。

load-balance {dst-ip | dst-mac | src-ip | src-mac | src-dst-ip | src-dst-mac}:Eth-Trunk 接口视图命令,配置 Eth-Trunk 的负载分担方式。

例 6-27   在 S1 上配置 LACP 负载分担

```
[S1]interface Eth-Trunk 1
[S1-Eth-Trunk1]load-balance src-mac
```

我们可以在命令 **display eth-trunk** 中查看 Hash arithmetic 字段,确认负载分担方式。从例 6-28 的命令输出中,我们可以看出,当前的散列计算是基于源 MAC 地址(SA)进行的。

例 6-28   验证 LACP 负载分担配置

```
[S1]display eth-trunk 1
Apr 28 2021 06:02:05-08:00 S1 DS/4/DATASYNC_CFGCHANGE:OID 1.3.6.1.4.1.2011.5.25.
191.3.1 configurations have been changed. The current change number is 26, the c
hange loop count is 0, and the maximum number of records is 4095.1
[S1]display eth-trunk 1
Eth-Trunk1's state information is:
Local:
LAG ID: 1                     WorkingMode: STATIC
Preempt Delay Time: 10        Hash arithmetic: According to SA
System Priority: 30000        System ID: 4c1f-cc29-4724
Least Active-linknumber: 2    Max Active-linknumber: 2
Operate status: up            Number Of Up Port In Trunk: 2
--------------------------------------------------------------------------------
ActorPortName            Status      PortType PortPri PortNo PortKey PortState Weight
GigabitEthernet0/0/10    Unselect    1GE      40000   11     305     10100000  1
GigabitEthernet0/0/11    Selected    1GE      32768   12     305     10111100  1
GigabitEthernet0/0/12    Selected    1GE      32768   13     305     10111100  1

Partner:
--------------------------------------------------------------------------------
ActorPortName            SysPri  SystemID          PortPri PortNo PortKey PortState
GigabitEthernet0/0/10    32768   4c1f-cc6c-03de    40000   11     305     10100000
GigabitEthernet0/0/11    32768   4c1f-cc6c-03de    32768   12     305     10111100
GigabitEthernet0/0/12    32768   4c1f-cc6c-03de    32768   13     305     10111100
```

华为交换机还支持增强模式的负载分担方式,需要使用增强模板进行设置,对此感兴趣的读者可以参考华为相应设备的配置指南。

第7章
实现VLAN间通信实验

本章主要内容

7.1　实验介绍

7.2　实验配置任务

通过前文的实验，读者已经了解了 IP 路由的基础配置及以太网交换的基础配置。本章将重点介绍如何在网络中将以太网与 IP 网络相结合，以及广播域隔离的不同 VLAN 之间该如何通信。

实现两个 VLAN 之间的通信，需要三层路由功能的介入，有两种方法可以实现这个目标。第一种方法是在网络中添加路由器，路由器作为三层设备为 VLAN 间的流量执行路由。第二种方法是在交换机上启用 VLANIF 接口，使用这个虚拟接口执行三层路由功能。能够支持三层路由功能的多层交换机才能使用这种方法，普通的二层接入交换机没有路由功能，因此无法借助这种方法实现 VLAN 间的通信。

在进行实验之前，我们先讲解一下这两种方法的工作原理。

方法一：添加路由器，为 VLAN 间的通信执行路由。

在这种方法中，交换机上的每个 VLAN 都要与路由器上的三层接口对应，看起来路由器需要消耗大量的三层接口，但在实际应用中，我们使用一个路由器接口就可以实现多个 VLAN 之间的路由转发。这种做法也被称为通过 Dot1q 终结子接口以实现 VLAN 间的通信。此时，路由器与交换机之间只有一条链路，通过交换机连接到网络中的主机分别属于不同的 VLAN，在这里，假设当前网络中使用的是 VLAN 10 和 VLAN 20，如图 7-1 所示。

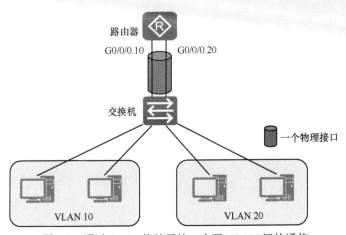

图 7-1 通过 Dot1q 终结子接口实现 VLAN 间的通信

路由器使用 G0/0/0 接口与交换机相连，G0/0/0 接口上创建了两个子接口，分别对应交换机上创建的两个 VLAN。子接口的格式为"物理接口 ID.子接口 ID"，比如 G0/0/0.10 和 G0/0/0.20，我们使用 VLAN ID 作为子接口 ID，由此可以较清晰地阐明子接口与 VLAN 的对应关系，读者在实际工作中可以采用其他子接口编号方式。每个子接口都是相应 VLAN 的终结，即子接口作为 VLAN 中终端设备的默认网关，将终端设备去往 VLAN 外的流量路由到相应的目的地。因此，对于路由器接口，我们需要进行如下配置：

① 创建子接口；
② 配置相应 VLAN 的 Dot1q 终结；
③ 配置 IP 地址，作为终端设备的默认网关。

交换机与路由器相连的端口需要传输两个 VLAN 的流量，因此我们需要将其配置为

Trunk，并放行相应 VLAN 的流量。

当 VLAN 10 主机向 VLAN 20 主机发送流量时，数据帧的处理过程大致如下：

① 交换机从连接 VLAN 10 主机的接口接收去往 VLAN 20 主机的数据帧，将其标记为 VLAN 10，并从连接路由器的 Trunk 接口发送出去；

② 路由器从 G0/0/0.10 子接口接收从 VLAN 10 发来的数据帧，标记为 VLAN 10，将其作为普通 IP 数据包进行路由，路由器根据目的 IP 地址查找路由表，并将数据包路由到 G0/0/0.20 子接口。G0/0/0.20 子接口向数据帧中添加 VLAN 20 标记并将其发送给交换机；

③ 交换机接收标记为 VLAN 20 的数据帧，并将其转发给相应的终端主机。

方法二：配置 VLANIF 接口，实现 VLAN 间通信。

VLANIF 是具有三层路由功能的逻辑接口，读者在前文实验中已经接触过两种虚拟接口：环回接口和 Eth-Trunk 接口。在支持三层路由功能的交换机上，管理员可以通过为每个 VLAN 配置相应的 VLANIF 接口，来实现 VLAN 间的通信。在配置 VLANIF 接口之前，首先要在交换机上创建相应的 VLAN，然后再创建 VLANIF 接口并配置 IP 地址，VLANIF 接口的 IP 地址会作为 VLAN 内终端设备的默认网关，当属于 VLAN 的物理端口中有 UP 状态的端口时，VLANIF 接口就会变为 UP 状态，具体的配置详见后文的实验配置任务。

7.1 实验介绍

7.1.1 关于本实验

本实验将通过上述两种方法（配置 Dot1q 终结子接口和 VLANIF 接口）实现不同 VLAN 间的通信，其中涵盖了与这两种方法相关的原理、配置命令和验证方法。

7.1.2 实验目的

- 掌握配置 Dot1q 终结子接口的方法，实现 VLAN 间通信。
- 掌握配置 VLANIF 接口的方法，实现 VLAN 间通信。
- 理解 VLAN 间通信的工作流程。

7.1.3 实验组网介绍

表 7-1 列出了本章使用的网络地址。

表 7-1 本章使用的网络地址

设备	接口	IP 地址	子网掩码	默认网关
AR1	G0/0/1.10	172.16.10.254	255.255.255.0	—
	G0/0/1.20	172.16.20.254	255.255.255.0	—
PC10	E0/0/1	172.16.10.10	255.255.255.0	172.16.10.254
PC20	E0/0/1	172.16.20.10	255.255.255.0	172.16.20.254

7.1.4 实验任务列表

配置任务 1：配置 Dot1q 终结子接口，实现 VLAN 间通信。
配置任务 2：配置 VLANIF 接口，实现 VLAN 间通信。
图 7-2 为实现 VLAN 间通信的实验拓扑。

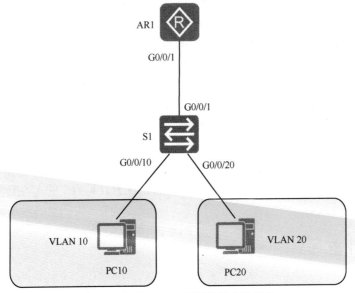

图 7-2　实现 VLAN 间通信的实验拓扑

7.2 实验配置任务

7.2.1 配置 Dot1q 终结子接口，实现 VLAN 间通信

在这个实验中，我们需要在路由器 AR1 的 G0/0/1 接口上创建两个子接口，这两个子接口分别为 VLAN 10 和 VLAN 20 提供默认网关。读者可以先按照之前实验中学习的配置方法和命令，为 PC10 和 PC20 配置 IP 地址、子网掩码和默认网关，再在交换机 S1 上创建 VLAN 10 和 VLAN 20，为相应的端口配置链路类型。

步骤 1　配置 PC 和交换机

读者可以参考图 7-3 来配置 PC10 和 PC20，需要注意的是，填写完 IP 参数后，需要单击"确定"按键才能使设置生效。

例 7-1 中展示了交换机 S1 的配置，读者已经学习过相关命令，建议读者根据前期学习过的相关命令自行进行配置。配置提示：将连接 PC 的两个端口设置为 Access 端口，并加入相应的 VLAN；将连接路由器 AR1 的端口设置为 Trunk 端口，并放行相应的 VLAN。

第 7 章 实现 VLAN 间通信实验

图 7-3 PC 配置参考

例 7-1 交换机 S1 上的配置

```
[S1]interface GigabitEthernet 0/0/10
[S1-GigabitEthernet0/0/10]port link-type access
[S1-GigabitEthernet0/0/10]port default vlan 10
[S1-GigabitEthernet0/0/10]quit
[S1]interface GigabitEthernet 0/0/20
[S1-GigabitEthernet0/0/20]port link-type access
[S1-GigabitEthernet0/0/20]port default vlan 20
[S1-GigabitEthernet0/0/20]quit
[S1]interface GigabitEthernet 0/0/1
[S1-GigabitEthernet0/0/1]port link-type trunk
[S1-GigabitEthernet0/0/1]port trunk allow-pass vlan 10 20
```

步骤 2 配置 Dot1q 终结子接口

在路由器上创建子接口的命令与进入物理接口的命令相同，在子接口下需要执行以下 4 条命令来启用 Dot1q 终结功能，例 7-2 展示了路由器 AR1 的配置。

① **interface** *interface-type interface-number.subinterface-number*：系统视图命令，创建子接口并进入子接口配置视图。

② **ip address** *ip-address mask-length*：子接口视图命令，配置子接口的 IP 地址，并将子接口的 IP 地址作为相应 VLAN 的默认网关。

③ **dot1q termination vid** *low-pe-vid*：子接口视图命令，指定该子接口终结的 VLAN。每个子接口只能关联并终结一个 VLAN；同一主接口下的不同子接口不能关联相同的 VLAN，但不同主接口下的子接口可以关联相同的 VLAN。

④ **arp broadcast enable**：子接口视图命令，在该子接口上启用 ARP 广播功能。如果没有启用 ARP 广播功能，路由器将不会从子接口发出 ARP 广播，也就无法根据 IP 地址查询其对应的 MAC 地址，从而无法为 VLAN 间的流量执行转发功能。

例 7-2　AR1 上的 Dot1q 终结子接口配置

```
[AR1]interface GigabitEthernet 0/0/1.10
[AR1-GigabitEthernet0/0/1.10]ip address 172.16.10.254 24
[AR1-GigabitEthernet0/0/1.10]dot1q termination vid 10
[AR1-GigabitEthernet0/0/1.10]arp broadcast enable
[AR1-GigabitEthernet0/0/1.10]quit
[AR1]interface GigabitEthernet 0/0/1.20
[AR1-GigabitEthernet0/0/1.20]ip address 172.16.20.254 24
[AR1-GigabitEthernet0/0/1.20]dot1q termination vid 20
[AR1-GigabitEthernet0/0/1.20]arp broadcast enable
```

配置完成后，我们可以从 PC10 向 PC20 发起 ping 测试，具体见例 7-3。

例 7-3　测试 VLAN 间通信

```
PC10>ping 172.16.20.10

Ping 172.16.20.10: 32 data bytes, Press Ctrl_C to break
Request timeout!
From 172.16.20.10: bytes=32 seq=2 ttl=127 time=32 ms
From 172.16.20.10: bytes=32 seq=3 ttl=127 time=63 ms
From 172.16.20.10: bytes=32 seq=4 ttl=127 time=46 ms
From 172.16.20.10: bytes=32 seq=5 ttl=127 time=47 ms

--- 172.16.20.10 ping statistics ---
  5 packet(s) transmitted
  4 packet(s) received
  20.00% packet loss
  round-trip min/avg/max = 0/47/63 ms
```

从例 7-3 的测试结果中我们可以看出，ping 测试成功，但第 1 次 ping 测试超时了，这是因为网络设备在进行第一次通信时需要先通过 ARP 查询 IP 地址所对应的 MAC 地址。PC10 向 PC20 发起 ping 测试的过程中，涉及两次 ARP 请求，如图 7-4 所示。

① PC10 查询默认网关的 MAC 地址：当我们在 PC10 上执行命令 **ping 172.16.20.10** 后，PC10 根据本地 IP 地址与子网掩码的计算，得出这个 IP 地址不是本地子网中的地址，因此 PC10 需要将默认网关的 MAC 地址封装为以太网数据帧的目的 MAC 地址。在此之前 PC10 没有与默认网关进行过通信，因此它会请求默认网关的 MAC 地址。

② AR1 查询 PC20 的 MAC 地址：AR1 为了向 PC20 发送 ping 请求，需要知道 PC20 的 MAC 地址，因此需要向 G0/0/1.20 子接口发出 ARP 广播，来查询 PC20 的 MAC 地址，子接口视图命令 **arp broadcast enable** 就是为了这个行为而配置的。若子接口没有启用 ARP 广播功能，AR1 便不会从子接口发出 ARP 广播，也无法请求 PC20 的 MAC 地址，更无法处理后续的数据流。

我们在路由器 AR1 的 G0/0/1 接口开启抓包，读者可以在 Wireshark 的抓包结果中执行过滤，图 7-4 仅显示了与 ARP 相关的数据帧，这几个数据帧包括以下内容。

① 序号 9：PC10 发出的 ARP 请求广播，询问其默认网关 172.16.10.254 的 MAC 地址。

② 序号 10：AR1 返回给 PC10 的 ARP 应答，告知其自己的 MAC 地址。

③ 序号 11（未显示）：PC10 获得默认网关的 MAC 地址后，封装并发出 ping 请求。

④ 序号 12：AR1 通过 G0/0/1.20 子接口发出的 ARP 请求广播，询问 PC20 的 MAC 地址。图 7-4 中将这个数据帧进一步展开，从阴影行我们可以看到，AR1 在这个数据帧中标记了 VLAN ID 20（ID: 20）。读者在进行实验练习时也可以进行抓包，观察 AR1 返回给 PC10 的 ARP 应答数据帧，其中标记了 VLAN ID 10。

⑤ 序号 13：PC20 返回给 AR1 的 ARP 应答，告知其自己的 MAC 地址。

第 7 章　实现 VLAN 间通信实验　　123

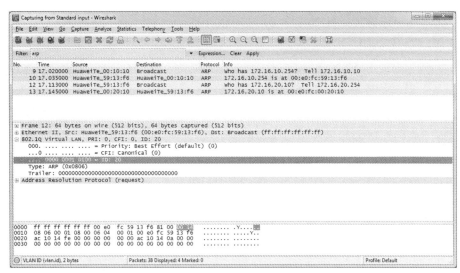

图 7-4　两次 ARP 请求和应答

我们通过抓包信息，来了解成功的 ping 测试中涉及的 4 个数据包。图 7-5 展示了 PC10 发出的 ping 请求，从中我们可以看到以下内容。

① 源 IP 地址：PC10（172.16.10.10）。

② 目的 IP 地址：PC20（172.16.20.10）。

③ 源 MAC 地址：PC10（00:e0:fc:00:10:10）。

④ 目的 MAC 地址：AR1（00:e0:fc:59:13:f6）。

⑤ VLAN ID：10。

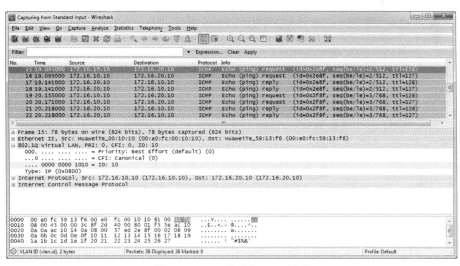

图 7-5　PC10 发出的 ping 请求

图 7-6 展示了 AR1 转发的 ping 请求，从中我们可以看到以下内容。

① 源 IP 地址：PC10（172.16.10.10）。

② 目的 IP 地址：PC20（172.16.20.10）。

③ 源 MAC 地址：AR1（00:e0:fc:59:13:f6）。

④ 目的 MAC 地址：PC20（00:e0:fc:00:20:10）。
⑤ VLAN ID：20。

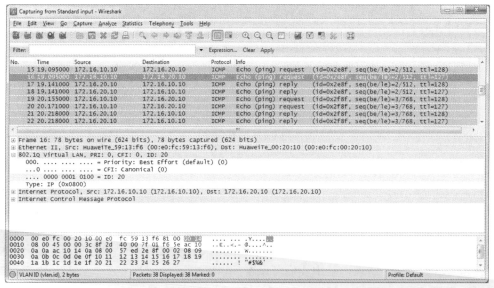

图 7-6　AR1 转发的 ping 请求

图 7-7 展示了 PC20 发出的 ping 应答，从中我们可以看到以下内容。
① 源 IP 地址：PC20（172.16.20.10）。
② 目的 IP 地址：PC10（172.16.10.10）。
③ 源 MAC 地址：PC20（00:e0:fc:00:20:10）。
④ 目的 MAC 地址：AR1（00:e0:fc:59:13:f6）。
⑤ VLAN ID：20。

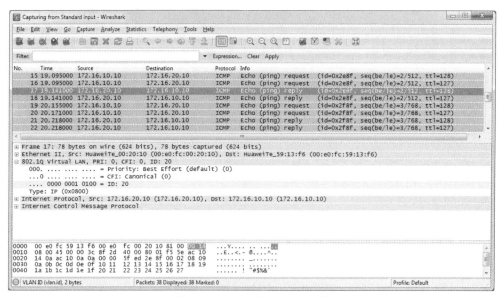

图 7-7　PC20 发出的 ping 应答

图 7-8 展示了 AR1 转发的 ping 应答,从中我们可以看到以下内容。
① 源 IP 地址:PC20(172.16.20.10)。
② 目的 IP 地址:PC10(172.16.10.10)。
③ 源 MAC 地址:AR1(00:e0:fc:59:13:f6)。
④ 目的 MAC 地址:PC10(00:e0:fc:00:10:10)。
⑤ VLAN ID:10。

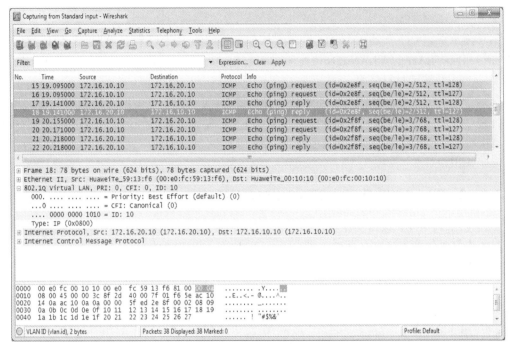

图 7-8　AR1 转发的 ping 应答

读者可以使用命令 **display arp** 在路由器 AR1 上查看 MAC 地址与 VLAN 的对应关系,具体见例 7-4。阴影部分表示 VLAN ID,以 PC10 为例,它的 IP 地址 172.16.10.10 对应的 MAC 地址是 00e0-fc00-1010,此 MAC 地址是从 GE0/0/1.10 子接口动态学习到的,属于 VLAN 10。

例 7-4　在 AR1 上查看 MAC 地址与 VLAN 的对应关系

```
[AR1]display arp
IP ADDRESS      MAC ADDRESS     EXPIRE(M) TYPE        INTERFACE      VPN-INSTANCE
                                          VLAN/CEVLAN PVC
------------------------------------------------------------------------------
172.16.10.254   00e0-fc59-13f6            I -         GE0/0/1.10
172.16.10.10    00e0-fc00-1010  20        D-0         GE0/0/1.10
                                          10/-
172.16.20.254   00e0-fc59-13f6            I -         GE0/0/1.20
172.16.20.10    00e0-fc00-2010  20        D-0         GE0/0/1.20
                                          20/-
------------------------------------------------------------------------------
Total:4         Dynamic:2       Static:0  Interface:2
```

在交换机 S1 上,读者可以使用命令 **display mac-address** 查看 MAC 地址与 VLAN 的对应关系,具体见例 7-5。仍以 PC10 为例,MAC 地址 00e0-fc00-1010 是从 GE0/0/10 端口动态学习到的,属于 VLAN 10。交换机 S1 作为二层交换机,不具备 IP 地址信息。

例 7-5 在 S1 上查看 MAC 地址与 VLAN 的对应关系

```
[S1]display mac-address
MAC address table of slot 0:
-------------------------------------------------------------------------------
MAC Address    VLAN/       PEVLAN CEVLAN Port         Type        LSP/LSR-ID
               VSI/SI                                             MAC-Tunnel
-------------------------------------------------------------------------------
00e0-fc59-13f6 20          -      -      GE0/0/1      dynamic     0/-
00e0-fc59-13f6 10          -      -      GE0/0/1      dynamic     0/-
00e0-fc00-1010 10          -      -      GE0/0/10     dynamic     0/-
00e0-fc00-2010 20          -      -      GE0/0/20     dynamic     0/-
-------------------------------------------------------------------------------
Total matching items on slot 0 displayed = 4
```

总体来说，AR1 在通过 Dot1q 终结子接口向相应的 VLAN 转发数据时，会在数据帧中标记相应 VLAN 的 VLAN ID，从而使对端的交换机 Trunk 端口能够识别出数据帧所属的 VLAN，并继续执行转发。

尽管这种方法能够实现 VLAN 间的通信，但其实际在中大型网络环境中并不常见。因为在中大型二层汇聚场景中，每个 VLAN 中的终端设备数量较为庞大，启用子接口下的 ARP 广播功能可能会因大量 ARP 请求而造成链路资源的浪费，影响网关设备的性能，从而影响用户业务的正常运行。

7.2.2 配置 VLANIF 接口，实现 VLAN 间通信

在使用 VLANIF 接口实现 VLAN 间通信的场景中，如果连接终端设备的交换机具有三层路由功能，则无须额外的路由器。要想通过 VLANIF 接口实现 VLAN 间通信，我们可以使用如下命令。

① **interface vlanif** *vlan-id*：系统视图命令，创建 VLANIF 接口并进入 VLANIF 接口视图。

② **ip address** *ip-address mask-length*：接口视图命令，配置 VLANIF 接口的 IP 地址，即 VLAN 中终端设备的默认网关。

我们在上一个实验配置任务的基础上继续本实验，为了做最少的变更，我们将 S1 的 G0/0/1 端口关闭，然后在 S1 上创建 VLANIF 10 和 VLANIF 20，并将其 IP 地址配置为 VLAN 10 和 VLAN 20 的网关地址，具体见例 7-6。

例 7-6 在 S1 上配置 VLANIF 接口

```
[S1]interface GigabitEthernet 0/0/1
[S1-GigabitEthernet0/0/1]shutdown
[S1-GigabitEthernet0/0/1]quit
[S1]interface Vlanif 10
[S1-Vlanif10]
Apr 30 2021 04:15:09-08:00 S1 %%01IFNET/4/IF_STATE(l)[3]:Interface Vlanif10 has
turned into UP state.
[S1-Vlanif10]ip address 172.16.10.254 24
[S1-Vlanif10]
Apr 30 2021 04:15:19-08:00 S1 %%01IFNET/4/LINK_STATE(l)[4]:The line protocol IP
on the interface Vlanif10 has entered the UP state.
[S1-Vlanif10]quit
[S1]interface Vlanif 20
[S1-Vlanif20]
Apr 30 2021 04:15:25-08:00 S1 %%01IFNET/4/IF_STATE(l)[5]:Interface Vlanif20 has
turned into UP state.
[S1-Vlanif20]ip address 172.16.20.254 24
[S1-Vlanif20]
Apr 30 2021 04:15:35-08:00 S1 %%01IFNET/4/LINK_STATE(l)[6]:The line protocol IP
on the interface Vlanif20 has entered the UP state.
```

我们从穿插在命令中间的系统日志可以看出，VLANIF 接口是逻辑接口，因此创建后它的状态会直接变为 UP，同时在配置了 IP 地址后，它的线路协议状态也会进入 UP。

此时我们在 PC10 上向 PC20 发起 ping 测试，测试成功，具体见例 7-7。

例 7-7　测试成功

```
PC10>ping 172.16.20.10

Ping 172.16.20.10: 32 data bytes, Press Ctrl_C to break
From 172.16.20.10: bytes=32 seq=1 ttl=127 time=62 ms
From 172.16.20.10: bytes=32 seq=2 ttl=127 time=46 ms
From 172.16.20.10: bytes=32 seq=3 ttl=127 time=47 ms
From 172.16.20.10: bytes=32 seq=4 ttl=127 time=62 ms
From 172.16.20.10: bytes=32 seq=5 ttl=127 time=62 ms

--- 172.16.20.10 ping statistics ---
  5 packet(s) transmitted
  5 packet(s) received
  0.00% packet loss
  round-trip min/avg/max = 46/55/62 ms
```

因为交换机 S1 启用了三层功能，所以我们现在可以使用 **display arp** 命令来查看它学习到的 MAC 地址，具体见例 7-8。

例 7-8　查看 S1 学习到的 MAC 地址

```
[S1]display arp
IP ADDRESS       MAC ADDRESS     EXPIRE(M) TYPE INTERFACE        VPN-INSTANCE
                                           VLAN
------------------------------------------------------------------------------
172.16.10.254    4c1f-cc02-52a4            I  -  Vlanif10
172.16.10.10     00e0-fc00-1010  20        D-0  GE0/0/10
                                           10
172.16.20.254    4c1f-cc02-52a4            I  -  Vlanif20
172.16.20.10     00e0-fc00-2010  20        D-0  GE0/0/20
                                           20
------------------------------------------------------------------------------
Total:4         Dynamic:2        Static:0     Interface:2
```

使用 **display interface vlanif** *vlan-id* 命令可以查看 VLANIF 接口的信息，具体见例 7-9。

例 7-9　查看 VLANIF 接口的信息

```
[S1]display interface Vlanif 10
Vlanif10 current state : UP
Line protocol current state : UP
Last line protocol up time : 2021-04-30 04:15:19 UTC-08:00
Description:
Route Port,The Maximum Transmit Unit is 1500
Internet Address is 172.16.10.254/24
IP Sending Frames' Format is PKTFMT_ETHNT_2, Hardware address is 4c1f-cc02-52a4
Current system time: 2021-04-30 04:38:02-08:00
    Input bandwidth utilization  : --
    Output bandwidth utilization : --
```

第 8 章
ACL 配置实验

本章主要内容

8.1　实验介绍

8.2　实验配置任务

ACL 是每位操作网络设备的工程师都必须熟练使用的工具,通过本实验,读者不仅可以熟悉配置 ACL 的命令,还可以练习如何根据需求来设计 ACL。我们先来学习一下与 ACL 相关的基础知识。

ACL 是由 permit 或 deny 规则组成的一系列有序规则的集合,分为基本 ACL、高级 ACL、二层 ACL 和用户自定义 ACL。本实验会展示如何使用基本 ACL 来实现基于源 IPv4 地址的匹配,以及如何使用高级 ACL 来实现基于源 IP 地址、目的 IP 地址、源端口号和目的端口号的匹配。

在做实验之前,读者需要明确 ACL 的方向。ACL 需要针对数据流中的数据包执行过滤,数据流是有方向的,因此 ACL 也是有方向的,数据流的方向是从数据源去往目的地,而 ACL 的方向是以网络设备为中心进行判断的,ACL 的方向如图 8-1 所示。

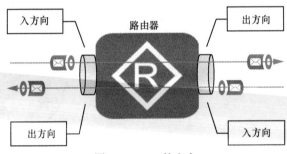

图 8-1　ACL 的方向

在图 8-1 中,路由器的左侧和右侧各有一个接口,每个接口都可以接收和发送数据包。从路由器的角度来看:从接口进入的数据包是入方向的数据包,如果对这些数据包进行过滤,则需要在接口上应用入方向的 ACL;与此类似,如果对接口转发出去的数据包进行过滤,需要在接口上应用出方向的 ACL。因此,每个接口上可以应用两个 IPv4 ACL,一个应用于出方向,另一个应用于入方向。

8.1 实验介绍

8.1.1 关于本实验

在本实验中,我们会提出两个通信需求,读者需要对 ACL 的部署位置和具体的命令进行设计,并且在应用后对效果进行测试。在此过程中,希望读者能够理解 ACL 的设计思路和配置命令,掌握 ACL 的应用。

8.1.2 实验目的

- 熟悉基本 ACL 和高级 ACL 的配置命令。
- 掌握基本 ACL 的设计思路。
- 掌握高级 ACL 的设计思路。

8.1.3 实验组网介绍

本章实验组网拓扑如图 8-2 所示。

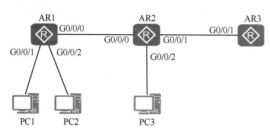

图 8-2 本章实验组网拓扑

表 8-1 列出了本章使用的网络地址。

表 8-1 本章使用的网络地址

设备	接口	IP 地址	子网掩码	默认网关
AR1	G0/0/0	10.10.12.1	255.255.255.0	—
	G0/0/1	10.10.10.254	255.255.255.0	—
	G0/0/2	10.10.20.254	255.255.255.0	—
	Loopback 0	1.1.1.1	255.255.255.255	—
AR2	G0/0/0	10.10.12.2	255.255.255.0	—
	G0/0/1	10.10.23.2	255.255.255.0	—
	G0/0/2	10.10.30.254	255.255.255.0	—
	Loopback 0	2.2.2.2	255.255.255.255	—
AR3	G0/0/1	10.10.23.3	255.255.255.0	—
	Loopback 0	3.3.3.3	255.255.255.255	—
PC1	E0/0/1	10.10.10.10	255.255.255.0	10.10.10.254
PC2	E0/0/1	10.10.20.10	255.255.255.0	10.10.20.254
PC3	E0/0/1	10.10.30.10	255.255.255.0	10.10.30.254

8.1.4 实验任务列表

配置任务 1：搭建拓扑，并实现全网互通。
配置任务 2：基本 ACL 的配置。
配置任务 3：高级 ACL 的配置。
配置任务 4：ACL 规则的顺序。

8.2 实验配置任务

8.2.1 搭建拓扑，并实现全网互通

首先，在使用 ACL 对流量进行限制之前，我们要实现全网所有网段的互通，即

表 8-1 中的所有设备及网段之间都能够相互 ping 通。本实验通过动态路由协议 OSPF 来实现全网互通，读者可以自行配置 OSPF 并进行验证，本节以 OSPF 为例提供相应的基础配置，读者可以进行参考，当然也可以使用其他方法来实现全网互通。

在 OSPF 的配置中，我们的规则是手动指定每台路由器的环回接口 IP 地址作为 OSPF 路由器 ID，同时也令环回接口加入 OSPF 域。例 8-1～例 8-3 分别展示了路由器 AR1、AR2 和 AR3 的初始配置。

例 8-1 AR1 的初始配置

```
[AR1]interface GigabitEthernet 0/0/0
[AR1-GigabitEthernet0/0/0]ip address 10.10.12.1 24
[AR1-GigabitEthernet0/0/0]quit
[AR1]interface GigabitEthernet 0/0/1
[AR1-GigabitEthernet0/0/1]ip address 10.10.10.254 24
[AR1-GigabitEthernet0/0/1]quit
[AR1]interface GigabitEthernet 0/0/2
[AR1-GigabitEthernet0/0/2]ip address 10.10.20.254 24
[AR1-GigabitEthernet0/0/2]quit
[AR1]interface LoopBack 0
[AR1-LoopBack0]ip address 1.1.1.1 32
[AR1-LoopBack0]quit
[AR1]ospf 1 router-id 1.1.1.1
[AR1-ospf-1]area 0
[AR1-ospf-1-area-0.0.0.0]network 10.10.12.1 0.0.0.0
[AR1-ospf-1-area-0.0.0.0]network 1.1.1.1 0.0.0.0
[AR1-ospf-1-area-0.0.0.0]network 10.10.10.254 0.0.0.0
[AR1-ospf-1-area-0.0.0.0]network 10.10.20.254 0.0.0.0
```

例 8-2 AR2 的初始配置

```
[AR2]interface GigabitEthernet 0/0/0
[AR2-GigabitEthernet0/0/0]ip address 10.10.12.2 24
[AR2-GigabitEthernet0/0/0]quit
[AR2]interface GigabitEthernet 0/0/1
[AR2-GigabitEthernet0/0/1]ip address 10.10.23.2 24
[AR2-GigabitEthernet0/0/1]quit
[AR2]interface GigabitEthernet 0/0/2
[AR2-GigabitEthernet0/0/2]ip address 10.10.30.254 24
[AR2-GigabitEthernet0/0/2]quit
[AR2]interface LoopBack 0
[AR2-LoopBack0]ip address 2.2.2.2 32
[AR2-LoopBack0]quit
[AR2]ospf 1 router-id 2.2.2.2
[AR2-ospf-1]area 0
[AR2-ospf-1-area-0.0.0.0]network 10.10.12.2 0.0.0.0
[AR2-ospf-1-area-0.0.0.0]network 10.10.23.2 0.0.0.0
[AR2-ospf-1-area-0.0.0.0]network 2.2.2.2 0.0.0.0
[AR2-ospf-1-area-0.0.0.0]network 10.10.30.254 0.0.0.0
```

例 8-3 AR3 的初始配置

```
[AR3]interface GigabitEthernet 0/0/1
[AR3-GigabitEthernet0/0/1]ip address 10.10.23.3 24
[AR3-GigabitEthernet0/0/1]quit
[AR3]interface LoopBack 0
[AR3-LoopBack0]ip address 3.3.3.3 32
[AR3-LoopBack0]quit
[AR3]ospf 1 router-id 3.3.3.3
[AR3-ospf-1]area 0
[AR3-ospf-1-area-0.0.0.0]network 10.10.23.3 0.0.0.0
[AR3-ospf-1-area-0.0.0.0]network 3.3.3.3 0.0.0.0
```

配置完成后，读者可以通过 IP 路由表查看非直连和环回接口的网段，例 8-4～例 8-6 分别展示了 3 台路由器的 IP 路由表。

例 8-4 AR1 上的 IP 路由表

```
[AR1]display ip routing-table
Route Flags: R - relay, D - download to fib
------------------------------------------------------------------------------
Routing Tables: Public
         Destinations : 18       Routes : 18

Destination/Mask    Proto   Pre  Cost       Flags NextHop         Interface

        1.1.1.1/32  Direct  0    0            D   127.0.0.1       LoopBack0
        2.2.2.2/32  OSPF    10   1            D   10.10.12.2      GigabitEthernet0/0/0
        3.3.3.3/32  OSPF    10   2            D   10.10.12.2      GigabitEthernet0/0/0
     10.10.10.0/24  Direct  0    0            D   10.10.10.254    GigabitEthernet0/0/1
   10.10.10.254/32  Direct  0    0            D   127.0.0.1       GigabitEthernet0/0/1
   10.10.10.255/32  Direct  0    0            D   127.0.0.1       GigabitEthernet0/0/1
     10.10.12.0/24  Direct  0    0            D   10.10.12.1      GigabitEthernet0/0/0
     10.10.12.1/32  Direct  0    0            D   127.0.0.1       GigabitEthernet0/0/0
   10.10.12.255/32  Direct  0    0            D   127.0.0.1       GigabitEthernet0/0/0
     10.10.20.0/24  Direct  0    0            D   10.10.20.254    GigabitEthernet0/0/2
   10.10.20.254/32  Direct  0    0            D   127.0.0.1       GigabitEthernet0/0/2
   10.10.20.255/32  Direct  0    0            D   127.0.0.1       GigabitEthernet0/0/2
     10.10.23.0/24  OSPF    10   2            D   10.10.12.2      GigabitEthernet0/0/0
     10.10.30.0/24  OSPF    10   2            D   10.10.12.2      GigabitEthernet0/0/0
      127.0.0.0/8   Direct  0    0            D   127.0.0.1       InLoopBack0
      127.0.0.1/32  Direct  0    0            D   127.0.0.1       InLoopBack0
 127.255.255.255/32 Direct  0    0            D   127.0.0.1       InLoopBack0
 255.255.255.255/32 Direct  0    0            D   127.0.0.1       InLoopBack0
```

例 8-5 AR2 上的 IP 路由表

```
[AR2]display ip routing-table
Route Flags: R - relay, D - download to fib
------------------------------------------------------------------------------
Routing Tables: Public
         Destinations : 18       Routes : 18

Destination/Mask    Proto   Pre  Cost       Flags NextHop         Interface

        1.1.1.1/32  OSPF    10   1            D   10.10.12.1      GigabitEthernet0/0/0
        2.2.2.2/32  Direct  0    0            D   127.0.0.1       LoopBack0
        3.3.3.3/32  OSPF    10   1            D   10.10.23.3      GigabitEthernet0/0/1
     10.10.10.0/24  OSPF    10   2            D   10.10.12.1      GigabitEthernet0/0/0
     10.10.12.0/24  Direct  0    0            D   10.10.12.2      GigabitEthernet0/0/0
     10.10.12.2/32  Direct  0    0            D   127.0.0.1       GigabitEthernet0/0/0
   10.10.12.255/32  Direct  0    0            D   127.0.0.1       GigabitEthernet0/0/0
     10.10.20.0/24  OSPF    10   2            D   10.10.12.1      GigabitEthernet0/0/0
     10.10.23.0/24  Direct  0    0            D   10.10.23.2      GigabitEthernet0/0/1
     10.10.23.2/32  Direct  0    0            D   127.0.0.1       GigabitEthernet0/0/1
   10.10.23.255/32  Direct  0    0            D   127.0.0.1       GigabitEthernet0/0/1
     10.10.30.0/24  Direct  0    0            D   10.10.30.254    GigabitEthernet0/0/2
   10.10.30.254/32  Direct  0    0            D   127.0.0.1       GigabitEthernet0/0/2
   10.10.30.255/32  Direct  0    0            D   127.0.0.1       GigabitEthernet0/0/2
      127.0.0.0/8   Direct  0    0            D   127.0.0.1       InLoopBack0
      127.0.0.1/32  Direct  0    0            D   127.0.0.1       InLoopBack0
 127.255.255.255/32 Direct  0    0            D   127.0.0.1       InLoopBack0
 255.255.255.255/32 Direct  0    0            D   127.0.0.1       InLoopBack0
```

例 8-6 AR3 上的 IP 路由表

```
[AR3]display ip routing-table
Route Flags: R - relay, D - download to fib
------------------------------------------------------------------------------
Routing Tables: Public
         Destinations : 14       Routes : 14

Destination/Mask    Proto   Pre  Cost       Flags NextHop         Interface

        1.1.1.1/32  OSPF    10   2            D   10.10.23.2      GigabitEthernet0/0/1
        2.2.2.2/32  OSPF    10   1            D   10.10.23.2      GigabitEthernet0/0/1
        3.3.3.3/32  Direct  0    0            D   127.0.0.1       LoopBack0
     10.10.10.0/24  OSPF    10   3            D   10.10.23.2      GigabitEthernet0/0/1
     10.10.12.0/24  OSPF    10   2            D   10.10.23.2      GigabitEthernet0/0/1
     10.10.20.0/24  OSPF    10   3            D   10.10.23.2      GigabitEthernet0/0/1
     10.10.23.0/24  Direct  0    0            D   10.10.23.3      GigabitEthernet0/0/1
     10.10.23.3/32  Direct  0    0            D   127.0.0.1       GigabitEthernet0/0/1
   10.10.23.255/32  Direct  0    0            D   127.0.0.1       GigabitEthernet0/0/1
     10.10.30.0/24  OSPF    10   2            D   10.10.23.2      GigabitEthernet0/0/1
      127.0.0.0/8   Direct  0    0            D   127.0.0.1       InLoopBack0
      127.0.0.1/32  Direct  0    0            D   127.0.0.1       InLoopBack0
 127.255.255.255/32 Direct  0    0            D   127.0.0.1       InLoopBack0
 255.255.255.255/32 Direct  0    0            D   127.0.0.1       InLoopBack0
```

我们也可以使用 ping 工具对连通性进行测试，例 8-7 在 PC1 上对 PC 和环回地址进行了测试。

例 8-7　测试连通性

```
PC1>ping 10.10.20.10

Ping 10.10.20.10: 32 data bytes, Press Ctrl_C to break
Request timeout!
From 10.10.20.10: bytes=32 seq=2 ttl=127 time=16 ms
From 10.10.20.10: bytes=32 seq=3 ttl=127 time=31 ms
From 10.10.20.10: bytes=32 seq=4 ttl=127 time=15 ms
From 10.10.20.10: bytes=32 seq=5 ttl=127 time<1 ms

--- 10.10.20.10 ping statistics ---
  5 packet(s) transmitted
  4 packet(s) received
  20.00% packet loss
  round-trip min/avg/max = 0/15/31 ms

PC1>ping 10.10.30.10

Ping 10.10.30.10: 32 data bytes, Press Ctrl_C to break
Request timeout!
From 10.10.30.10: bytes=32 seq=2 ttl=126 time=31 ms
From 10.10.30.10: bytes=32 seq=3 ttl=126 time=31 ms
From 10.10.30.10: bytes=32 seq=4 ttl=126 time=16 ms
From 10.10.30.10: bytes=32 seq=5 ttl=126 time=31 ms

--- 10.10.30.10 ping statistics ---
  5 packet(s) transmitted
  4 packet(s) received
  20.00% packet loss
  round-trip min/avg/max = 0/27/31 ms

PC1>ping 1.1.1.1

Ping 1.1.1.1: 32 data bytes, Press Ctrl_C to break
From 1.1.1.1: bytes=32 seq=1 ttl=255 time=47 ms
From 1.1.1.1: bytes=32 seq=2 ttl=255 time=16 ms
From 1.1.1.1: bytes=32 seq=3 ttl=255 time=16 ms
From 1.1.1.1: bytes=32 seq=4 ttl=255 time=31 ms
From 1.1.1.1: bytes=32 seq=5 ttl=255 time=15 ms

--- 1.1.1.1 ping statistics ---
  5 packet(s) transmitted
  5 packet(s) received
  0.00% packet loss
  round-trip min/avg/max = 15/25/47 ms

PC1>ping 2.2.2.2

Ping 2.2.2.2: 32 data bytes, Press Ctrl_C to break
From 2.2.2.2: bytes=32 seq=1 ttl=254 time=47 ms
From 2.2.2.2: bytes=32 seq=2 ttl=254 time=16 ms
From 2.2.2.2: bytes=32 seq=3 ttl=254 time=31 ms
From 2.2.2.2: bytes=32 seq=4 ttl=254 time=32 ms
From 2.2.2.2: bytes=32 seq=5 ttl=254 time=31 ms

--- 2.2.2.2 ping statistics ---
  5 packet(s) transmitted
  5 packet(s) received
  0.00% packet loss
  round-trip min/avg/max = 16/31/47 ms

PC1>ping 3.3.3.3

Ping 3.3.3.3: 32 data bytes, Press Ctrl_C to break
From 3.3.3.3: bytes=32 seq=1 ttl=253 time=47 ms
From 3.3.3.3: bytes=32 seq=2 ttl=253 time=62 ms
From 3.3.3.3: bytes=32 seq=3 ttl=253 time=47 ms
From 3.3.3.3: bytes=32 seq=4 ttl=253 time=47 ms
From 3.3.3.3: bytes=32 seq=5 ttl=253 time=31 ms

--- 3.3.3.3 ping statistics ---
  5 packet(s) transmitted
  5 packet(s) received
  0.00% packet loss
  round-trip min/avg/max = 31/46/62 ms
```

现在我们已经确保实验环境中的每个网段之间都能够进行通信，接着按照以下需求来设计并配置 ACL。

8.2.2 基本 ACL 的配置

基本 ACL 只能基于 IP 数据包的源 IP 地址、是否分片及时间戳信息来进行过滤，本实验仅介绍基于源 IP 地址进行匹配的方式。基本 ACL 配置语句的命令语法如下，完整的命令语法可参考华为设备配置指南。

rule [*rule-id*] { **deny** | **permit** } [**source** { *source-address source-wildcard* | **any** }]

① *rule-id*：指定 ACL 规则编号。这是一个可选参数，若不指定 ID，系统会自动分配 ID 值。在自动分配 ID 值时，系统会使用"步长"的概念，步长指每两个自动生成的 ID 之间的间隔。缺省的步长为 5，系统自动生成的 ID 为 5，并按照 5 的倍数自动生成序号，即 5、10、15，以此类推。

② **deny** 或 **permit**：必选参数，指定处理行为，拒绝或允许匹配数据包。

③ **source** { *source-address source-wildcard* | **any** }：如果根据源 IP 地址匹配数据包，就需要使用这一组关键字和参数。*source-address* 指定源 IP 地址；*source-wildcard* 指定通配符掩码，在子网掩码中，0 表示不关心，1 表示匹配，而在通配符掩码中，0 表示匹配，1 表示不关心，因此通配符掩码也被称为反掩码，比如，如果匹配子网 10.10.0.0/24，需要使用的通配符掩码为 0.0.0.255；**any** 匹配任意源 IP 地址。

在使用命令创建了基本 ACL 后，我们还需要将它应用在接口上，以使 ACL 的功能生效。要想在接口上应用 ACL，需要先进入相应接口，然后使用以下命令进行应用。

traffic-filter {**outbound** | **inbound**} **acl** *acl-number*：接口视图命令，关键字 **outbound** 指定将 ACL 应用于出方向的流量，关键字 **inbound** 指定将 ACL 应用于入方向的流量。

了解了上述规则后，我们就可以设计并应用基本 ACL。

步骤 1 选择基本 ACL 的最佳部署位置

在这个实验中，我们通过使用基本 ACL，禁止 PC1 所属网段的设备访问 PC3 所属网段的设备，其他流量不受影响。

我们从 ACL 的设计进行分析，首先，基本 ACL 只能根据源 IP 地址来对数据包进行匹配。因为流量是双向的，所以根据本实验的需求，我们有两种选择：PC1 所属网段的 IP 地址（10.10.10.0/24）和 PC3 所属网段的 IP 地址（10.10.30.0/24）。图 8-3 展示了 PC1 与 PC3 之间的数据包流。

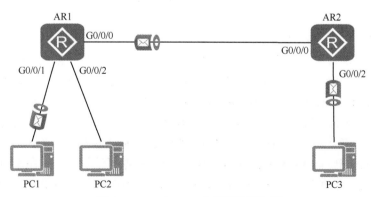

图 8-3　PC1 与 PC3 之间的数据包流

若以 PC1 所属子网作为源 IP 地址进行过滤，我们可以在以下接口的方向上应用基本 ACL。

① AR1 接口 G0/0/1 的入方向（但同时也会导致 PC1 无法访问 PC2）。
② AR1 接口 G0/0/0 的出方向。
③ AR2 接口 G0/0/0 的入方向。
④ AR2 接口 G0/0/2 的出方向。

若以 PC3 所属子网作为源 IP 地址进行过滤，我们可以在以下接口的方向上应用基本 ACL。

① AR1 接口 G0/0/1 的出方向。
② AR1 接口 G0/0/0 的入方向。
③ AR2 接口 G0/0/0 的出方向。
④ AR2 接口 G0/0/2 的入方向（但同时也会导致 PC3 无法访问 PC2）。

在选择基本 ACL 的应用位置时，应遵循的原则是：在尽可能靠近目的地的位置应用基本 ACL。基于此，若以 PC1 子网为源，最佳应用位置为 AR2 接口 G0/0/2 的出方向；若以 PC3 子网为源，最佳应用位置为 AR1 接口 G0/0/1 的出方向。这样做是为了避免无意间扩大阻塞范围。比如：以 PC1 子网为源，若在接近源的 AR2 接口 G0/0/0 的入方向阻塞流量，则 PC1 将无法访问 AR2 所连接的任何子网；若 AR2 不仅连接了 PC3 子网，还连接了其他子网，甚至连接了互联网，则 PC1 不仅无法访问 PC3 子网，还无法访问其他子网和互联网；若在距离源最近的 AR1 接口 G0/0/1 的入方向阻塞流量，AR1 就会丢弃 PC1 的流量，使流量无法到达子网之外。

步骤 2　创建并应用基本 ACL

根据上述分析，我们选择在 AR2 的出方向上应用以 PC1 子网为源的基本 ACL，具体配置命令见例 8-8。

例 8-8　创建并应用基本 ACL

```
[AR2]acl 2000
[AR2-acl-basic-2000]rule deny source 10.10.10.0 0.0.0.255
[AR2-acl-basic-2000]rule permit source any
[AR2-acl-basic-2000]quit
[AR2]interface GigabitEthernet 0/0/2
[AR2-GigabitEthernet0/0/2]traffic-filter outbound acl 2000
```

在例 8-8 中，读者可以看到我们创建了包含两条规则的基本 ACL，具体命令语法解析如下。

① **acl 2000** 为系统视图命令，创建 ACL，其中基本 ACL 的编号范围是 2000～2999。在缺省情况下，设备中没有任何 ACL，设备首次使用该命令时会创建相应的 ACL，并进入 ACL 视图；设备已有相应编号的 ACL 时，使用该命令会进入 ACL 视图。

② **rule deny source 10.10.10.0 0.0.0.255** 为 ACL 视图命令，根据语法规则，我们可以看出，它拒绝了源 IP 地址为 10.10.10.0～10.10.10.255 的数据包。需要注意的是，用来限定子网范围的参数是通配符掩码（0.0.0.255），而不是子网掩码（255.255.255.0）。

③ **rule permit source any** 为 ACL 视图命令，根据语法规则，这条命令放行了所有数据包。理论上，这种情况下无须配置这条命令，因为在 ACL 末尾会隐含一条放行所有数据包的规则。也就是说，若一个数据包与 ACL 中的所有规则都不匹配，就会被判定为

"不匹配",而 ACL 不会对"不匹配"的数据包进行任何处理,即设备会照常转发"不匹配"数据包。需要注意的是,并不是所有厂商的设备都使用相同的末尾隐含规则,因此读者在实际工作中,为避免无意间的错误配置或 ACL 解读模糊,可以在每个 ACL 的末尾都明确配置放行或拒绝所有数据包的语句,另外,这样做还可以让 ACL 针对末尾放行或拒绝的数据包进行计数。

④ **traffic-filter outbound acl 2000** 为接口视图命令,在接口上启用 ACL。关键字 **outbound** 指定了针对哪个方向的数据包进行过滤,**outbound** 过滤出方向的数据包,**inbound** 过滤入方向的数据包(本例未显示)。

在创建了 ACL 后,我们可以使用命令 **display acl all** 来查看设备中配置的所有 ACL。注意,命令输出的第一行会显示当前设备中 ACL 的总数量(本例为 1),具体见例 8-9。在例 8-9 中,读者还可以关注 ACL 2000 的具体信息,比如,该 ACL 使用的步长为默认值 5(Acl's step is 5),并且根据默认步长,设备为我们配置的两条规则进行了自动编号。

例 8-9　查看配置中的所有 ACL

```
[AR2]display acl all
 Total quantity of nonempty ACL number is 1

Basic ACL 2000, 2 rules
Acl's step is 5
 rule 5 deny source 10.10.10.0 0.0.0.255
 rule 10 permit
```

此时,我们可以从 PC1 上发起 ping 测试,测试 PC1 与 PC3 之间的连通性是否受到影响。例 8-10 中展示了 PC1 与 PC3 之间的通信测试,测试结果为无法通信,说明 ACL 发挥了作用。同时我们通过测试 PC1 与 AR3 环回接口之间的通信,验证了其他流量没有受到影响。

例 8-10　PC1 与 PC3 之间的通信测试

```
PC1>ping 10.10.30.10

Ping 10.10.30.10: 32 data bytes, Press Ctrl_C to break
Request timeout!
Request timeout!
Request timeout!
Request timeout!
Request timeout!

--- 10.10.30.10 ping statistics ---
  5 packet(s) transmitted
  0 packet(s) received
  100.00% packet loss

PC1>ping 3.3.3.3

Ping 3.3.3.3: 32 data bytes, Press Ctrl_C to break
From 3.3.3.3: bytes=32 seq=1 ttl=253 time=47 ms
From 3.3.3.3: bytes=32 seq=2 ttl=253 time=16 ms
From 3.3.3.3: bytes=32 seq=3 ttl=253 time=47 ms
From 3.3.3.3: bytes=32 seq=4 ttl=253 time=47 ms
From 3.3.3.3: bytes=32 seq=5 ttl=253 time=15 ms

--- 3.3.3.3 ping statistics ---
  5 packet(s) transmitted
  5 packet(s) received
  0.00% packet loss
  round-trip min/avg/max = 15/34/47 ms
```

测试后再次查看 ACL 2000，我们可以将 **display acl all** 中的关键字 **all** 替换为具体的 ACL 编号，此时设备就会只展示这一个 ACL 的信息。在例 8-11 的命令输出内容中，每个规则后面都出现了与该规则相匹配的数据包数量，从中可以发现，在进行测试时，ping 命令发出 5 个 ICMP Echo 请求数据包，这 5 个数据包都与规则 5 相匹配，在被拒绝的同时也会被计入匹配次数。

例 8-11　查看 ACL 2000

```
[AR2]display acl 2000
Basic ACL 2000, 2 rules
Acl's step is 5
 rule 5 deny source 10.10.10.0 0.0.0.255 (5 matches)
 rule 10 permit (4 matches)
```

另外，在例 8-11 中我们还可以看到，有 4 个数据包与规则 10 相匹配，这是由于 AR2 接口 G0/0/2 加入了 OSPF 路由协议，所以该接口会周期性地发送 OSPF Hello 数据包。

8.2.3　高级 ACL 的配置

高级 ACL 可以根据源 IP 地址、目的 IP 地址、IP 类型、传输层源端口、传输层目的端口、是否分片及时间戳信息来进行过滤。高级 ACL 规则的命令语法需要根据 IP 数据包承载的协议类型来选择不同的参数组合，比如，当协议类型为 TCP（Transmission Control Protocol，传输控制协议）时，高级 ACL 的语法格式为（完整的命令语法可参考华为设备配置指南）：

rule [*rule-id*] { **deny** | **permit** } { *protocol-number* | **tcp** } [**destination** { *destination-address destination-wildcard* | **any** } | **destination-port** { **eq** *port* | **gt** *port* | **lt** *port* | **range** *port-start port-end* } | **source** { *source-address source-wildcard* | **any** } | **source-port** { **eq** *port* | **gt** *port* | **lt** *port* | **range** *port-start port-end* } | **tcp-flag** { **ack** | **established** | **fin** | **psh** | **rst** | **syn** | **urg** }]

① *rule-id*：指定 ACL 规则编号。这是一个可选参数，若不指定 ID，系统会自动分配 ID 值。在自动分配 ID 值时，系统会使用"步长"的概念，步长指每两个自动生成的 ID 之间的间隔。缺省的步长为 5，系统自动生成的 ID 为 5，并按照 5 的倍数自动生成序号，即 5、10、15，以此类推。

② **deny** 或 **permit**：必选参数，指定处理行为，拒绝或允许匹配数据包。

③ *protocol-number* | **tcp**：必选参数，通过 *protocol-number* 指定具体的 TCP 端口，或者通过关键字 **tcp** 来笼统地指定全部 TCP 端口。

④ **destination** { *destination-address destination-wildcard* | **any** }：如果根据目的 IP 地址匹配数据包，就需要使用这一组关键字和参数。*destination-address* 指定目的 IP 地址；*destination-wildcard* 指定通配符掩码；**any** 匹配任意源 IP 地址。

⑤ **destination-port** { **eq** *port* | **gt** *port* | **lt** *port* | **range** *port-start port-end* }：如果根据目的端口来匹配数据包，就需要使用这一组关键字和参数。其中 **eq** 关键字表示等于，**gt** 关键字表示大于，**lt** 关键字表示小于，**range** 关键字表示匹配一个目的端口号范围。

⑥ **source** { *source-address source-wildcard* | **any** }：如果根据源 IP 地址匹配数据包，就需要使用这一组关键字和参数。*source-address* 指定源 IP 地址；*source-wildcard* 指定通

配符掩码；**any** 匹配任意源 IP 地址。

⑦ **source-port** { **eq** *port* | **gt** *port* | **lt** *port* | **range** *port-start port-end* }：如果根据源端口来匹配数据包，就需要使用这一组关键字和参数。其中 **eq** 关键字表示等于，**gt** 关键字表示大于，**lt** 关键字表示小于，**range** 关键字表示匹配一个源端口号范围。

⑧ **tcp-flag** { **ack** | **established** | **fin** | **psh** | **rst** | **syn** | **urg** }：如果根据 TCP 标记来匹配数据包，就需要使用这一组关键字组合。**ack** 关键字匹配的 TCP 标记为 ACK（010000），**established** 关键字匹配的 TCP 标记为 ACK（010000）或 RST（000100），**fin** 关键字匹配的 TCP 标记为 FIN（000001），**psh** 关键字匹配的 TCP 标记为 PSH（001000），**rst** 关键字匹配的 TCP 标记为 RST（000100），**syn** 关键字匹配的 TCP 标记为 SYN（000010），**urg** 关键字匹配的 TCP 标记为 URG（100000）。

当协议类型为 UDP 时，高级 ACL 的语法格式为（完整的命令语法可参考华为设备配置指南）：

rule [*rule-id*] { **deny** | **permit** } { *protocol-number* | **udp** } [**destination** { *destination-address destination-wildcard* | **any** } | **destination-port** { **eq** *port* | **gt** *port* | **lt** *port* | **range** *port-start port-end* } | **source** { *source-address source-wildcard* | **any** } | **source-port** { **eq** *port* | **gt** *port* | **lt** *port* | **range** *port-start port-end* }]

通过观察匹配 TCP 和匹配 UDP 数据包的命令语法可以发现，UDP 命令语法中没有与标记相关的关键字，这是因为 TCP 与 UDP 本身存在的差异。UDP 所使用的其他匹配参数与 TCP 相同，在此不做赘述。

当协议类型为 ICMP 时，高级 ACL 的语法格式为（完整的命令语法可参考华为设备配置指南）：

rule [*rule-id*] { **deny** | **permit** } { *protocol-number* | **icmp** } [**destination** { *destination-address destination-wildcard* | **any** } | **icmp-type** { *icmp-name* | *icmp-type* [*icmp-code*] } | **source** { *source-address source-wildcard* | **any** }]

① 在针对 ICMP 数据包进行匹配时，我们可以将 *protocol-number* 参数设置为 1，或者直接使用关键字 **icmp**。

② 除了源和目的 IP 地址，我们还可以针对 ICMP 数据包类型进行匹配，此时需要使用 **icmp-type** 关键字及其相关的必选参数。

在这里我们仅列出部分高级 ACL 规则的命令语法，完整的语法请参考华为设备配置指南。

在使用命令创建了高级 ACL 后，我们还需要将它应用在接口上，才能使 ACL 的功能生效。与基本 ACL 相同，要想在接口上应用 ACL，首先需要进入相应接口，然后使用以下命令进行应用：

traffic-filter {**outbound** | **inbound**} **acl** *acl-number*：接口视图命令，关键字 **outbound** 指定将 ACL 应用于出方向的流量，关键字 **inbound** 指定将 ACL 应用于入方向的流量。

了解了上述规则后，我们就可以设计并应用高级 ACL。

步骤 1　选择高级 ACL 的最佳部署位置

在这个实验中，我们要通过使用高级 ACL 的方式，禁止 PC2 所属网段的设备访问 PC3 所属网段的设备，其他流量不受影响。

高级 ACL 能够针对数据包的多重参数进行匹配，因此它的应用位置比较灵活，PC2 与 PC3 之间的数据包流如图 8-4 所示。

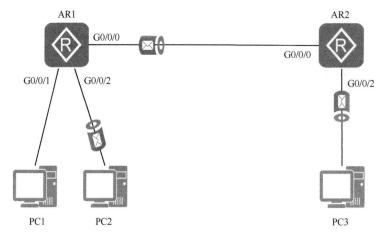

图 8-4　PC2 与 PC3 之间的数据包流

假设我们只考虑从 PC2 去往 PC3 这个方向，即源 IP 地址为 PC2 所属网段，目的 IP 地址为 PC3 所属网段，则可以在以下接口应用高级 ACL：

① AR1 接口 G0/0/2 的入方向；
② AR1 接口 G0/0/0 的出方向；
③ AR2 接口 G0/0/0 的入方向；
④ AR2 接口 G0/0/2 的出方向。

高级 ACL 可以同时匹配源和目的 IP 地址，无论在哪个接口上应用 ACL，都不会扩大阻塞的范围，因此我们可以择优选择高级 ACL 的应用位置。为了节省网络带宽资源和网络设备处理资源，选择高级 ACL 的应用位置时，应遵循的原则是：在尽可能靠近源的位置应用高级 ACL。也就是说，应用高级 ACL 的最佳位置是 AR1 接口 G0/0/2 的入方向。

步骤 2　创建并应用高级 ACL

根据上述分析，我们选择在 AR1 的入方向上应用以 PC2 子网为源、以 PC3 子网为目的的高级 ACL。高级 ACL 的编号范围是 3000～3999，例 8-12 使用编号 3000 创建高级 ACL，具体命令如下。

例 8-12　创建并应用高级 ACL

```
[AR1]acl 3000
[AR1-acl-adv-3000]rule deny ip source 10.10.20.0 0.0.0.255 destination 10.10.30.0 0.0.0.255
[AR1-acl-adv-3000]quit
[AR1]interface GigabitEthernet 0/0/2
[AR1-GigabitEthernet0/0/2]traffic-filter inbound acl 3000
```

与基本 ACL 的配置示例不同，此时我们只设置了一条 **deny** 规则，没有在其之后再设置 **permit** 规则，我们通过以下测试来观察它们的区别。

例 8-13 通过 PC2 对 PC3 执行 ping 测试，从命令的输出可以看出，ping 请求超时，高级 ACL 生效。同时 PC2 对 AR3 的环回接口执行了 ping 测试，且通信成功，说明高级 ACL 的生效并没有影响到其他流量。

例 8-13 通过 PC2 对 PC3 执行 ping 测试

```
PC2>ping 10.10.30.10

Ping 10.10.30.10: 32 data bytes, Press Ctrl_C to break
Request timeout!
Request timeout!
Request timeout!
Request timeout!
Request timeout!

--- 10.10.30.10 ping statistics ---
  5 packet(s) transmitted
  0 packet(s) received
  100.00% packet loss

PC2>ping 3.3.3.3

Ping 3.3.3.3: 32 data bytes, Press Ctrl_C to break
From 3.3.3.3: bytes=32 seq=1 ttl=253 time=31 ms
From 3.3.3.3: bytes=32 seq=2 ttl=253 time=32 ms
From 3.3.3.3: bytes=32 seq=3 ttl=253 time=47 ms
From 3.3.3.3: bytes=32 seq=4 ttl=253 time=16 ms
From 3.3.3.3: bytes=32 seq=5 ttl=253 time=32 ms

--- 3.3.3.3 ping statistics ---
  5 packet(s) transmitted
  5 packet(s) received
  0.00% packet loss
  round-trip min/avg/max = 16/31/47 ms
```

在测试后，我们来查看 ACL 的配置和状态。从例 8-14 中可以看出，当前已有 5 个数据包匹配了 ACL 3000 的规则 5，与基本 ACL 不同的是，我们没有配置 **permit** 规则，因此无法知道在 ACL 应用期间，该接口放行了多少个数据包。读者可以根据自己的需求选择是否配置 ACL 末尾的 **permit** 规则。

例 8-14 查看高级 ACL

```
[AR1]display acl 3000
Advanced ACL 3000, 1 rule
Acl's step is 5
 rule 5 deny ip source 10.10.20.0 0.0.0.255 destination 10.10.30.0 0.0.0.255 (5 matches)
```

一个 ACL 通常由多条规则构成，这些规则是有顺序的，设备首先执行编号最小的规则，一旦匹配成功，设备便退出 ACL 匹配逻辑，为数据包执行该匹配规则所定义的行为。因此在设计一个包含多条规则的 ACL 时，我们需要综合考虑，制订适当的规则并将其合理排列。下面我们通过一些更复杂的需求来讨论 ACL 规则的顺序。

8.2.4 ACL 规则的顺序

ACL 是由多条 **deny** 或 **permit** 规则构成的一组有序的规则列表，有时这些规则之间会存在重叠或行为矛盾的地方，比如以下两条规则：

rule deny ip destination 10.1.0.0 0.0.255.255

rule permit ip destination 10.1.1.0 0.0.0.255

第一条规则拒绝了目的网段为 10.1.0.0/16 的数据包通过，第二条规则允许了目的网段为 10.1.1.0/24 的数据包通过。通过对比这两条规则，我们可以发现，10.1.0.0/16 网段中包含 10.1.1.0/24 网段，这两条规则对这个网段的处理行为是矛盾的。

若这两条规则以上述顺序出现在 ACL 中，则目的网段为 10.1.1.0/24 的数据包永远没有机会与第二条规则相匹配。这是因为 ACL 是按顺序来处理每个规则的，一旦发现与数据包相匹配的规则，设备便会执行规则中指定的行为并退出 ACL 匹配进程。因此目的网段为 10.1.1.0/24 的数据包会与第一条规则相匹配，从而被丢弃。

在这个实验任务中,我们需要通过高级 ACL 禁止 AR1 的 G0/0/0 接口 ping 通 AR3 的环回接口,但 AR1 能够 Telnet 到 AR3 的环回接口,同时能够禁止其他设备 Telnet 到 AR3 的环回接口。ACL 规则的顺序实验拓扑如图 8-5 所示。地址、接口均与此前配置任务相同且互联配置已经完成,仅不含 PC1-PC3 及连接 PC 的接口。

图 8-5 ACL 规则的顺序实验拓扑

步骤 1 设计 ACL

上述需求中涉及两种协议:ICMP 和 Telnet。

① 对于 ICMP 来说,需求中规定禁止 AR1 的 G0/0/0 接口 ping 通 AR3 的环回接口,但允许其他 IP 地址 ping 通 AR3 的环回接口。要想满足这个需求,只需要一条规则即可,为了便于参考,我们列出以下匹配 ICMP 的命令语法:

rule [*rule-id*] { **deny** | **permit** } { *protocol-number* | **icmp** } [**destination** { *destination-address destination-wildcard* | **any** } | **icmp-type** { *icmp-name* | *icmp-type* [*icmp-code*] } | **source** { *source-address source-wildcard* | **any** }]

根据命令语法,我们需要在规则中将协议定义为 ICMP(或协议号 1),将源 IP 地址定义为 10.10.12.1(AR1 的 G0/0/0 接口 IP 地址),将目的 IP 地址定义为 3.3.3.3(AR3 的环回接口 IP 地址)。AR1 发来的 ICMP 数据包的类型是 ICMP 请求(Echo-request),因此我们可以将 ICMP 指定为 *echo*(在该命令语法中,*echo* 参数代表 ICMP 请求)。最终命令为:**rule deny icmp source 10.10.12.1 0 destination 3.3.3.3 0 icmp-type echo**。注意,我们在这里只匹配了一个 IP 地址,使用的通配符掩码为 0.0.0.0,这个通配符掩码可以简写为 0。

② 对于 Telnet 来说,需求中规定只有 AR1 能够 Telnet 到 AR3 的环回接口,同时禁止其他设备对其进行 Telnet。我们需要使用两条命令来满足这个需求:放行 AR1 去往 AR3 的 Telnet 流量,拒绝其他设备去往 AR3 的 Telnet 流量。Telnet 使用 TCP(端口 23),因此我们将匹配 TCP 的命令语法列出:

rule [*rule-id*] { **deny** | **permit** } { *protocol-number* | **tcp** } [**destination** { *destination-address destination-wildcard* | **any** } | **destination-port** { **eq** *port* | **gt** *port* | **lt** *port* | **range** *port-start port-end* } | **source** { *source-address source-wildcard* | **any** } | **source-port** { **eq** *port* | **gt** *port* | **lt** *port* | **range** *port-start port-end* } | **tcp-flag** { **ack** | **established** | **fin** | **psh** | **rst** | **syn** | **urg** }]

根据命令语法,在放行 AR1 流量的规则中,我们需要将协议定义为 TCP,将源 IP 地址定义为 10.10.12.1(AR1 的 G0/0/0 接口 IP 地址),将目的 IP 地址定义为 3.3.3.3(AR3 的环回接口 IP 地址),同时指定目的端口 23(Telnet 协议使用的端口),最终命令为:**rule permit tcp source 10.10.12.1 0 destination 3.3.3.3 0 destination-port eq 23**。接着我们还需要配置一条命令来禁止其他设备对 IP 地址 3.3.3.3 发起 Telnet,此时的源 IP 地址要定义为 any,因此最终命令为:**rule deny tcp source any destination 3.3.3.3 0 destination-port eq 23**。

步骤 2 创建并应用 ACL

在 AR3 接口 G0/0/1 的入方向上应用 ACL,具体见例 8-15。

例 8-15　创建并应用高级 ACL

```
[AR3]acl 3000
[AR3-acl-adv-3000]rule deny icmp source 10.10.12.1 0 destination 3.3.3.3 0 icmp-type echo
[AR3-acl-adv-3000]rule permit tcp source 10.10.12.1 0 destination 3.3.3.3 0 destination-port
eq 23
[AR3-acl-adv-3000]rule deny tcp source any destination 3.3.3.3 0 destination-port eq 23
[AR3-acl-adv-3000]quit
[AR3]interface GigabitEthernet 0/0/1
[AR3-GigabitEthernet0/0/1]traffic-filter inbound acl 3000
```

例 8-16 展示了配置完成的 ACL 3000。需要注意的是，在规则 10 和规则 15 的配置中，我们指定的目的端口 23 已被自动转换为对应的协议 Telnet，详见阴影部分。读者在配置时也可以使用关键字 **telnet** 来代替 **23**。另外，规则 15 中并没有出现配置中的关键字 **source any**，这是因为 **any** 表示不对源 IP 地址进行限制，因此设备在将命令写入运行配置之前，通过对命令的"解析"，省略了这部分配置。

例 8-16　ACL 3000 配置完成

```
[AR3]display acl 3000
Advanced ACL 3000, 3 rules
Acl's step is 5
 rule 5 deny icmp source 10.10.12.1 0 destination 3.3.3.3 0 icmp-type echo
 rule 10 permit tcp source 10.10.12.1 0 destination 3.3.3.3 0 destination-port eq telnet
 rule 15 deny tcp destination 3.3.3.3 0 destination-port eq telnet
```

步骤 3　在 AR3 上启用 Telnet

在进行测试之前，我们需要先在 AR3 上启用 Telnet 功能，读者可以按照例 8-17 中的操作，在 AR3 上配置几条基本的命令，允许网络设备通过发起 Telnet 的方式，使用密码 Huawei@123 登录 AR3。

例 8-17　在 AR3 上启用 Telnet

```
[AR3]user-interface vty 0 4
[AR3-ui-vty0-4]authentication-mode password
Please configure the login password (maximum length 16):Huawei@123
```

步骤 4　验证配置结果

我们通过 AR1 和 AR2 分别向 AR3 的环回接口 3.3.3.3 发起 ping 和 Telnet 测试，以验证 ACL 的配置是否满足需求。例 8-18 展示了 AR1 上的 ping 和 Telnet 测试，结果与需求匹配：ping 测试失败，Telnet 测试成功。需要注意的是，我们需要在用户视图中使用 **Telnet** 命令。

例 8-18　从 AR1 上进行验证

```
<AR1>ping 3.3.3.3
  PING 3.3.3.3: 56  data bytes, press CTRL_C to break
    Request time out
    Request time out
    Request time out
    Request time out
    Request time out

  --- 3.3.3.3 ping statistics ---
    5 packet(s) transmitted
    0 packet(s) received
    100.00% packet loss

<AR1>telnet 3.3.3.3
  Press CTRL_] to quit telnet mode
  Trying 3.3.3.3 ...
  Connected to 3.3.3.3 ...
Login authentication

Password:Huawei@123
<AR3>
<AR3>quit
  Configuration console exit, please retry to log on
  The connection was closed by the remote host
<AR1>
```

例 8-19 展示了测试后 ACL 3000 的计数器状态,从中我们可以发现规则 5 和规则 10 的计数器都增长了。

例 8-19 查看 ACL 3000 的计数器状态

```
[AR3]display acl 3000
Advanced ACL 3000, 3 rules
Acl's step is 5
 rule 5 deny icmp source 10.10.12.1 0 destination 3.3.3.3 0 icmp-type echo (5 matches)
 rule 10 permit tcp source 10.10.12.1 0 destination 3.3.3.3 0 destination-port eq telnet (35 matches)
 rule 15 deny tcp destination 3.3.3.3 0 destination-port eq telnet
```

最后我们再从 AR2 上发起相同的测试,所得到的结果与 AR1 上的结果相反,即 AR2 能够 ping 通 3.3.3.3,但无法对其发起 Telnet 命令,具体见例 8-20。在测试 Telnet 时,读者可以使用组合键 Ctrl+]中断 Telnet 进程。

例 8-20 从 AR2 上进行验证

```
<AR2>ping 3.3.3.3
  PING 3.3.3.3: 56  data bytes, press CTRL_C to break
    Reply from 3.3.3.3: bytes=56 Sequence=1 ttl=255 time=60 ms
    Reply from 3.3.3.3: bytes=56 Sequence=2 ttl=255 time=20 ms
    Reply from 3.3.3.3: bytes=56 Sequence=3 ttl=255 time=40 ms
    Reply from 3.3.3.3: bytes=56 Sequence=4 ttl=255 time=20 ms
    Reply from 3.3.3.3: bytes=56 Sequence=5 ttl=255 time=30 ms

  --- 3.3.3.3 ping statistics ---
    5 packet(s) transmitted
    5 packet(s) received
    0.00% packet loss
    round-trip min/avg/max = 20/34/60 ms

<AR2>telnet 3.3.3.3
  Press CTRL_] to quit telnet mode
  Trying 3.3.3.3 ...
  Error: Can't connect to the remote host
<AR2>
```

再次查看 ACL 3000 的状态,并观察计数器的变化,具体见例 8-21,此时规则 15 的计数器增加,证明 ACL 的配置符合需求。

例 8-21 查看 ACL 3000 的状态

```
[AR3]display acl 3000
Advanced ACL 3000, 3 rules
Acl's step is 5
 rule 5 deny icmp source 10.10.12.1 0 destination 3.3.3.3 0 icmp-type echo (5 matches)
 rule 10 permit tcp source 10.10.12.1 0 destination 3.3.3.3 0 destination-port eq telnet (35 matches)
 rule 15 deny tcp destination 3.3.3.3 0 destination-port eq telnet (3 matches)
```

步骤 5 使用自动排序

在本实验配置的 3 条规则中,规则 10 和规则 15 是有序的,它们的顺序不能颠倒,否则将会导致 AR1 无法对 3.3.3.3 发起 Telnet 命令,因为数据包会先与 **deny** 规则相匹配,导致数据包被丢弃。

规则 5 的顺序与另外两个规则无关,我们可以将它配置在两个 Telnet 规则之后,因为它与这两条规则不存在重叠和冲突。

华为设备提供了 ACL 自动排序功能,设备会根据规则之间的重叠和冲突,自动为规则进行排序。设备默认按照管理员配置的顺序进行排序,如需让设备自行判断规则的顺序,就要在创建 ACL 时进行指定,否则,当 ACL 中存在规则时,通过 **match-order auto** 更改排序规则会产生错误提示,具体见例 8-22。

例 8-22　更改 ACL 规则的顺序

```
[AR3]acl 3000 match-order auto
Error: Cannot execute match order command because this ACL is not empty.
[AR3-acl-adv-3000]
```

我们仍以 ACL 3000 中的 3 条规则为例，展示自动排序的效果，首先使用命令 **undo acl 3000** 删除 ACL 3000 并重新创建。例 8-23 删除了 ACL 3000 并创建了第一条 ping 规则。**display this** 命令可以根据管理员当前所处的配置视图给出不同的显示信息，检查当前配置。从本例的 **display this** 命令输出中，我们可以看到当前 ACL 3000 的规则顺序已更改为自动排序，具体见例 8-23 中的阴影部分。

例 8-23　删除并重新创建 ACL 3000

```
[AR3]undo acl 3000
[AR3]acl 3000 match-order auto
[AR3-acl-adv-3000]rule deny icmp source 10.10.12.1 0 destination 3.3.3.3 0 icmp-type echo
[AR3-acl-adv-3000]display this
[V200R003C00]
#
acl number 3000  match-order auto
 rule 5 deny icmp source 10.10.12.1 0 destination 3.3.3.3 0 icmp-type echo
#
return
```

为了测试自动排序的功能，我们将两条 Telnet 规则的顺序调换并进行配置，先配置禁止所有设备对 3.3.3.3 发起 Telnet 的 deny 规则，具体见例 8-24。从 **display this** 命令的输出内容中，我们可以看出 Telnet deny 规则与 ICMP deny 规则不存在重叠与冲突，因此 Telnet deny 规则按序被排列为规则 10。

例 8-24　配置 deny 规则

```
[AR3-acl-adv-3000]rule deny tcp source any destination 3.3.3.3 0 destination-port eq telnet
[AR3-acl-adv-3000]display this
[V200R003C00]
#
acl number 3000  match-order auto
 rule 5 deny icmp source 10.10.12.1 0 destination 3.3.3.3 0 icmp-type echo
 rule 10 deny tcp destination 3.3.3.3 0 destination-port eq telnet
#
return
```

接着添加与 Telnet deny 规则存在重叠与冲突的 Telnet permit 规则，具体见例 8-25。当我们添加了与 deny 规则存在重叠且匹配的 IP 地址范围更小的 permit 规则后，ACL 自动将匹配范围更精确的 permit 规则放到 deny 规则前面。此时，规则 10 变为了 permit 规则，而 deny 规则的编号则变为了 15。

例 8-25　添加 permit 规则

```
[AR3-acl-adv-3000]rule permit tcp source 10.10.12.1 0 destination 3.3.3.3 0 destination-port eq telnet
[AR3-acl-adv-3000]display this
[V200R003C00]
#
acl number 3000  match-order auto
 rule 5 deny icmp source 10.10.12.1 0 destination 3.3.3.3 0 icmp-type echo
 rule 10 permit tcp source 10.10.12.1 0 destination 3.3.3.3 0 destination-port eq telnet
 rule 15 deny tcp destination 3.3.3.3 0 destination-port eq telnet
#
return
```

我们按照之前的测试命令对这个可自动调整顺序的高级 ACL 进行测试，测试结果与需求相符合，具体见例 8-26。

例 8-26   从 AR1 上进行验证

```
<AR1>ping 3.3.3.3
  PING 3.3.3.3: 56  data bytes, press CTRL_C to break
    Request time out
    Request time out
    Request time out
    Request time out
    Request time out

  --- 3.3.3.3 ping statistics ---
    5 packet(s) transmitted
    0 packet(s) received
    100.00% packet loss

<AR1>telnet 3.3.3.3
  Press CTRL_] to quit telnet mode
  Trying 3.3.3.3 ...
  Connected to 3.3.3.3 ...

Login authentication

Password:Huawei@123
<AR3>
<AR3>quit

  Configuration console exit, please retry to log on

  The connection was closed by the remote host
<AR1>
```

例 8-27 展示了从 AR1 上进行验证后的 ACL 3000，从中我们可以看到规则 5 已命中 5 次，规则 10 已命中 26 次。

例 8-27   AR1 验证后的 ACL 3000

```
[AR3]display acl 3000
Advanced ACL 3000, 3 rules, match-order is auto
Acl's step is 5
 rule 5 deny icmp source 10.10.12.1 0 destination 3.3.3.3 0 icmp-type echo (5 matches)
 rule 10 permit tcp source 10.10.12.1 0 destination 3.3.3.3 0 destination-port eq telnet (26 matches)
 rule 15 deny tcp destination 3.3.3.3 0 destination-port eq telnet
```

再次在 AR2 上进行测试，测试结果与需求相吻合，具体见例 8-28。

例 8-28   从 AR2 上进行验证

```
<AR2>ping 3.3.3.3
  PING 3.3.3.3: 56  data bytes, press CTRL_C to break
    Reply from 3.3.3.3: bytes=56 Sequence=1 ttl=255 time=20 ms
    Reply from 3.3.3.3: bytes=56 Sequence=2 ttl=255 time=30 ms
    Reply from 3.3.3.3: bytes=56 Sequence=3 ttl=255 time=40 ms
    Reply from 3.3.3.3: bytes=56 Sequence=4 ttl=255 time=20 ms
    Reply from 3.3.3.3: bytes=56 Sequence=5 ttl=255 time=20 ms

  --- 3.3.3.3 ping statistics ---
    5 packet(s) transmitted
    5 packet(s) received
    0.00% packet loss
    round-trip min/avg/max = 20/26/40 ms

<AR2>telnet 3.3.3.3
  Press CTRL_] to quit telnet mode
  Trying 3.3.3.3 ...
  Error: Can't connect to the remote host
<AR2>
```

在例 8-29 中，我们可以确认此时规则 15 有 3 次命中。

例 8-29 AR2 上验证后的 ACL 3000

```
[AR3]display acl 3000
Advanced ACL 3000, 3 rules, match-order is auto
Acl's step is 5
 rule 5 deny icmp source 10.10.12.1 0 destination 3.3.3.3 0 icmp-type echo (5 matches)
 rule 10 permit tcp source 10.10.12.1 0 destination 3.3.3.3 0 destination-port eq telnet (26
matches)
 rule 15 deny tcp destination 3.3.3.3 0 destination-port eq telnet (3 matches)
```

读者在实验环境或生产环境中对 ACL 进行配置和检查时，不仅要关注 ACL 中需要匹配的各个参数是否正确，还要关注每条规则的顺序。

第 9 章
本地 AAA 配置实验

本章主要内容

9.1 实验介绍

9.2 实验配置任务

AAA 是指认证（Authentication）、授权（Authorization）和计费（Accounting），是网络管理中被广泛部署的安全控制机制。具体来说：认证所要保障的是用户身份是合法的，也就是说用户是被允许进行访问的人员/实体；授权所要保障的是用户行为是合法的，也就是说用户的行为是在被允许范围内的行为；计费所要保障的是用户行为是可追溯的，也就是说用户使用网络资源的情况会被记录并保留。

本实验旨在通过 AAA 来确保只有合法用户能够登录华为数通设备，并且用户只能够执行授权范围内的命令。华为数通设备支持以下 3 种方式的 AAA 部署。

① 不认证：任何用户都可以成功登录，设备不会对登录用户的身份进行认证。

② 本地认证：设备会使用本地存储的用户信息对登录用户进行认证和授权。

③ 远程认证：设备会将登录用户的身份信息发送给单独部署的 AAA 服务器进行认证和授权，设备与 AAA 服务器之间使用 RADIUS（Remote Authentication Dial In User Service，远程用户拨号认证服务）协议或 HWTACACS（Huawei Terminal Access Controller Access Control System，华为终端访问控制器访问控制系统）协议进行通信。

在本实验中，我们会展示如何将华为数通设备配置为 AAA 服务器并对 Telnet 登录用户进行认证和授权。在 AAA 本地认证的环境中，用户登录信息都被配置并存储在网络设备本地，其中包括用户的登录名、密码和权限等信息。这种方式适用于人员稳定的小型网络环境。

9.1 实验介绍

9.1.1 关于本实验

本实验要求将路由器 AR1 配置为 AAA 服务器，以本地认证方式对尝试登录 AR1 的用户进行身份认证和授权。路由器 AR2 作为登录用户（AAA 客户端），以 Telnet 的方式登录 AR1。读者需要在 AR1 中创建一个名为 datacom 的管理员域，并创建两个本地用户，允许这两个本地用户进行 Telnet 登录，并且用户 1（hcia-admin）能够进入系统视图实施配置和调试，用户 2（hcia-user）不能进入系统视图。

9.1.2 实验目的

- 掌握 AAA 本地认证的配置方法。
- 掌握 AAA 本地授权的配置方法。
- 掌握 AAA 维护的方法。

9.1.3 实验组网介绍

本地 AAA 配置实验拓扑如图 9-1 所示。

图 9-1 本地 AAA 配置实验拓扑

第 9 章　本地 AAA 配置实验

表 9-1 列出了本章使用的网络地址。

表 9-1　本章使用的网络地址

设备	接口	IP 地址	子网掩码	默认网关
AR1	G0/0/0	10.10.12.1	255.255.255.0	—
AR2	G0/0/0	10.10.12.2	255.255.255.0	—

9.1.4　实验任务列表

配置任务 1：配置 hcia-admin 用户 AAA 本地认证。
配置任务 2：配置 hcia-operator 用户 AAA 本地认证。

9.2　实验配置任务

9.2.1　配置 hcia-admin 用户 AAA 本地认证

在本实验中，我们要在 AR1 的 datacom 域中配置 AAA 本地认证，允许管理员用户 hcia-admin 通过 Telnet 的方式登录 AR1 的 CLI 并进行远程管理，管理员可以获得完全管理权限。在此过程中，管理员需要使用 AR1 中本地配置的用户账户信息进行登录，用户名为 hcia-admin，密码为 HCIA-Datacom。

步骤 1　基本配置

在配置 AAA 本地认证和授权之前，需要完成基本联通性配置。例 9-1 和例 9-2 分别展示了路由器 AR1 和 AR2 的相关配置。

例 9-1　完成 AR1 的基础配置

```
<Huawei>system-view
Enter system view, return user view with Ctrl+Z.
[Huawei]sysname AR1
[AR1]interface GigabitEthernet 0/0/0
[AR1-GigabitEthernet0/0/0]ip address 10.10.12.1 24
```

例 9-2　完成 AR2 的基础配置

```
<Huawei>system-view
Enter system view, return user view with Ctrl+Z.
[Huawei]sysname AR2
[AR2]interface GigabitEthernet 0/0/0
[AR2-GigabitEthernet0/0/0]ip address 10.10.12.2 24
```

配置完成后，读者可以使用 ping 命令来验证两台设备之间的连通性，具体见例 9-3。

例 9-3　验证连通性

```
[AR2]ping 10.10.12.1
  PING 10.10.12.1: 56  data bytes, press CTRL_C to break
    Reply from 10.10.12.1: bytes=56 Sequence=1 ttl=255 time=170 ms
    Reply from 10.10.12.1: bytes=56 Sequence=2 ttl=255 time=30 ms
    Reply from 10.10.12.1: bytes=56 Sequence=3 ttl=255 time=40 ms
    Reply from 10.10.12.1: bytes=56 Sequence=4 ttl=255 time=30 ms
    Reply from 10.10.12.1: bytes=56 Sequence=5 ttl=255 time=40 ms

  --- 10.10.12.1 ping statistics ---
    5 packet(s) transmitted
    5 packet(s) received
    0.00% packet loss
    round-trip min/avg/max = 30/62/170 ms
```

在使用 AAA 本地认证的环境中，我们需要在基于域的配置中设置 AAA 方案和本地用户。

在配置 AAA 本地认证之前，可以先使用命令 **display aaa configuration** 来确认设备上是否有足够的资源，具体见例 9-4。

例 9-4　查看 AAA 资源

```
[AR1]display aaa configuration
  Domain Name Delimiter              : @
  Domainname parse direction         : Left to right
  Domainname location                : After-delimiter
  Administrator user default domain: default_admin
  Normal user default domain         : default
  Domain                             : total: 32        used: 2
  Authentication-scheme              : total: 32        used: 1
  Accounting-scheme                  : total: 32        used: 1
  Authorization-scheme               : total: 32        used: 1
  Service-scheme                     : total: 256       used: 0
  Recording-scheme                   : total: 32        used: 0
  Local-user                         : total: 512       used: 1
```

从命令的输出内容中我们可以看到一些默认设置。

① 域名分隔符：缺省为"@"，可以进行更改。

② 用户名和域名的解析方向：缺省为从左到右进行解析，可以进行更改。

③ 域名的位置：缺省为域名位于分隔符之后，可以进行更改。

④ 缺省的管理员域和缺省的普通域，以及各种方案的可创建方案总数和已创建方案数量。

步骤 2　配置 AAA 方案

在 AAA 方案中，我们需要配置认证方案和授权方案，本实验要求两者都使用本地方案。读者可以使用以下命令来配置 AAA 方案。

① 在系统视图中，使用命令 **aaa** 进入 AAA 视图。在 AAA 视图中，管理员可以设置与用户接入相关的安全参数。

② 在 AAA 视图中，使用命令 **authentication-scheme** *authentication-scheme-name* 创建认证方案，并进入认证方案视图。缺省情况下设备中存在一个名为 default 的认证方案，管理员无法删除这个认证方案，但可以对其进行修改。

③ 在认证方案视图中，使用命令 **authentication-mode local** 将认证模式指定为本地认证，这也是缺省的认证模式。

④ 在 AAA 视图中，使用命令 **authorization-scheme** *authorization-scheme-name* 创建授权方案，并进入授权方案视图。缺省情况下设备中存在一个名为 default 的授权方案，管理员无法删除这个授权方案，但可以对其进行修改。

⑤ 在授权方案视图中，使用命令 **authorization-mode local** 将授权模式指定为本地授权，这也是缺省的授权模式。

例 9-5 展示了 AR1 中配置的 AAA 方案。

例 9-5　配置 AAA 方案

```
[AR1]aaa
[AR1-aaa]authentication-scheme datacom-authentication
Info: Create a new authentication scheme.
[AR1-aaa-authen-datacom]authentication-mode local
[AR1-aaa-authen-datacom]quit
[AR1-aaa]authorization-scheme datacom-authorization
Info: Create a new authorization scheme.
[AR1-aaa-author-datacom]authorization-mode local
```

配置完成后，我们可以使用命令 **display authentication-scheme** 来查看配置中的 AAA 认证方案，具体见例 9-6。阴影部分显示出我们刚配置的认证方案 datacom-authentication 和认证方式（本地认证）。

例 9-6　查看系统中的认证方案

```
[AR1]display authentication-scheme
------------------------------------------------------------
 Authentication-scheme-name        Authentication-method
------------------------------------------------------------
 default                           Local
 datacom-authentication            Local
------------------------------------------------------------
 Total of authentication scheme: 2
```

为了查看这个认证方案的详细信息，我们可以在这条命令后面加上认证方案名称，具体见例 9-7。

例 9-7　查看认证方案详情

```
[AR1]display authentication-scheme datacom-authentication
------------------------------------------------------------
  Authentication-scheme-name    : datacom-authentication
  Authentication-method         : Local
  Authentication-super method   : Super
------------------------------------------------------------
```

同样我们可以使用命令 **display authorization-scheme** 来查看配置中的授权方案，具体见例 9-8。阴影部分显示出我们刚配置的授权方案 datacom-authorization 和授权方式（本地授权）。

例 9-8　查看系统中的授权方案

```
[AR1]display authorization-scheme
------------------------------------------------------------
 Authorization-scheme-name         Authorization-method
------------------------------------------------------------
 default                           Local
 datacom-authorization             Local
------------------------------------------------------------
 Total of authortication-scheme: 2
```

为了查看这个授权方案的详细信息，我们可以在这条命令后面加上授权方案名称，具体见例 9-9。授权命令级别指是否为这个级别的用户启用按命令授权的功能。Authorization-cmd no-response-policy 指当按命令授权失败时，采取何种应对策略。本例采取的应对策略为允许用户上线。

例 9-9　查看授权方案详情

```
[AR1]display authorization-scheme datacom-authorization
------------------------------------------------------------
  Authorization-scheme-name     : datacom-authorization
  Authorization-method          : Local
  Authorization-cmd level   0   : Disabled
  Authorization-cmd level   1   : Disabled
  Authorization-cmd level   2   : Disabled
  Authorization-cmd level   3   : Disabled
  Authorization-cmd level   4   : Disabled
  Authorization-cmd level   5   : Disabled
  Authorization-cmd level   6   : Disabled
  Authorization-cmd level   7   : Disabled
  Authorization-cmd level   8   : Disabled
  Authorization-cmd level   9   : Disabled
  Authorization-cmd level  10   : Disabled
  Authorization-cmd level  11   : Disabled
  Authorization-cmd level  12   : Disabled
  Authorization-cmd level  13   : Disabled
  Authorization-cmd level  14   : Disabled
  Authorization-cmd level  15   : Disabled
  Authorization-cmd no-response-policy     : Online
------------------------------------------------------------
```

步骤 3 配置业务方案

除了 AAA 方案，我们可以在业务方案中设置与管理员用户相关的一些参数，比如管理员用户登录后的用户级别。华为设备对设备 CLI 中的命令执行分级管理，即每条命令都有其各自对应的级别，只有用户的优先级大于等于命令行的优先级，用户才可以执行这条命令。

缺省情况下，命令级别分为 0、1、2、3 级，具体说明如下。

① 级别 0（参观级）：能够使用 ping 等网络连通性诊断工具，可以从本设备向其他设备发出访问命令（比如 Telnet 等），不允许对设备配置进行更改，也无法保存配置文件。

② 级别 1（监控级）：在级别 0 的基础上，允许用户使用部分 display 命令，不允许用户保存配置文件。

③ 级别 2（配置级）：用户可以使用配置命令，比如路由配置等，也可以保存配置文件。

④ 级别 3（管理级）：能够使用所有配置命令，同时还可以使用用于业务故障诊断的 debugging 命令。

管理员还可以将某些命令的级别扩展到更高级别，以实现更精细化的管理。我们可以使用以下命令在业务方案中配置管理员的用户级别。

① 在 AAA 视图中，使用命令 **service-scheme** *service-scheme-name* 配置业务方案并进入业务方案视图。缺省情况下，设备中没有配置业务方案。

② 在业务方案视图中，使用命令 **admin-user privilege level** *level* 指定管理员用户登录后的用户级别。*level* 参数的取值范围是 0~15。

例 9-10 展示了 AR2 中业务方案的配置命令。

例 9-10 配置业务方案

```
[AR1]aaa
[AR1-aaa]service-scheme datacom-service
Info: Create a new service scheme.
[AR1-aaa-service-datacom-service]admin-user privilege level 3
```

步骤 4 创建自定义管理域

基于域的配置是为了提供更精细化和差异化的 AAA 服务，华为数通设备中默认有以下两个域。

① default_admin：全局默认管理域，管理员用户会被划分到这个域中，比如通过 Telnet、SSH、FTP、HTTP（Hyper Text Transfer Protocol，超文本传输协议）等方式登录到设备本地的用户。

② default：全局默认普通域，接入用户会被划分到这个域中，比如 PPP 用户、NAC（Network Admission Control，网络接入控制）用户（通过 802.1x 认证、MAC 认证和 Portal 认证）等。

在未配置自定义域时，用户会被匹配到默认域中，并使用默认域中的各种设置。本实验需要为管理员用户创建一个名为 datacom 的域，并在其中调用指定的 AAA 方案和业务方案。在此步骤中，我们可以使用以下命令进行配置。

① 在 AAA 视图中，使用命令 **domain** *domain-name* 来创建域并进入域视图，本实验使用 datacom 作为域名。

② 在域视图中，使用命令 **authentication-scheme** *authentication-scheme-name* 应用认证方案。

③ 在域视图中，使用命令 **authorization-scheme** *authorization-scheme-name* 应用授权方案。

④ 在域视图中，使用命令 **service-scheme** *service-scheme-name* 应用业务方案。

例 9-11 展示了 datacom 域的配置命令。

例 9-11 配置 datacom 域

```
[AR1]aaa
[AR1-aaa]domain datacom
Info: Success to create a new domain.
[AR1-aaa-domain-datacom]authentication-scheme datacom-authentication
[AR1-aaa-domain-datacom]authorization-scheme datacom-authorization
[AR1-aaa-domain-datacom]service-scheme datacom-service
```

创建完成后，我们可以使用命令 **display domain** 来查看配置中的域，具体见例 9-12，阴影部分显示了我们刚配置的域 datacom。

例 9-12 查看系统中的域

```
[AR1]display domain
---------------------------------------------------------------
 index    DomainName
---------------------------------------------------------------
 0        default
 1        default_admin
 2        datacom
---------------------------------------------------------------
Total: 3
```

我们还可以使用命令 **display domain name** *domain-name* 来查看某个域的详细信息，具体见例 9-13，阴影部分显示了我们在这个域中调用的 AAA 方案和业务方案。

例 9-13 查看域的详细信息

```
[AR1]display domain name datacom
  Domain-name                    : datacom
  Domain-state                   : Active
  Authentication-scheme-name     : datacom-authentication
  Accounting-scheme-name         : default
  Authorization-scheme-name      : datacom-authorization
  Service-scheme-name            : datacom-service
  RADIUS-server-template         : -
  HWTACACS-server-template       : -
  User-group                     : -
```

步骤 5 创建本地用户 hcia-admin

在配置本地用户时，管理员可以设置多种参数，比如用户级别、密码复杂度、空闲切断时间等。本实验要求使用的用户名为 hcia-admin，密码为 HCIA-Datacom，允许接入类型为 Telnet。

我们可以使用以下命令来配置本地用户。

① 在 AAA 视图中，使用命令 **local-user** *user-name* **password cipher** *password* 创建本地账号和登录密码。需要注意的是，本实验要求在 datacom 域中设置管理员用户，因此在配置本地用户的命令中，使用的用户名格式应该为"用户名@域名"。本实验的管理员用户名为 hcia-admin，域名为 datacom，因此完整用户名为 hcia-admin@datacom。

② 在 AAA 视图中，使用命令 **local-user** *user-name* **service-type telnet** 指定本地用户的接入类型为 Telnet。在这条命令中，管理员还可以指定其他接入类型，一个用户可以设置多种接入类型。缺省情况下未设置本地用户的接入类型。

例 9-14 展示了 AR1 中本地用户 hcia-admin 的创建命令。

例 9-14　创建本地用户

```
[AR1]aaa
[AR1-aaa]local-user hcia-admin@datacom password cipher HCIA-Datacom
Info: Add a new user.
[AR1-aaa]local-user hcia-admin@datacom service-type telnet
```

创建本地用户后，我们可以使用一些命令来查看本地设备配置的用户信息。首先通过命令 **display local-user** 查看设备上有哪些用户，以及这些用户的摘要信息，具体见例 9-15。

例 9-15　查看设备中的本地用户

```
[AR1]display local-user
--------------------------------------------------------------------------
User-name                    State   AuthMask   AdminLevel
--------------------------------------------------------------------------
admin                        A       H          -
hcia-admin@datacom           A       T          -
--------------------------------------------------------------------------
Total 2 user(s)
```

从命令的输出内容中，我们可以看到刚配置的用户 hcia-admin@datacom。AuthMask 指的是本地用户的接入类型，具体如下。

① T：Telnet 用户。

② M：终端用户，通常指的是 Console 用户。

③ S：SSH 用户。

④ F：FTP 用户。

⑤ W：Web（World Wide Web，全球广域网）用户。

⑥ X：802.1x 用户。

⑦ A：用户可以使用所有接入类型。

⑧ H：HTTP 用户。

⑨ 组合类型：比如 TH，表示这个用户可以使用 Telnet 和 HTTP 方式接入。

读者还可以使用命令 **display local-user username hcia-admin@datacom** 查看本地用户 hcia-admin@datacom 的详细信息，具体见例 9-16。从命令输出内容中，我们可以看到密码是以符号"*"显示的，可增强安全性。

例 9-16　查看本地用户的详细信息

```
[AR1]display local-user username hcia-admin@datacom
  The contents of local user(s):
  Password                : ****************
  State                   : active
  Service-type-mask       : T
  Privilege level         : -
  Ftp-directory           : -
  Access-limit            : -
  Accessed-num            : 0
  Idle-timeout            : -
  User-group              : -
```

步骤 6　配置 Telnet 功能

在本步骤中，我们需要启用路由器 AR1 的 Telnet 功能，并且将 Telnet 访问的用户设置成执行本地 AAA 认证和授权。读者可以使用以下命令启用 Telnet 功能。

telnet server enable：系统视图命令，启用 Telnet 服务器功能。缺省情况下设备的 Telnet 功能是否启用与设备型号相关，管理员可以先使用命令 **display telnet server status** 来查看设备 Telnet 服务器功能的状态。

例 9-17 展示了 **telnet server enable** 命令的功能，系统显示错误提示，表示 Telnet 服务器功能已启用。我们使用命令 **display telnet server status** 查看 Telnet 服务器的状态，命令输出内容中显示 Telnet 服务已启用。

例 9-17　启用 Telnet 服务并查看 Telnet 服务状态

```
[AR1]telnet server enable
 Error: TELNET server has been enabled
[AR1]display telnet server status
 TELNET IPV4 server                       :Enable
 TELNET IPV6 server                       :Enable
 TELNET server port                       :23
```

通过 Telnet 服务进行登录的用户会根据 VTY 用户界面的配置进行认证，按照本实验的要求，我们需要在 VTY 线路上配置 AAA 认证。

在 VTY 用户界面上能够配置以下认证方式。

① AAA 认证：用户在登录时需要输入用户名和密码。设备根据本地配置的 AAA 方案来对用户进行认证，认证成功后允许用户登录，认证失败则断开连接。

② 密码认证：用户在登录时需要输入认证密码。设备根据本地配置的认证密码来判断是否允许用户登录。

③ 不认证：用户在登录时不需要输入任何认证信息，就可以成功登录设备。

我们可以使用以下命令在 VTY 用户界面上设置用 AAA 认证对 Telnet 用户执行认证。

① **user-interface vty** *first-ui-number* [*last-ui-number*]：进入 VTY 用户界面视图。

② **authentication-mode aaa**：VTY 视图命令，指定 VTY 用户界面的认证模式为 AAA 认证。

例 9-18 展示了 AR1 中 VTY 用户界面的配置。

例 9-18　VTY 用户界面的配置

```
[AR1]user-interface vty 0 4
[AR1-ui-vty0-4]authentication-mode aaa
```

步骤 7　验证配置

首先在路由器 AR2 上使用 Telnet 命令，以管理员用户 hcia-admin@datacom 的身份登录 AR1，具体见例 9-19。从示例输出内容中可以看到，执行了 **telnet 10.10.12.1** 命令后，设备弹出让管理员输入用户名的提示，此时我们需要输入完整的用户名@域名，否则认证将会失败。从示例阴影行的命令提示符可以看出，管理员用户已经成功登录了 AR1。

例 9-19　从 AR2 登录 AR1

```
<AR2>telnet 10.10.12.1
  Press CTRL_] to quit telnet mode
  Trying 10.10.12.1 ...
  Connected to 10.10.12.1 ...

Login authentication


Username:hcia-admin@datacom
Password:
<AR1>
```

登录后我们可以通过命令 **display users** 来查看当前已经登录设备的用户，具体见例 9-20。从命令输出内容中可以看到，阴影行显示的用户前端有一个加号，它表示当前用户所在的用户界面。我们从命令输出内容中还可以看到用户 hcia-admin@datacom 的 IP 地址是 10.10.12.2，表示用户是通过 Telnet 登录的。

例 9-20　查看登录设备的用户

```
<AR1>display users
  User-Intf    Delay      Type    Network Address       AuthenStatus    AuthorcmdFlag
  0  CON 0    00:00:20                                      pass
  Username : Unspecified

+ 129 VTY 0   00:00:00    TEL     10.10.12.2                pass
  Username : hcia-admin@datacom
```

我们也可以使用命令 **display access-user** 来查看设备上已存在的用户链接，具体见例 9-21。

例 9-21　查看已存在的用户链接

```
[AR1]display access-user
------------------------------------------------------------------------
  UserID Username                     IP address               MAC
------------------------------------------------------------------------
  8      hcia-admin@datacom           2.12.10.10               -

------------------------------------------------------------------------
Total 1,1 printed
```

根据上一条命令展示出的用户 ID 8，我们还可以使用命令 **display access-user user-id 8** 来查看该用户的详细信息，具体见例 9-22。

例 9-22　查看用户的详细信息

```
[AR1]display access-user user-id 8
Basic:
  User id                        : 8
  User name                      : hcia-admin@datacom
  Domain-name                    : datacom
  User MAC                       : -
  User IP address                : 2.12.10.10
  User access time               : 2021/07/19 10:18:27
  User accounting session ID     : AR10025525500000000000519262000008
  User access type               : Telnet
  Idle Timeout                   : 4294967236(s)

AAA:
  User authentication type       : Administrator authentication
  Current authentication method  : Local
  Current authorization method   : Local
  Current accounting method      : None
```

从上述命令输出内容中，我们可以确认登录用户的用户名、所属的域，以及 IP 地址，还可以看到它所触发的 AAA 认证类型和 AAA 方案。

当管理员希望断开其他连接在设备上的用户时，可以使用用户视图命令 **free user-interface ui-type ui-number**，具体见例 9-23。示例中先使用命令 **display users** 确认有一个用户是通过 VTY 用户界面 0 登录的，接着使用命令将这个用户断开连接。

例 9-23　断开其他用户

```
<AR1>display users
  User-Intf    Delay      Type    Network Address       AuthenStatus    AuthorcmdFlag
+ 0  CON 0    00:00:00                                      pass
  Username : Unspecified

  129 VTY 0   00:00:29    TEL     10.10.12.2                pass
  Username : hcia-admin@datacom


<AR1>free user-interface vty 0
Warning: User interface vty0 will be freed. Continue? [Y/N]:y
 [OK]
<AR1>
Jul 19 2021 09:16:22-08:00 AR1 %%01LINE/3/CLR_ONELINE(l)[0]:The user chose Y whe
n deciding whether to disconnect the specified user interface.
<AR1>
```

管理员可以使用命令 **display aaa statistics offline-reason** 来查看用户下线的原因计数统计，具体见例 9-24。从命令输出内容中，我们可以看到当前设备上的用户触发了两种下线原因：编码 19 表示用户主动下线，这种下线事件发生了 6 次；编码 146 表示认证失败，这种事件发生了 4 次。

例 9-24 查看用户下线原因计数统计

```
[AR1]display aaa statistics offline-reason
 19  user request to offline          :6
 146 Authenticate fail                :4
```

9.2.2 配置 hcia-operator 用户 AAA 本地认证

在上一个实验的基础上，我们需要再创建一个用户，并且为这个用户设置监控级的权限，这个用户使用的用户名为 hcia-operator，密码为 Huawei@123。

例 9-25 展示了 AR1 上新用户的配置命令。需要注意的是，实验要求该用户的权限级别为监控级，因此命令中为其配置的级别为 1。

例 9-25 配置 hcia-operator 用户

```
[AR1]aaa
[AR1-aaa]local-user hcia-operator@datacom password cipher Huawei@123
Info: Add a new user.
[AR1-aaa]local-user hcia-operator@datacom service-type telnet
[AR1-aaa]local-user hcia-operator@datacom privilege level 1
```

配置完成后，通过 AR2 向 AR1 发起 Telnet 连接，并使用新用户进行登录，具体见例 9-26，从示例中可以看出用户登录成功。

例 9-26 使用 hcia-operator 进行登录

```
<AR2>telnet 10.10.12.1
  Press CTRL_] to quit telnet mode
  Trying 10.10.12.1 ...
  Connected to 10.10.12.1 ...

Login authentication

Username:hcia-operator@datacom
Password:
<AR1>
```

当用户成功登录 AR1 后，尝试输入命令 **system-view** 进入系统视图，结果失败了，具体见例 9-27，因为这个用户获得的授权级别为 1 级，无权进入系统视图。

例 9-27 尝试进入系统视图

```
<AR1>system-view
         ^
Error: Unrecognized command found at '^' position.
<AR1>
```

但用户仍可以使用部分 **display** 命令，比如可以使用命令 **display access-user** 来查看当前登录的用户信息，具体见例 9-28。

例 9-28 查看当前登录的用户信息

```
<AR1>display access-user
------------------------------------------------------------------------------
UserID Username                  IP address               MAC
------------------------------------------------------------------------------
10     hcia-operator@datacom     2.12.10.10               -

------------------------------------------------------------------------------
Total 1,1 printed
```

根据从命令输出内容中获得的用户 ID，我们可以查看登录用户的详细信息，具体见例 9-29。

例 9-29 查看登录用户的详细信息

```
<AR1>display access-user user-id 10
Basic:
  User id                        : 10
  User name                      : hcia-operator@datacom
  Domain-name                    : datacom
  User MAC                       : -
  User IP address                : 2.12.10.10
  User access time               : 2021/07/19 12:11:13
  User accounting session ID     : AR10025525500000000073d444000010
  User access type               : Telnet
  Idle Timeout                   : 4294967236(s)

AAA:
  User authentication type       : Administrator authentication
  Current authentication method  : Local
  Current authorization method   : Local
  Current accounting method      : None
```

第10章
网络地址转换配置实验

本章主要内容

10.1 实验介绍

10.2 实验配置任务

NAT 几乎是每个园区网都会部署的非常重要的工具。IP 数据包在网络中进行路由时，源 IP 地址和目的 IP 地址是网络设备路由数据包的依据。通常来说，这些地址信息是不会发生变化的。网络地址转换功能会使路由设备在转发 IP 数据包时，将 IP 数据包头部的源 IP 地址或目的 IP 地址进行相应改写。

一般来说，我们会在网络的出口/入口处使用 NAT，比如在企业网络的网关路由器上使用 NAT。当内部主机访问 Internet 时，网关路由器将内部主机使用的私有 IP 地址转换为 ISP 给企业分配的公有 IP 地址，此时转换的是源 IP 地址。当内部主机/服务器使用私有 IP 地址，并且需要向 Internet 提供服务(比如 Web 服务)时，网关路由器会将来自 Internet 数据包的公有 IP 地址转换为内部服务器的私有 IP 地址，此时转换的是目的 IP 地址。

在这两种应用场景中，NAT 提供了以下功能。

① 在使用私有 IP 地址以节省全球 IP 地址空间的环境中，为使用私有 IP 地址的主机提供了连接外部网络的机会。

② 对外隐藏了网络内部 IP 地址结构，能够有效避免来自外部的诸多攻击，提高网络安全性。

10.1 实验介绍

10.1.1 关于本实验

当使用私有 IP 地址的内部主机访问外网时，需要使用 NAT 将其私有 IP 地址转换为公有 IP 地址，此时需要在网关路由器上配置 NAT 来提供相应的地址转换服务。当网关路由器连接 ISP 的接口上未使用固定 IP 地址，而是动态地从 ISP 获取 IP 地址时，为了向内部使用私有 IP 地址的主机提供访问外网的服务，需要在网关路由器上配置 Easy IP。当使用私有 IP 地址的内部主机/服务器需要向外部主机提供服务时，需要在网关路由器上配置 NAT Server 特性，将来自外部的目的 IP 地址转换为相应的内部主机的私有 IP 地址。本实验通过 4 种不同的需求场景分别演示 4 种 NAT 配置。

另外需要注意的是，我们无法通过 ping 来展示 NAT Server 实验的结果，因此本实验选择使用 HTTP 服务来展示 NAT Server 实验的结果。

10.1.2 实验目的

- 掌握静态 NAT 的配置。
- 掌握动态 NAT 的配置。
- 掌握 Easy IP 的配置。
- 掌握 NAT Server 的配置。

10.1.3 实验组网介绍

NAT 配置实验拓扑如图 10-1 所示。

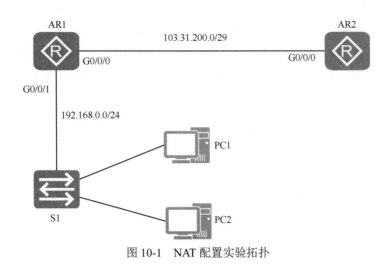

图 10-1　NAT 配置实验拓扑

本章使用的网络地址见表 10-1。

表 10-1　本章使用的网络地址

设备	接口	IP 地址	子网掩码	默认网关
AR1	G0/0/0	103.31.200.5	255.255.255.248	—
	G0/0/1	192.168.0.254	255.255.255.0	—
AR2	G0/0/0	103.31.200.6	255.255.255.248	—
	Loopback 0	8.8.8.8	255.255.255.255	—
PC1	E0/0/1	192.168.0.1	255.255.255.0	192.168.0.254
PC2	E0/0/1	192.168.0.2	255.255.255.0	192.168.0.254

10.1.4　实验任务列表

配置任务 1：静态 NAT 的配置。
配置任务 2：动态 NAT 的配置。
配置任务 3：Easy IP 的配置。
配置任务 4：NAT Server 的配置。

10.2　实验配置任务

10.2.1　静态 NAT 的配置

在这个实验中，AR1 是企业网关路由器，G0/0/0 接口连接 ISP，G0/0/1 接口连接内网主机；AR2 是 ISP 路由器。我们要实现内网主机 PC1 和 PC2 访问 Internet，AR2 的环回接口地址用来模拟 Internet 中的服务器，也就是要使 PC1 和 PC2 能够访问 AR2 的环回接口地址 8.8.8.8。AR2 连接 AR1，并且 ISP 为这个连接分配了 6 个 IP 地址，其中 103.31.200.5/29 是 AR1 接口 G0/0/0 使用的 IP 地址，103.31.200.6/29 是 AR2 接口

G0/0/0 使用的 IP 地址。其余 IP 地址可以由企业自行分配使用。

要使内网主机能够访问 Internet，AR1 上可以配置一条默认路由，指向本地网关 AR2。按照表 10-1 在每台设备上配置的接口 IP 地址，例 10-1 和例 10-2 分别展示了 AR1 和 AR2 的基础配置。S1 无须配置。

例 10-1　AR1 的基础配置

```
[AR1]interface GigabitEthernet 0/0/0
[AR1-GigabitEthernet0/0/0]ip address 103.31.200.5 29
[AR1-GigabitEthernet0/0/0]quit
[AR1]interface GigabitEthernet 0/0/1
[AR1-GigabitEthernet0/0/1]ip address 192.168.0.254 24
[AR1-GigabitEthernet0/0/1]quit
[AR1]ip route-static 0.0.0.0 0.0.0.0 103.31.200.6
```

例 10-2　AR2 的基础配置

```
[AR2]interface GigabitEthernet 0/0/0
[AR2-GigabitEthernet0/0/0]ip address 103.31.200.6 29
[AR2-GigabitEthernet0/0/0]quit
[AR2]interface LoopBack 0
[AR2-LoopBack0]ip address 8.8.8.8 32
```

我们没有在 AR2 中配置任何路由信息，这是因为一般情况下 ISP 无须了解客户内网的 IP 规划，客户会使用 ISP 分配的 IP 地址发送数据包，即 103.31.200.0/29。在这个网段中，103.31.200.5 和 103.31.200.6 用于 AR1 和 AR2 的直连链路，因此内网主机可以使用 4 个 IP 地址来访问 Internet（8.8.8.8）：103.31.200.1～103.31.200.4。

要想配置静态 NAT，我们需要进入执行 NAT 的接口视图，即 AR1 的 G0/0/0 接口，在接口视图中使用以下命令，完整的命令语法请参考华为设备配置指南。

nat static global *global-address* **inside** *host-address*：在配置静态 NAT 时，需要保证 *global-address*（外部地址）和 *host-address*（内部地址）是唯一且没有重复的，要避免使用设备接口地址。

我们为 PC1 配置静态 NAT，使它能够使用 103.31.200.1 来访问 Internet，具体见例 10-3。

例 10-3　为 PC1 配置静态 NAT

```
[AR1]interface GigabitEthernet 0/0/0
[AR1-GigabitEthernet0/0/0]nat static global 103.31.200.1 inside 192.168.0.1
```

通过在 PC1 和 PC2 上分别对 8.8.8.8 发起 ping 测试，验证一下配置效果，具体见例 10-4 和例 10-5。

例 10-4　验证 PC1 可以访问 Internet

```
PC1>ping 8.8.8.8

Ping 8.8.8.8: 32 data bytes, Press Ctrl_C to break
From 8.8.8.8: bytes=32 seq=1 ttl=254 time=125 ms
From 8.8.8.8: bytes=32 seq=2 ttl=254 time=47 ms
From 8.8.8.8: bytes=32 seq=3 ttl=254 time=63 ms
From 8.8.8.8: bytes=32 seq=4 ttl=254 time=156 ms
From 8.8.8.8: bytes=32 seq=5 ttl=254 time=94 ms

--- 8.8.8.8 ping statistics ---
  5 packet(s) transmitted
  5 packet(s) received
  0.00% packet loss
  round-trip min/avg/max = 47/97/156 ms
```

例 10-5 验证 PC2 无法访问 Internet

```
PC2>ping 8.8.8.8

Ping 8.8.8.8: 32 data bytes, Press Ctrl_C to break
Request timeout!
Request timeout!
Request timeout!
Request timeout!
Request timeout!

--- 8.8.8.8 ping statistics ---
  5 packet(s) transmitted
  0 packet(s) received
  100.00% packet loss
```

从上述验证结果可以看出，配置了静态 NAT 的 PC1 能够正常访问 Internet。图 10-2 所示为静态 NAT 工作原理。

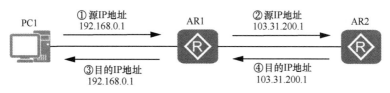

图 10-2 静态 NAT 工作原理

静态 NAT 基本原理具体如下。

① PC1 以 192.168.0.1 为源 IP 地址，将数据包发送出去。AR1 从 G0/0/1 接口接收该数据包。

② AR1 根据路由表判断出应将数据包从 G0/0/0 接口转发出去。在执行转发之前，它根据接口上配置的静态 NAT，将源 IP 地址 192.168.0.1 转换为 103.31.200.1，然后将数据包从 G0/0/0 接口转发出去。

③ AR2 收到源 IP 地址为 103.31.200.1 的数据包并进行响应。由于 103.31.200.0/29 是 AR2 的直连网络，因此 AR2 将该响应数据包从 G0/0/0 接口转发出去。AR1 从 G0/0/0 接口收到目的 IP 地址为 103.31.200.1 的数据包，根据接口配置的静态 NAT，先将目的 IP 地址 103.31.200.1 转换为 192.168.0.1，然后根据路由表将转换后的数据包从 G0/0/1 接口转发出去。

④ PC1 成功接收响应数据包，ping 测试成功。

我们可以使用命令 **display nat static** 来查看静态 NAT 配置信息，具体见例 10-6。

例 10-6 查看静态 NAT 配置信息

```
[AR1]display nat static
 Static Nat Information:
 Interface : GigabitEthernet0/0/0
    Global IP/Port     : 103.31.200.1/----
    Inside IP/Port     : 192.168.0.1/----
    Protocol : ----
    VPN instance-name  : ----
    Acl number         : ----
    Netmask : 255.255.255.255
    Description : ----

 Total :    1
```

静态 NAT 转换方式有两个缺点：第一个是无法实现公有 IP 地址的复用，一个公有 IP 地址只能支持一台内网 PC 访问 Internet；第二个是将内部主机"暴露"在公网中，使公网中的主机能够主动访问内部主机，这种做法增加了内部主机的安全风险。

接着我们为 PC2 配置静态 NAT，使它能够访问 Internet，具体见例 10-7。

例 10-7 为 PC2 配置静态 NAT

```
[AR1]interface GigabitEthernet 0/0/0
[AR1-GigabitEthernet0/0/0]nat static global 103.31.200.2 inside 192.168.0.2
[AR1-GigabitEthernet0/0/0]quit
[AR1]display nat static
 Static Nat Information:
 Interface : GigabitEthernet0/0/0
  Global IP/Port     : 103.31.200.1/----
  Inside IP/Port     : 192.168.0.1/----
  Protocol : ----
  VPN instance-name  : ----
  Acl number         : ----
  Netmask : 255.255.255.255
  Description : ----

  Global IP/Port     : 103.31.200.2/----
  Inside IP/Port     : 192.168.0.2/----
  Protocol : ----
  VPN instance-name  : ----
  Acl number         : ----
  Netmask : 255.255.255.255
  Description : ----

 Total :    2
```

在 PC2 对 8.8.8.8 发起 ping 测试的同时，我们在 AR1 与 AR2 连接的链路上开启 Wireshark 抓包，以此来确认 AR1 转发数据包所使用的 IP 地址，如图 10-3 所示。从抓包信息我们可以看出，AR1 以 103.31.200.2 作为源 IP 地址转发了 PC2 发出的 ping 消息。

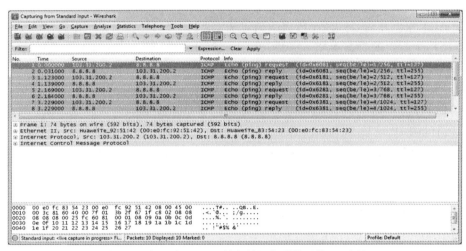

图 10-3 抓包确认 AR1 转发数据包所使用的 IP 地址

10.2.2 动态 NAT 的配置

如果访问 Internet 的内网主机数量多于企业所拥有的公网 IP 地址数量，就可以使用 NAPT（Network Access Port Translation，网络地址端口转换）来实现多对一的地址转换。NAPT 使用"IP 地址+端口号"的形式执行转换，使多个内网主机能够共用一个公有 IP 地址访问 Internet。

要想配置动态 NAT，我们需要执行以下几个步骤。

① 使用 ACL 对需要进行转换的私有 IP 地址进行匹配。管理员可以按照实际需求来选择使用基本 ACL（编号范围为 2000～2999）或高级 ACL（编号范围为 3000～3999）。当 ACL 的规则为 permit 时，表示设备需要对相匹配的源 IP 地址执行地址转换。当 ACL 没有设置 permit 规则时，与 ACL 关联的 NAT 功能将不生效，也就是不会对任何私有 IP 地址执行地址转换。

② 配置动态 NAT 使用的地址池，也就是转换后的公有 IP 地址范围。使用的命令为：**nat address-group** *group-index start-address end-address*。

③ 进入接口视图配置 NAT，将 ACL 与 IP 地址池进行关联，其中 ACL 定义了需要转换的私有 IP 地址，IP 地址池中定义了转换后的公有 IP 地址。使用的命令语法为：**nat outbound** *acl-number* **address-group** *group-index*。

如果读者是在上一个实验任务完成的基础上执行本实验任务，那么需要先删除 AR1 G0/0/0 接口配置的静态 NAT，具体见例 10-8。删除完成后可以在接口视图中使用命令 **display this** 来查看接口当前的配置信息，确认 NAT 配置已清除，也可以使用 **display nat static** 命令查看已清空的 NAT 配置信息。

例 10-8 清除多余配置

```
[AR1]interface GigabitEthernet 0/0/0
[AR1-GigabitEthernet0/0/0]undo nat static global 103.31.200.1 inside 192.168.0.1
[AR1-GigabitEthernet0/0/0]undo nat static global 103.31.200.2 inside 192.168.0.2
[AR1-GigabitEthernet0/0/0]display this
[V200R003C00]
#
interface GigabitEthernet0/0/0
 ip address 103.31.200.5 255.255.255.248
#
Return
[AR1-GigabitEthernet0/0/0]quit
[AR1]display nat static
 Static Nat Information:
Total :    0
```

例 10-9 中展示了 AR1 中 NAPT 的配置命令。从命令中我们可以看出，ACL 2000 定义了需要转换的私有 IP 地址范围为 192.168.0.0/24，NAT 地址池 1 中指定了 4 个可以用于转换的公有 IP 地址：103.31.200.1～103.31.200.4。在 G0/0/0 接口视图中，我们调用了 ACL 2000 和 NAT 地址池 1。

例 10-9 配置 NAPT

```
[AR1]acl 2000
[AR1-acl-basic-2000]rule permit source 192.168.0.0 0.0.0.255
[AR1-acl-basic-2000]quit
[AR1]nat address-group 1 103.31.200.1 103.31.200.4
[AR1]interface GigabitEthernet 0/0/0
[AR1-GigabitEthernet0/0/0]nat outbound 2000 address-group 1
```

我们可以通过一些 **display** 命令来查看配置信息，在 **display nat address-group** 中我们可以查看当前设备的 NAT 地址池信息，具体见例 10-10。设备中包含一个 NAT 地址池，编号为 1，其中包含了 103.31.200.1~103.31.200.4 的 IP 地址。

例 10-10 查看 NAT 地址池

```
[AR1]display nat address-group

 NAT Address-Group Information:
 --------------------------------------
 Index   Start-address      End-address
 --------------------------------------
 1        103.31.200.1      103.31.200.4
 --------------------------------------
  Total : 1
```

我们还可以使用命令 **display nat outbound** 来查看 NAT 转换表，具体见例 10-11。从命令输出内容中我们可以看到，G0/0/0 接口启用了 NAT 功能，将 ACL 2000 中定义的源 IP 地址转换为地址池 1 中的 IP 地址。注意最后一列的类型显示为 pat，表示我们配置的是 NAPT，即设备会使用"IP 地址 + 端口号"的组合进行地址转换。

例 10-11　查看 NAT 转换表

```
[AR1]display nat outbound
 NAT Outbound Information:
--------------------------------------------------------------------------
  Interface                 Acl      Address-group/IP/Interface     Type
--------------------------------------------------------------------------
  GigabitEthernet0/0/0      2000                              1     pat
--------------------------------------------------------------------------
  Total : 1
```

我们分别在 PC1 和 PC2 上对 8.8.8.8 发起 ping 测试，例 10-12 和例 10-13 分别展示了测试结果。

例 10-12　在 PC1 上发起 ping 测试

```
PC1>ping 8.8.8.8

Ping 8.8.8.8: 32 data bytes, Press Ctrl_C to break
From 8.8.8.8: bytes=32 seq=1 ttl=254 time=63 ms
From 8.8.8.8: bytes=32 seq=2 ttl=254 time=63 ms
From 8.8.8.8: bytes=32 seq=3 ttl=254 time=78 ms
From 8.8.8.8: bytes=32 seq=4 ttl=254 time=31 ms
From 8.8.8.8: bytes=32 seq=5 ttl=254 time=78 ms

--- 8.8.8.8 ping statistics ---
  5 packet(s) transmitted
  5 packet(s) received
  0.00% packet loss
  round-trip min/avg/max = 31/62/78 ms
```

例 10-13　在 PC2 上发起 ping 测试

```
PC2>ping 8.8.8.8

Ping 8.8.8.8: 32 data bytes, Press Ctrl_C to break
From 8.8.8.8: bytes=32 seq=1 ttl=254 time=78 ms
From 8.8.8.8: bytes=32 seq=2 ttl=254 time=78 ms
From 8.8.8.8: bytes=32 seq=3 ttl=254 time=47 ms
From 8.8.8.8: bytes=32 seq=4 ttl=254 time=63 ms
From 8.8.8.8: bytes=32 seq=5 ttl=254 time=63 ms

--- 8.8.8.8 ping statistics ---
  5 packet(s) transmitted
  5 packet(s) received
  0.00% packet loss
  round-trip min/avg/max = 47/65/78 ms
```

在进行上述测试的同时，我们可以在 AR1 上查看 NAT 转发表，具体见例 10-14。从命令 **display nat session all** 的第一部分输出内容中我们可以看出，NAT 路由器将 PC1 的源 IP 地址 192.168.0.1 转换为 103.31.200.1，同时把 ICMP ID 从 38441 转换为 10267。在第二部分输出内容中，NAT 路由器将 PC2 的源 IP 地址 192.168.0.2 也转换为 103.31.200.1，ICMP ID 从 38442 转换为 10268。由此可见，NAT 路由器正在使用"IP 地址+端口号"的组合进行地址转换。

例 10-14　查看 NAT 转发表

```
[AR1]display nat session all
 NAT Session Table Information:
    Protocol          : ICMP(1)
    SrcAddr    Vpn    : 192.168.0.1
    DestAddr   Vpn    : 8.8.8.8
    Type Code IcmpId  : 0    8    38441
    NAT-Info
      New SrcAddr     : 103.31.200.1
      New DestAddr    : ----
      New IcmpId      : 10267

    Protocol          : ICMP(1)
    SrcAddr    Vpn    : 192.168.0.2
    DestAddr   Vpn    : 8.8.8.8
    Type Code IcmpId  : 0    8    38442
    NAT-Info
      New SrcAddr     : 103.31.200.1
      New DestAddr    : ----
      New IcmpId      : 10268

    Protocol          : ICMP(1)
    SrcAddr    Vpn    : 192.168.0.1
    DestAddr   Vpn    : 8.8.8.8
    Type Code IcmpId  : 0    8    38440
    NAT-Info
      New SrcAddr     : 103.31.200.1
      New DestAddr    : ----
      New IcmpId      : 10265

    Protocol          : ICMP(1)
    SrcAddr    Vpn    : 192.168.0.2
    DestAddr   Vpn    : 8.8.8.8
    Type Code IcmpId  : 0    8    38441
    NAT-Info
      New SrcAddr     : 103.31.200.1
      New DestAddr    : ----
      New IcmpId      : 10266

 Total : 4
```

在一些小型企业网络环境中，企业内网中需要访问 Internet 的设备数量很少，因此企业不会向 ISP 购买多个公有 IP 地址，而是使用同一个公有 IP 地址来连接 ISP，以支持内网主机访问 Internet。同时，企业也可能会通过动态的方式，从 ISP 那里获取公有 IP 地址，因此这个公有 IP 地址并不是固定的，而是会随着 ISP 的每次连接而发生变化。在这种情况下，我们无法使用固定的 NAT 地址池来配置 NAT，此时可以使用 Easy IP 的方式进行配置。

10.2.3　Easy IP 的配置

在这个实验中，我们要实现的效果是：使用 AR1 接口 G0/0/0 的 IP 地址作为 NAT 的 IP 地址，帮助内网主机访问 Internet。我们依然使用 AR1 G0/0/0 接口上配置的固定 IP 地址 103.31.200.5 作为企业从 ISP 获取的公有 IP 地址，忽略自动获得 IP 地址的配置。对于实验结果来说，只要内网主机能够访问 Internet（8.8.8.8），并且使用的公有 IP 地址是 103.31.200.5，则实验即为成功。

如果读者在上一个实验的基础上进行本实验，那么需要删除多余配置，具体见例 10-15。

例 10-15　删除多余配置

```
[AR1]interface GigabitEthernet 0/0/0
[AR1-GigabitEthernet0/0/0]undo nat outbound 2000 address-group 1
[AR1-GigabitEthernet0/0/0]quit
[AR1]undo nat address-group 1
```

Easy IP 的配置步骤是动态 NAT 配置步骤的一部分。

① 使用 ACL 对需要进行转换的私有 IP 地址进行匹配。管理员可以按照实际需求来选择使用基本 ACL（编号范围为 2000～2999）或高级 ACL（编号范围为 3000～3999）。当 ACL 的规则为 permit 时，表示设备需要对相匹配的源 IP 地址执行地址转换。当 ACL 中没有设置 permit 规则时，与 ACL 关联的 NAT 功能将不生效，也就是不会对任何私有地址执行地址转换。

② 进入接口视图来配置 NAT，通过 ACL 来指定哪些私有 IP 地址需要转换。使用的命令语法为：**nat outbound** *acl-number*。

从上述步骤可以看出，在 Easy IP 的配置环境中，我们只需要限定哪些私有 IP 地址需要进行转换，无须配置公有 IP 地址池，这是因为我们会使用出接口的 IP 地址来作为转换后的 IP 地址。

例 10-16 展示了 Easy IP 的配置命令，为了保证配置的清晰和完整，我们仍重复展示了 ACL 的配置。

例 10-16　配置 Easy IP

```
[AR1]acl 2000
[AR1-acl-basic-2000]rule permit source 192.168.0.0 0.0.0.255
[AR1-acl-basic-2000]quit
[AR1]interface GigabitEthernet 0/0/0
[AR1-GigabitEthernet0/0/0]nat outbound 2000
```

从中可以看出，Easy IP 的配置比动态 NAT 的配置简单，这也是其称为 Easy 的原因。配置完成后，我们可以使用 **display nat outbound** 查看 NAT 状态。例 10-17 展示了命令输出内容。可以看到，当前 G0/0/0 接口上的公有 IP 地址不再是例 10-11 中的组 1，而是具体的 IP 地址 103.31.200.5，同时类型展示为 easyip。

例 10-17　查看 NAT 转换表

```
[AR1]display nat outbound
 NAT Outbound Information:
 --------------------------------------------------------------------
   Interface              Acl       Address-group/IP/Interface    Type
 --------------------------------------------------------------------
   GigabitEthernet0/0/0   2000                    103.31.200.5   easyip
 --------------------------------------------------------------------
   Total : 1
```

我们在 PC1 上对 Internet（8.8.8.8）发起 ping 测试，具体见例 10-18。

例 10-18　在 PC1 发起 ping 测试

```
PC1>ping 8.8.8.8

Ping 8.8.8.8: 32 data bytes, Press Ctrl_C to break
From 8.8.8.8: bytes=32 seq=1 ttl=254 time=47 ms
From 8.8.8.8: bytes=32 seq=2 ttl=254 time=62 ms
From 8.8.8.8: bytes=32 seq=3 ttl=254 time=46 ms
From 8.8.8.8: bytes=32 seq=4 ttl=254 time=47 ms
From 8.8.8.8: bytes=32 seq=5 ttl=254 time=62 ms

--- 8.8.8.8 ping statistics ---
  5 packet(s) transmitted
  5 packet(s) received
  0.00% packet loss
  round-trip min/avg/max = 46/52/62 ms
```

与此同时，我们在 AR1 上查看当前的 NAT 会话，具体见例 10-19。可以看到，ping 消息的源 IP 地址被转换为 103.31.200.5，即 AR1 G0/0/0 接口的 IP 地址。

例 10-19 在 AR1 上查看 NAT 会话

```
[AR1]display nat session all
 NAT Session Table Information:
    Protocol            : ICMP(1)
    SrcAddr    Vpn      : 192.168.0.1
    DestAddr   Vpn      : 8.8.8.8
    Type Code IcmpId    : 0    8    17943
    NAT-Info
       New SrcAddr      : 103.31.200.5
       New DestAddr     : ----
       New IcmpId       : 10250

    Protocol            : ICMP(1)
    SrcAddr    Vpn      : 192.168.0.1
    DestAddr   Vpn      : 8.8.8.8
    Type Code IcmpId    : 0    8    17944
    NAT-Info
       New SrcAddr      : 103.31.200.5
       New DestAddr     : ----
       New IcmpId       : 10251

    Protocol            : ICMP(1)

    SrcAddr    Vpn      : 192.168.0.1
    DestAddr   Vpn      : 8.8.8.8
    Type Code IcmpId    : 0    8    17945
    NAT-Info
       New SrcAddr      : 103.31.200.5
       New DestAddr     : ----
       New IcmpId       : 10252

    Protocol            : ICMP(1)
    SrcAddr    Vpn      : 192.168.0.1
    DestAddr   Vpn      : 8.8.8.8
    Type Code IcmpId    : 0    8    17946
    NAT-Info
       New SrcAddr      : 103.31.200.5
       New DestAddr     : ----
       New IcmpId       : 10253

    Protocol            : ICMP(1)
    SrcAddr    Vpn      : 192.168.0.1
    DestAddr   Vpn      : 8.8.8.8
    Type Code IcmpId    : 0    8    17947
    NAT-Info
       New SrcAddr      : 103.31.200.5
       New DestAddr     : ----
       New IcmpId       : 10254

 Total : 5
```

在上述 3 个实验任务中,我们介绍了 3 种方法实现内网主机访问 Internet。这 3 种方式适用于不同的网络环境,可以满足不同需求。接下来我们要介绍如何允许 Internet 主机主动访问内网主机。

10.2.4　NAT Server 的配置

在本实验任务中,我们需要在企业内网中添加一台 HTTP 服务器,在 Internet 中添加一台 HTTP 客户端,并且在 ISP 不了解企业内网 IP 结构的情况下,使 Internet 客户端能够主动访问位于内网的 HTTP 服务器。在本实验中,拓扑如图 10-4 所示。

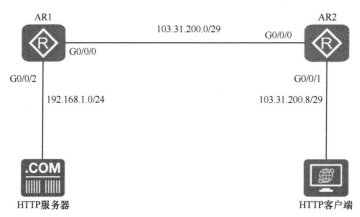

图 10-4 添加 HTTP 服务器和 HTTP 客户端的实验拓扑

新增设备和接口的 IP 地址规划见表 10-2。

表 10-2 新增设备和接口的 IP 地址规划

设备	接口	IP 地址	子网掩码	默认网关
AR1	G0/0/2	192.168.1.254	255.255.255.0	—
AR2	G0/0/1	103.31.200.14	255.255.255.248	—
HTTP 服务器	E0/0/0	192.168.1.1	255.255.255.0	192.168.1.254
HTTP 客户端	E0/0/0	103.31.200.9	255.255.255.248	103.31.200.14

我们需要配置服务器，使其提供 HTTP 服务，并选择提供内容的目录。我们在这里保留默认的端口号 80，读者可以更改提供 HTTP 服务的端口号，并注意在后续路由器的配置中进行相应的设置。在设置 HTTP 服务器时还需要选择提供内容的路径，并在其中添加.html 文件，在后续的测试中，HTTP 客户端就会向服务器请求获取这个文件。

我们在路由器上完成基础配置，具体见例 10-20 和例 10-21。

例 10-20 补全 AR1 的基础配置

```
[AR1]interface GigabitEthernet 0/0/2
[AR1-GigabitEthernet0/0/2]ip address 192.168.1.254 24
```

例 10-21 补全 AR2 的基础配置

```
[AR2]interface GigabitEthernet 0/0/1
[AR2-GigabitEthernet0/0/1]ip address 103.31.200.14 29
```

NAT Server 的配置与静态 NAT 的配置类似，会将"私有 IP 地址 + 端口号"与"公有 IP 地址 + 端口号"进行映射，使 NAT 设备能够根据这个映射进行地址转换。与静态 NAT 配置一样，我们需要在出接口视图中执行映射配置，命令语法如下。

nat server protocol {**tcp** | **udp**} **global** *global-address global-port* **inside** *host-address host-port*：在 IP 地址映射的基础上，添加了端口号的映射，使外网主机能够主动访问内网服务器。本书简化了命令语法，以突出重点，完整的命令语法可以参考华为设备配置指南。

在本实验中，我们需要匹配的是 HTTP，该协议使用 TCP 80 端口。若读者在配置 HTTP 服务器时更改了端口号，此时也需要在路由器的映射命令中指定服务器实际使用的端口号。例 10-22 展示了 AR1 上的 NAT Server 配置。

例 10-22 配置 NAT Server

```
[AR1]interface GigabitEthernet 0/0/0
[AR1-GigabitEthernet0/0/0]nat    server    protocol    tcp    global    103.31.200.4    www    inside
192.168.1.1 www
```

配置完成后,我们可以使用命令 **display nat server** 查看生成的配置,具体见例 10-23。从命令的输出内容我们可以看出,接口 G0/0/0 启用了 NAT Server,将全局 IP 地址和端口号 103.31.200.4:80 与内部 IP 地址 192.168.1.1:80 进行映射。

例 10-23 查看 NAT Server 设置

```
[AR1]display nat server
 Nat Server Information:
 Interface   : GigabitEthernet0/0/0
   Global IP/Port      : 103.31.200.4/80(www)
   Inside IP/Port      : 192.168.1.1/80(www)
   Protocol : 6(tcp)
   VPN instance-name   : ----
   Acl number          : ----
   Description : ----

 Total :    1
```

现在我们可以在 HTTP 客户端上进行最终测试,验证是否可以通过 http://103.31.200.4 来访问企业的内部服务器。注意:能够获取的文件会因读者搭建的实验环境而有所不同。

在这几种 NAT 配置中,读者需要根据实际情况进行选择。要慎重选择静态 NAT,这是因为这种做法相当于将内网主机暴露于 Internet 之中,会为内部网络带来安全隐患。

第11章
FTP 基础配置实验

本章主要内容

11.1 实验介绍

11.2 实验配置任务

华为网络设备支持使用多种方式对文件系统进行管理，管理员可以根据具体情况选择合适的文件管理方式。比如管理员可以直接登录网络设备以进行文件管理，或者通过 FTP、TFTP（Trivial File Transfer Protocol，简单文件传输协议）和 SFTP（SSH File Transfer Protocol，安全文件传送协议）进行文件管理，包括将设备升级包上传到设备、从设备中下载日志和配置。

FTP 使用客户端/服务器模型，在对网络设备进行 FTP 连接和管理时，网络设备充当 FTP 服务器，管理员 PC 充当 FTP 客户端。管理员向网络设备主动发起 FTP 连接，在此之前，需要先对网络设备进行一些配置，使其成为 FTP 服务器。

11.1 实验介绍

11.1.1 关于本实验

在本实验中，我们通过两台路由器来展示通过 FTP 在两台路由器之间传输文件。其中一台路由器 AR2 作为 FTP 服务器，另一台路由器 AR1 以 FTP 的方式登录 AR2，并对 AR2 的文件系统进行一些更改。

在此之前，由于我们需要对文件系统进行操作，因此在开始 FTP 的基础配置实验之前，为读者展示 VRP 系统的文件系统基本操作。

11.1.2 实验目的

- 熟悉华为网络设备文件系统的管理。
- 掌握华为网络设备上 FTP 服务器的配置。
- 掌握通过 FTP 传输文件的方法。

11.1.3 实验组网介绍

FTP 基础配置实验拓扑如图 11-1 所示。

图 11-1　FTP 基础配置实验拓扑

表 11-1 中列出了本章使用的网络地址。

表 11-1　本章使用的网络地址

设备	接口	IP 地址	子网掩码	默认网关
AR1	G0/0/0	192.168.0.1	255.255.255.0	—
AR2	G0/0/0	192.168.0.2	255.255.255.0	—

11.1.4 实验任务列表

配置任务 1：熟悉文件系统命令。
配置任务 2：FTP 基础配置实验。

11.2 实验配置任务

11.2.1 熟悉文件系统命令

在本实验任务中，我们先介绍 VRP 系统中的一些文件管理命令，使读者能够通过命令行的方式管理华为网络设备中的文件系统。首先通过用户视图的命令 **dir** 查看存储器中的文件和目录的信息，具体见例 11-1。在关键字 **dir** 后面还可以加上具体的文件名或目录名称，完整的语法格式为 **dir** [/**all**] [*filename* | *directory*]。

例 11-1 查看当前目录下的文件

```
<AR1>dir
Directory of flash:/

  Idx  Attr     Size(Byte)   Date         Time(LMT)   FileName
    0  drw-              -   Jun 24 2021  05:31:34    dhcp
    1  -rw-        121,802   May 26 2014  09:20:58    portalpage.zip
    2  -rw-          2,263   Jun 24 2021  05:31:27    statemach.efs
    3  -rw-        828,482   May 26 2014  09:20:58    sslvpn.zip
    4  -rw-            249   Jun 24 2021  06:14:12    private-data.txt

1,090,732 KB total (784,452 KB free)
```

从例 11-1 命令输出内容的第一行可以看到当前目录为 flash:/，我们可以通过命令 **mkdir** 来创建新目录，具体见例 11-2。

例 11-2 创建新目录

```
<AR1>mkdir flash:/test
Info: Create directory flash:/test......Done
```

命令 **pwd** 可以用来查看当前管理员所处的目录，通过命令 **cd** 可以更换目录，例 11-3 展示了当前的目录为 flash:/。我们通过命令 **cd flash:/test** 进入新创建的目录，并通过 **pwd** 命令进行验证。

例 11-3 更改和查看目录

```
<AR1>pwd
flash:
<AR1>cd flash:/test
<AR1>pwd
flash:/test
```

若想要删除我们创建的目录，可以使用 **rmdir** 命令，具体见例 11-4。需要注意的是，只有空目录才能被删除。另外，当管理员处于该目录时，无法删除该目录，因此我们先使用 **cd** 命令返回到 flash:中。

例 11-4 删除目录

```
<AR1>cd flash:
<AR1>pwd
flash:
<AR1>rmdir flash:/test
Remove directory flash:/hcia-datacom? (y/n)[n]:y
%Removing directory flash:/test...Done!
```

我们可以使用 **copy** 命令复制文件，完整的语法格式为 **copy** *source-filename destination-filename*，具体见例 11-5。在指定 *destination-filename* 时，若指定了目标文件的路径但未指定文件名，则会使用与源文件相同的文件名；此时若目标文件与源文件在相同目录下，则必须指定不同的目标文件名，否则复制不成功。我们以路由器中的 private-data.txt 文件为例，private-data.txt 文件会在我们对路由器执行 **save**（保存配置）操作后自动生成，路由器中的文件数量和名称会因实验环境而异。我们要将原文件复制到新的目录中，并使用相同的文件名，因此本例首先创建了一个新目录 hcia-datacom。

例 11-5　复制文件

```
<AR1>mkdir flash:/hcia-datacom
Info: Create directory flash:/hcia-datacom......Done
<AR1>cd flash:/hcia-datacom
<AR1>pwd
flash:/hcia-datacom
<AR1>dir
Info: File can't be found in the directory
1,090,732 KB total (784,448 KB free)
<AR1>copy flash:/private-data.txt flash:/hcia-datacom/private-data.txt
Copy flash:/private-data.txt to flash:/hcia-datacom/private-data.txt? (y/n)[n]:y


100%   complete
Info: Copied file flash:/private-data.txt to flash:/hcia-datacom/private-data.txt...Done
<AR1>dir
Directory of flash:/hcia-datacom/

  Idx  Attr     Size(Byte)  Date        Time(LMT)   FileName
    0  -rw-            249  Jun 24 2021 08:44:52    private-data.txt

1,090,732 KB total (784,440 KB free)
```

使用 **move** 命令可以移动文件，具体见例 11-6。我们将 hcia-datacom 目录中的文件 private-data.txt 移动到 flash:中，并将其保存为 private-data-move.txt。文件移动后 hcia-datacom 目录为空，文件复制后仍存在于 flash:目录中，这就是 **copy** 与 **move** 的区别：**copy** 命令相当于复制并粘贴，源文件同时存在于源路径和目标路径中；**move** 命令相当于剪切并粘贴，源文件只存在于目标路径中。

例 11-6　移动文件

```
<AR1>move flash:/hcia-datacom/private-data.txt flash:/private-data-move.txt
Move flash:/hcia-datacom/private-data.txt to flash:/private-data-move.txt? (y/n)[n]:y
%Moved file flash:/hcia-datacom/private-data.txt to flash:/private-data-move.txt.
<AR1>dir
Info: File can't be found in the directory
1,090,732 KB total (784,444 KB free)
<AR1>cd flash:
<AR1>dir
Directory of flash:/

  Idx  Attr     Size(Byte)  Date        Time(LMT)   FileName
    0  drw-              -  Jun 24 2021 05:31:34    dhcp
    1  -rw-        121,802  May 26 2014 09:20:58    portalpage.zip
    2  drw-              -  Jun 24 2021 08:49:21    hcia-datacom
    3  -rw-          2,263  Jun 24 2021 05:31:27    statemach.efs
    4  -rw-        828,482  May 26 2014 09:20:58    sslvpn.zip
    5  -rw-            249  Jun 24 2021 06:14:12    private-data.txt
    6  -rw-            249  Jun 24 2021 08:44:52    private-data-move.txt

1,090,732 KB total (784,444 KB free)
```

使用 **rename** 命令可以将文件重命名，具体见例 11-7。我们将 private-data-move.txt 文件重命名为 private-data-rename.txt。由于此时管理员正处于 flash:目录中，因此在对 flash:目录中的文件进行重命名时，可以省略完整路径，只输入文件名即可。

例 11-7 重命名文件

```
<AR1>rename private-data-move.txt private-data-rename.txt
Rename flash:/private-data-move.txt to flash:/private-data-rename.txt? (y/n)[n]:y
Info: Rename file flash:/private-data-move.txt to flash:/private-data-rename.txt
 ......Done
<AR1>dir
Directory of flash:/

  Idx  Attr     Size(Byte)  Date        Time(LMT)  FileName
    0  drw-              -  Jun 24 2021 05:31:34   dhcp
    1  -rw-        121,802  May 26 2014 09:20:58   portalpage.zip
    2  -rw-            249  Jun 24 2021 08:44:52   private-data-rename.txt
    3  drw-              -  Jun 24 2021 08:49:21   hcia-datacom
    4  -rw-          2,263  Jun 24 2021 05:31:27   statemach.efs
    5  -rw-        828,482  May 26 2014 09:20:58   sslvpn.zip
    6  -rw-            249  Jun 24 2021 06:14:12   private-data.txt

1,090,732 KB total (784,444 KB free)
```

我们可以使用 **delete** 命令删除文件，完整的语法格式为 **delete** [/**unreserved**] [/**force**] {*filename* | *devicename*}，其中，可选关键词/**unreserved** 表示永久删除且无法恢复，在执行命令时系统会进行提示并让管理员进行确认；使用可选关键词/**force** 时，系统不会给出任何提示。我们分别介绍普通删除和永久删除。

在例 11-8 中，我们复制 3 个文件待删除。

例 11-8 复制 3 个文件

```
<AR1>copy private-data.txt del-1.txt
Copy flash:/private-data.txt to flash:/del-1.txt? (y/n)[n]:y
100% complete
Info: Copied file flash:/private-data.txt to flash:/del-1.txt...Done
<AR1>copy private-data.txt del-2.txt
Copy flash:/private-data.txt to flash:/del-2.txt? (y/n)[n]:y
100% complete
Info: Copied file flash:/private-data.txt to flash:/del-2.txt...Done
<AR1>copy private-data.txt del-3.txt
Copy flash:/private-data.txt to flash:/del-3.txt? (y/n)[n]:y
100% complete
Info: Copied file flash:/private-data.txt to flash:/del-3.txt...Done
<AR1>dir
Directory of flash:/

  Idx  Attr     Size(Byte)  Date        Time(LMT)  FileName
    0  drw-              -  Jun 24 2021 05:31:34   dhcp
    1  -rw-        121,802  May 26 2014 09:20:58   portalpage.zip
    2  -rw-            249  Jun 24 2021 08:44:52   private-data-rename.txt
    3  -rw-            249  Jun 24 2021 08:58:57   del-3.txt
    4  drw-              -  Jun 24 2021 08:49:21   hcia-datacom
    5  -rw-          2,263  Jun 24 2021 05:31:27   statemach.efs
    6  -rw-        828,482  May 26 2014 09:20:58   sslvpn.zip
    7  -rw-            249  Jun 24 2021 08:58:43   del-1.txt
    8  -rw-            249  Jun 24 2021 06:14:12   private-data.txt
    9  -rw-            249  Jun 24 2021 08:58:51   del-2.txt

1,090,732 KB total (784,420 KB free)
```

我们对 del-1.txt 文件执行普通删除，对 del-2.txt 文件使用关键词/**unreserved**，对 del-3.txt 文件使用关键词/**force**。例 11-9 展示了删除 del-1.txt 的过程和结果，在输入删除命令后，系统会再次询问是否删除。文件删除后，我们可以通过命令 **dir /all** 查看被放入回收站的文件 del-1.txt，[]表示该文件已被放入回收站中。

例 11-9 删除文件至回收站

```
<AR1>delete del-1.txt
Delete flash:/del-1.txt? (y/n)[n]:y
Info: Deleting file flash:/del-1.txt...succeed.
<AR1>dir /all
Directory of flash:/

  Idx  Attr   Size(Byte)  Date         Time(LMT)   FileName
    0  drw-            -  Jun 24 2021  05:31:34    dhcp
    1  -rw-      121,802  May 26 2014  09:20:58    portalpage.zip
    2  -rw-          249  Jun 24 2021  08:44:52    private-data-rename.txt
    3  -rw-          249  Jun 24 2021  08:58:57    del-3.txt
    4  drw-            -  Jun 24 2021  08:49:21    hcia-datacom
    5  -rw-        2,263  Jun 24 2021  05:31:27    statemach.efs
    6  -rw-      828,482  May 26 2014  09:20:58    sslvpn.zip
    7  -rw-          249  Jun 24 2021  06:14:12    private-data.txt
    8  -rw-          249  Jun 24 2021  08:58:51    del-2.txt
    9  -rw-          249  Jun 24 2021  09:03:42    [del-1.txt]

1,090,732 KB total (784,436 KB free)
```

例 11-10 展示了永久删除的过程和结果，在使用命令 **delete /unreserved del-2.txt** 后，系统进行了确认；在使用命令 **delete /unreserved /force del-3.txt** 后，系统直接删除了文件，没有进行任何询问。删除这两个文件后，我们通过命令 **dir /all** 可以确认这两个文件没有被放入回收站中，而是被彻底删除。

例 11-10 永久删除文件

```
<AR1>delete /unreserved del-2.txt
Warning: The contents of file flash:/del-2.txt cannot be recycled. Continue? (y/n)[n]:y
Info: Deleting file flash:/del-2.txt...
Deleting file permanently from flash will take a long time if needed...succeed.
<AR1>delete /unreserved /force del-3.txt
Info: Deleting file flash:/del-3.txt...
Deleting file permanently from flash will take a long time if needed...succeed.
<AR1>dir /all
Directory of flash:/

  Idx  Attr   Size(Byte)  Date         Time(LMT)   FileName
    0  drw-            -  Jun 24 2021  05:31:34    dhcp
    1  -rw-      121,802  May 26 2014  09:20:58    portalpage.zip
    2  -rw-          249  Jun 24 2021  08:44:52    private-data-rename.txt
    3  drw-            -  Jun 24 2021  08:49:21    hcia-datacom
    4  -rw-        2,263  Jun 24 2021  05:31:27    statemach.efs
    5  -rw-      828,482  May 26 2014  09:20:58    sslvpn.zip
    6  -rw-          249  Jun 24 2021  06:14:12    private-data.txt
    7  -rw-          249  Jun 24 2021  09:03:42    [del-1.txt]

1,090,732 KB total (784,444 KB free)
```

对于被删除到回收站的文件，我们可以使用命令 **undelete** 进行恢复，具体见例 11-11。

例 11-11 恢复被删除的文件

```
<AR1>undelete del-1.txt
Undelete flash:/del-1.txt? (y/n)[n]:y
%Undeleted file flash:/del-1.txt.
<AR1>dir
Directory of flash:/

  Idx  Attr   Size(Byte)  Date         Time(LMT)   FileName
    0  drw-            -  Jun 24 2021  05:31:34    dhcp
    1  -rw-      121,802  May 26 2014  09:20:58    portalpage.zip
    2  -rw-          249  Jun 24 2021  08:44:52    private-data-rename.txt
    3  drw-            -  Jun 24 2021  08:49:21    hcia-datacom
    4  -rw-        2,263  Jun 24 2021  05:31:27    statemach.efs
    5  -rw-      828,482  May 26 2014  09:20:58    sslvpn.zip
    6  -rw-          249  Jun 24 2021  08:58:43    del-1.txt
    7  -rw-          249  Jun 24 2021  06:14:12    private-data.txt

1,090,732 KB total (784,444 KB free)
```

对于被删除且放到回收站的文件，我们可以使用命令 **reset recycle-bin** 进行彻底删除，具体见例 11-12。我们再次删除 del-1.txt 文件至回收站，并将其从回收站中彻底删除。

例 11-12　彻底删除被删除的文件

```
<AR1>delete del-1.txt
Delete flash:/del-1.txt? (y/n)[n]:y
Info: Deleting file flash:/del-1.txt...succeed.
<AR1>dir /all
Directory of flash:/

  Idx  Attr     Size(Byte)  Date        Time(LMT)   FileName
    0  drw-              -  Jun 24 2021 05:31:34    dhcp
    1  -rw-        121,802  May 26 2014 09:20:58    portalpage.zip
    2  -rw-            249  Jun 24 2021 08:44:52    private-data-rename.txt
    3  drw-              -  Jun 24 2021 08:49:21    hcia-datacom
    4  -rw-          2,263  Jun 24 2021 05:31:27    statemach.efs
    5  -rw-        828,482  May 26 2014 09:20:58    sslvpn.zip
    6  -rw-            249  Jun 24 2021 06:14:12    private-data.txt
    7  -rw-            249  Jun 24 2021 09:44:46    [del-1.txt]

1,090,732 KB total (784,444 KB free)
<AR1>reset recycle-bin del-1.txt
Squeeze flash:/del-1.txt? (y/n)[n]:y
Clear file from flash will take a long time if needed...Done.
%Cleared file flash:/del-1.txt.
<AR1>dir /all
Directory of flash:/

  Idx  Attr     Size(Byte)  Date        Time(LMT)   FileName
    0  drw-              -  Jun 24 2021 05:31:34    dhcp
    1  -rw-        121,802  May 26 2014 09:20:58    portalpage.zip
    2  -rw-            249  Jun 24 2021 08:44:52    private-data-rename.txt
    3  drw-              -  Jun 24 2021 08:49:21    hcia-datacom
    4  -rw-          2,263  Jun 24 2021 05:31:27    statemach.efs
    5  -rw-        828,482  May 26 2014 09:20:58    sslvpn.zip
    6  -rw-            249  Jun 24 2021 06:14:12    private-data.txt

1,090,732 KB total (784,448 KB free)
```

11.2.2　FTP 基础配置实验

在这个实验中，AR1 通过 FTP 对 AR2 的文件系统进行管理，向 AR2 传输文件，并从 AR2 删除文件。因此我们需要将 AR2 配置为 FTP 服务器，将 AR1 作为 FTP 客户端。

步骤 1　完成准备工作

按照图 11-1 搭建实验环境，在 AR1 和 AR2 上完成基础配置，实现 AR1 与 AR2 之间的通信，具体见例 11-13 和例 11-14。

例 11-13　在 AR1 上完成基础配置

```
[AR1]interface GigabitEthernet 0/0/0
[AR1-GigabitEthernet0/0/0]ip address 192.168.0.1 24
```

例 11-14　在 AR2 上完成基础配置

```
[AR2]interface GigabitEthernet 0/0/0
[AR2-GigabitEthernet0/0/0]ip address 192.168.0.2 24
```

我们在两台路由器上将配置保存为指定文件，以它作为 FTP 传输的文件。将 AR1 的配置保存为 ar1-conf.cfg，将 AR2 的配置保存为 ar2-conf.cfg，具体见例 11-15 和例 11-16。

例 11-15 保存 AR1 的配置

```
<AR1>save ar1-conf.cfg
 Are you sure to save the configuration to ar1-conf.cfg? (y/n)[n]:y
  It will take several minutes to save configuration file, please wait........
  Configuration file had been saved successfully
  Note: The configuration file will take effect after being activated
<AR1>dir
Directory of flash:/

  Idx  Attr     Size(Byte)   Date         Time(LMT)   FileName
    0  drw-              -   Jun 25 2021  00:42:28    dhcp
    1  -rw-        121,802   May 26 2014  09:20:58    portalpage.zip
    2  -rw-            849   Jun 25 2021  00:54:06    ar1-conf.cfg
    3  -rw-          2,263   Jun 25 2021  00:42:21    statemach.efs
    4  -rw-        828,482   May 26 2014  09:20:58    sslvpn.zip
    5  -rw-            249   Jun 25 2021  00:54:07    private-data.txt

1,090,732 KB total (784,448 KB free)
```

例 11-16 保存 AR2 的配置

```
<AR2>save ar2-conf.cfg
 Are you sure to save the configuration to ar2-conf.cfg? (y/n)[n]:y
  It will take several minutes to save configuration file, please wait........
  Configuration file had been saved successfully
  Note: The configuration file will take effect after being activated
<AR2>dir
Directory of flash:/

  Idx  Attr     Size(Byte)   Date         Time(LMT)   FileName
    0  drw-              -   Jun 25 2021  00:36:08    dhcp
    1  -rw-        121,802   May 26 2014  09:20:58    portalpage.zip
    2  -rw-          2,263   Jun 25 2021  00:36:00    statemach.efs
    3  -rw-        828,482   May 26 2014  09:20:58    sslvpn.zip
    4  -rw-            849   Jun 25 2021  00:55:33    ar2-conf.cfg
    5  -rw-            249   Jun 25 2021  00:55:33    private-data.txt

1,090,732 KB total (784,448 KB free)
```

步骤 2 配置 FTP 服务器

我们需要将路由器 AR2 配置为 FTP 服务器，配置步骤如下，具体见例 11-17。

① 使用系统视图命令 **ftp server enable** 启用 FTP 功能。

② 创建本地用户，并为其赋予通过 FTP 访问路由器的能力。

a. 使用命令 **aaa** 进入 AAA 视图，此时可以对用户接入实施安全配置。

b. 在 AAA 视图中使用命令 **local-user** *user-name* **password cipher** *password* 创建本地账号和登录密码。本实验将用户名设置为 ftp-user，密码设置为 Huawei@123。

c. 在 AAA 视图中使用命令 **local-user** *user-name* **service-type ftp** 允许本地用户使用 FTP 进行接入。

d. 在 AAA 视图中使用命令 **local-user** *user-name* **privilege level** *level* 配置本地用户的级别。在配置 FTP 用户级别时，必须将级别配置为 3 级或 3 级以上，否则 FTP 无法成功连接。本实验将用户级别设置为 3。

e. 在 AAA 视图中使用命令 **local-user** *user-name* **ftp-directory** *directory* 设置允许 FTP 用户访问的目录。本实验将目录指定为 flash:。

例 11-17 配置 FTP 服务器

```
[AR2]ftp server enable
Info: Succeeded in starting the FTP server
[AR2]aaa
[AR2-aaa]local-user ftp-user password cipher Huawei@123
Info: Add a new user.
[AR2-aaa]local-user ftp-user service-type ftp
[AR2-aaa]local-user ftp-user privilege level 3
[AR2-aaa]local-user ftp-user ftp-directory flash:
```

我们在 AR1 上，通过用户名 ftp-user 和密码 Huawei@123 对 AR2 发起 FTP 连接，具体见例 11-18。从命令提示符[AR1-ftp]可以确认我们已经以 FTP 的方式登录到 AR2 中。

例 11-18　从 AR1 上以 FTP 的方式登录 AR2

```
<AR1>ftp 192.168.0.2
Trying 192.168.0.2 ...
Press CTRL+K to abort
Connected to 192.168.0.2.
220 FTP service ready.
User(192.168.0.2:(none)):ftp-user
331 Password required for ftp-user.
Enter password:Huawei@123
230 User logged in.

[AR1-ftp]
```

我们对 AR2 的文件系统进行以下更改：

① 将 AR1 的配置文件 ar1-conf.cfg 传输到 AR2；
② 将 AR2 的配置文件 ar2-conf.cfg 下载到 AR1；
③ 将 AR2 本地的配置文件 ar2-conf.cfg 删除。

操作后的结果将为：AR1 上拥有 ar1-conf.cfg 和 ar2-conf.cfg 两个配置文件，AR2 上拥有 ar1-conf.cfg 一个配置文件。

例 11-19 使用 **put** 命令，将 AR1 的配置文件 ar1-conf.cfg 上传到 AR2。

例 11-19　通过 FTP 上传文件

```
[AR1-ftp]put ar1-conf.cfg
200 Port command okay.
150 Opening ASCII mode data connection for ar1-conf.cfg.

 100%
226 Transfer complete.
FTP: 849 byte(s) sent in 0.250 second(s) 3.39Kbyte(s)/sec.
```

例 11-20 使用 **get** 命令，将 AR2 的配置文件 ar2-conf.cfg 下载到 AR1。

例 11-20　通过 FTP 下载文件

```
[AR1-ftp]get ar2-conf.cfg
200 Port command okay.
150 Opening ASCII mode data connection for ar2-conf.cfg.
226 Transfer complete.
FTP: 849 byte(s) received in 0.250 second(s) 3.39Kbyte(s)/sec.
```

例 11-21 使用 **delete** 命令，将 AR2 本地的配置文件 ar2-conf.cfg 删除。

例 11-21　通过 FTP 删除文件

```
[AR1-ftp]delete ar2-conf.cfg
Warning: The contents of file ar2-conf.cfg cannot be recycled. Continue? (y/n)[n]:y
250 DELE command successful.
```

我们可以使用命令 **bye** 断开 FTP 连接，返回到 AR1 的命令行，具体见例 11-22。

例 11-22　断开 FTP 连接

```
[AR1-ftp]bye
221 Server closing.

<AR1>
```

我们分别查看 AR1 和 AR2 的文件目录，验证操作结果。例 11-23 展示了 AR1 的文件目录，可以看到两个配置文件 ar1-conf.cfg 和 ar2-conf.cfg。

例 11-23　查看 AR1 的文件目录

```
<AR1>dir
Directory of flash:/

  Idx  Attr     Size(Byte)   Date        Time(LMT)  FileName
    0  drw-              -   Jun 25 2021 00:42:28   dhcp
    1  -rw-        121,802   May 26 2014 09:20:58   portalpage.zip
    2  -rw-            849   Jun 25 2021 00:54:06   ar1-conf.cfg
    3  -rw-          2,263   Jun 25 2021 00:42:21   statemach.efs
    4  -rw-        828,482   May 26 2014 09:20:58   sslvpn.zip
    5  -rw-            849   Jun 25 2021 01:30:34   ar2-conf.cfg
    6  -rw-            249   Jun 25 2021 00:54:07   private-data.txt

1,090,732 KB total (784,452 KB free)
```

例 11-24 展示了 AR2 的文件目录，可以看到一个配置文件 ar1-conf.cfg。

例 11-24　查看 AR2 的文件目录

```
<AR2>dir
Directory of flash:/

  Idx  Attr     Size(Byte)   Date        Time(LMT)  FileName
    0  drw-              -   Jun 25 2021 00:36:08   dhcp
    1  -rw-        121,802   May 26 2014 09:20:58   portalpage.zip
    2  -rw-            849   Jun 25 2021 01:30:03   ar1-conf.cfg
    3  -rw-          2,263   Jun 25 2021 00:36:00   statemach.efs
    4  -rw-        828,482   May 26 2014 09:20:58   sslvpn.zip
    5  -rw-            249   Jun 25 2021 00:55:33   private-data.txt

1,090,732 KB total (784,456 KB free)
```

第 12 章
DHCP 基础配置实验

本章主要内容

12.1 实验介绍

12.2 实验配置任务

所有接入网络的终端设备都需要拥有 IP 地址才能与其他网络设备进行通信。网络规模越大，为所有终端设备配置 IP 地址的工作越繁重。DHCP（Dynamic Host Configuration Protocol，动态主机配置协议）可以帮助管理员减轻这部分工作负担，DHCP 服务器可以自动向 DHCP 客户端分配 IP 地址信息，其中不仅包括终端设备的 IP 地址/子网掩码，还包含比如默认网关、DNS 地址等信息。

DHCP 采用客户端/服务器模式，由 DHCP 客户端发起 DHCP 请求，再由 DHCP 服务器进行应答并提供 IP 地址，具体包含以下阶段。

① **发现阶段**：DHCP 客户端通过广播形式发送 DHCP 发现报文（DISCOVER），子网中的所有网络设备都会接收到这个广播包，但只有 DHCP 服务器会对其进行响应。

② **提供阶段**：DHCP 服务器接收到 DHCP 请求包，从相应的 IP 地址池中选择一个可用的 IP 地址分配给客户端，以单播形式向 DHCP 客户端发送包含这个 IP 地址的 DHCP 提供报文（OFFER）。

③ **选择阶段**：DHCP 客户端会接受它接收到的第一个 DHCP 提供报文中提供的 IP 地址信息，并以广播形式向 DHCP 服务器请求使用这个 IP 地址。

④ **确认阶段**：DHCP 服务器在接收到 DHCP 客户端发来的 DHCP 请求报文（REQUEST）后，会以单播形式发送 DHCP 确认报文（ACK）来确认 DHCP 客户端能够使用这个 IP 地址。

为了简化环境部署，并不是每个子网中都有 DHCP 服务器，当一个子网中没有 DHCP 服务器时，DHCP 客户端就无法获得 IP 地址。为了实现这种集中式 DHCP 服务器的部署，我们可以在每个子网中设置一个 DHCP 中继，使它充当 DHCP 客户端和 DHCP 服务器之间的代理，对 DHCP 报文进行转发。

在本章中，我们会通过 3 个实验任务介绍在华为路由器上配置 DHCP 服务器、DHCP 客户端，以及 DHCP 中继。

12.1 实验介绍

12.1.1 关于本实验

在以华为路由器作为 DHCP 服务器的环境中，路由器在分配 IP 地址时，可以使用全局地址池，也可以使用接口地址池。同时，路由器既可以充当 DHCP 客户端，从其他 DHCP 服务器获取 IP 地址，也可以充当 DHCP 中继，为 DHCP 客户端和 DHCP 服务器传递 DHCP 报文。

本实验包含以下 3 个任务。

第一个是由 AR1 和 AR2 组成的部署环境，AR1 作为 DHCP 服务器，向其直连的 AR2 接口分配 IP 地址。

第二个是由 AR1、S1 和两台 PC（PC1 和 PC2）组成的部署环境，其中 S1 无须进行配置。AR1 作为 DHCP 服务器，为同 IP 网段中的 DHCP 客户端分配 IP 地址、默认网关、DNS 地址等信息。

第三个是由 AR1、AR2 和 PC3 组成的部署环境,其中 AR1 作为 DHCP 服务器,通过 AR2(DHCP 中继),向非直连的 DHCP 客户端分配 IP 地址、默认网关、DNS 地址等信息。

12.1.2 实验目的

- 掌握 DHCP 服务器的配置方法。
- 掌握 DHCP 客户端的配置方法。
- 掌握 DHCP 中继的配置方法。
- 掌握更改 DHCP 其他参数的方法。
- 掌握通过 DHCP 下发其他参数的方法。
- 掌握地址池的配置方法。

12.1.3 实验组网介绍

DHCP 基础配置实验拓扑如图 12-1 所示。

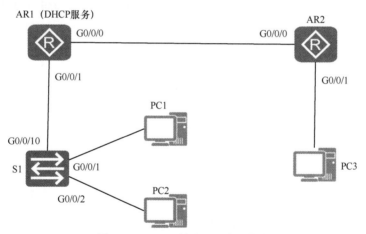

图 12-1　DHCP 基础配置实验拓扑

表 12-1 列出了本章使用的网络地址。

表 12-1　本章使用的网络地址

设备	接口	IP 地址	子网掩码	默认网关
AR1	G0/0/0	10.0.0.1	255.255.255.0	—
	G0/0/1	192.168.10.254	255.255.255.0	—
AR2	G0/0/0	DHCP	DHCP	—
	G0/0/1	192.168.20.254	255.255.255.0	—
PC1	E0/0/1	DHCP	DHCP	DHCP
PC2	E0/0/1	DHCP	DHCP	DHCP
PC3	E0/0/1	DHCP	DHCP	DHCP

12.1.4 实验任务列表

配置任务 1：配置基于全局地址池的 DHCP 服务器。
配置任务 2：配置基于接口的 DHCP 服务器。
配置任务 3：配置 DHCP 中继。

12.2 实验配置任务

12.2.1 配置基于全局地址池的 DHCP 服务器

在使用全局地址池的环境中，当接口接收到 DHCP 请求报文时，路由器会根据该接口的 IP 网段来匹配全局地址池中的 IP 网段，匹配后会从中选择可用的 IP 地址进行分配。因此，当路由器上有多个接口都充当其直连网段的 DHCP 服务器时，管理员只需要根据不同的 IP 网段配置多个全局地址池即可。同时，一个路由器接口也可以充当不同 IP 网段的 DHCP 服务器。

在这个实验任务中，我们要将 AR1 配置为 DHCP 服务器，通过 G0/0/1 接口为 PC1 和 PC2 分配 IP 地址。要求可分配的 IP 地址范围为 192.168.10.1～192.168.10.10，网关为 192.168.10.254，租期为 2 天，DNS 为 8.8.8.8。基于全局地址池的 DHCP 服务器配置实验拓扑如图 12-2 所示，具体步骤如下。

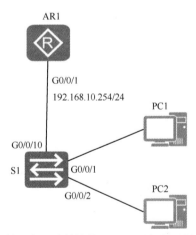

图 12-2 基于全局地址池的 DHCP 服务器配置实验拓扑

步骤 1 完成基础配置

我们需要在 AR1 上配置接口 G0/0/1 的 IP 地址，具体见例 12-1。

例 12-1 完成 AR1 上的基础配置

```
[AR1]interface GigabitEthernet 0/0/1
[AR1-GigabitEthernet0/0/1]ip address 192.168.10.254 24
```

步骤 2 创建全局地址池

在创建全局地址池时，需要注意的是所配置网段中有 3 个 IP 地址不会参与自动分配：

网络地址，即主机号二进制全为 0 的地址；广播地址，即主机号二进制全为 1 的地址；已配置了客户端的网关 IP 地址。我们可以使用以下命令来配置全局地址池。

① **ip pool** *ip-pool-name*：系统视图命令，创建全局地址池，并且进入全局地址池视图。在缺省情况下，设备上没有配置任何全局地址池。

② **network** *ip-address* [**mask** {*mask* | *mask-length*}]：全局地址池视图命令，指定可动态分配的 IP 地址范围，一个地址池中只能配置一个 IP 网段。在缺省情况下，地址池中没有任何 IP 地址。

在全局地址池中指定可动态分配的 IP 地址范围时，应该确保配置的 IP 地址范围与 DHCP 服务器接口的 IP 地址网段相同。因此，在地址池视图中，我们需要指定 G0/0/1 接口的 IP 网段，具体见例 12-2。

例 12-2 创建全局地址池

```
[AR1]ip pool hcia-pool
Info: It's successful to create an IP address pool.
[AR1-ip-pool-hcia-pool]network 192.168.10.0
```

在 **network** 命令中没有指定子网掩码时，设备会根据地址分类来自动应用默认掩码。由于 192.168.10.0 属于 C 类地址，其默认掩码为 24 位，即 255.255.255.0，因此设备会自动使用这个掩码。我们可以通过写入设备的配置来验证这一点，具体见例 12-3。

例 12-3 确认配置中的掩码

```
[AR1-ip-pool-hcia-pool]display this
[V200R003C00]
#
ip pool hcia-pool
network 192.168.10.0 mask 255.255.255.0
#
return
```

步骤 3 精确指定自动分配的 IP 地址范围

允许 DHCP 客户端获得的 IP 地址范围为 192.168.10.1～192.168.10.10，若想指定能够分配的 IP 地址，我们需要使用以下命令。

excluded-ip-address *start-ip-address* [*end-ip-address*]：全局地址池视图命令，用来指定地址池中不参与自动分配的 IP 地址。在缺省情况下，地址池中的所有可用 IP 地址参与自动分配。需要注意的是，系统会从地址池中最后一个可用地址开始进行分配，比如在本实验中，会将 192.168.10.10 和 192.168.10.9 分别分配给 PC1 和 PC2。

例 12-4 展示了 AR1 上的配置命令。

例 12-4 指定自动分配的 IP 地址范围

```
[AR1]ip pool hcia-pool
Info: It's successful to create an IP address pool.
[AR1-ip-pool-hcia-pool]excluded-ip-address 192.168.10.11 192.168.10.253
```

步骤 4 指定默认网关

从例 12-4 中可以看出，在指定的 IP 地址范围中不包含 G0/0/1 接口的 IP 地址 192.168.10.254，这是因为我们要将这个 IP 地址作为默认网关 IP 地址提供给 DHCP 客户端，所以这个 IP 地址不属于可用 IP 地址。在 DHCP 服务器上指定了默认网关后，客户端会根据默认网关地址自动生成到达该地址的缺省路由，使客户端能够访问本地网段之外的主机。

为了指定默认网关，我们需要使用以下命令。

gateway-list *ip-address*：全局地址池视图命令，用来配置 DHCP 客户端使用的默认网关地址。在缺省情况下，地址池中没有指定默认网关地址。

例 12-5 展示了 AR1 中默认网关的配置命令。

例 12-5　配置 AR1 中的默认网关

```
[AR1]ip pool hcia-pool
Info: It's successful to create an IP address pool.
[AR1-ip-pool-hcia-pool] gateway-list 192.168.10.254
```

步骤 5　配置自动分配的 IP 地址租期

租期是 DHCP 服务器自动分配的 IP 地址中的一个参数，我们可以使用以下命令来配置租期。

lease {**day** *day* [**hour** *hour* [**minute** *minute*]] | **unlimited**}：全局地址池视图命令，用来配置地址池中 IP 地址的租期。在缺省情况下，租期有效期为 1 天。

我们将 IP 地址有效期配置为 2 天，具体见例 12-6。

例 12-6　设置租期为 2 天

```
[AR1]ip pool hcia-pool
Info: It's successful to create an IP address pool.
[AR1-ip-pool-hcia-pool] lease day 2
```

步骤 6　配置 DNS 服务器地址

当管理员希望通过 DHCP 向主机配置 DNS 服务器信息时，可以使用以下命令来设置 DNS 地址。当园区网为内部服务器应用了 DNS，可以将内部 DNS 服务器的 IP 地址指定为 DNS 服务器地址；当园区网内部没有 DNS 服务器时，可以将公网可用的 DNS 服务器 IP 地址指定为 DNS 服务器地址。

dns-list *ip-address*：全局地址池视图命令，用来指定 DHCP 客户端所使用的 DNS 服务器 IP 地址。在缺省情况下，地址池中没有指定 DNS 服务器地址。

例 12-7 展示了 DNS 地址的配置命令。

例 12-7　配置 DNS 地址

```
[AR1]ip pool hcia-pool
Info: It's successful to create an IP address pool.
[AR1-ip-pool-hcia-pool]dns-list 8.8.8.8
```

至此，我们已经根据需求配置了 DHCP 服务器，例 12-8 展示了其完整的配置信息。

例 12-8　查看 DHCP 服务器的完整配置信息

```
[AR1-ip-pool-hcia-pool]display this
[V200R003C00]
#
ip pool hcia-pool
 gateway-list 192.168.10.254
 network 192.168.10.0 mask 255.255.255.0
 excluded-ip-address 192.168.10.11 192.168.10.253
 lease day 2 hour 0 minute 0
 dns-list 8.8.8.8
#
return
```

步骤 7　采用全局地址池方式提供 DHCP 服务

为了使 AR1 能够向 PC1 和 PC2 分配 IP 地址，我们需要将 AR1 配置为 DHCP 服务器，并使用基于全局地址池的方式分配 IP 地址。我们需要使用以下命令。

① **dhcp enable**：系统视图命令，在配置 DHCP 服务器功能之前，管理员必须先在系统视图下使用这条命令启用 DHCP 功能。与 DHCP 功能相关的其他特性在启用 DHCP

功能后才可以继续配置，比如 DHCP 客户端、DHCP 中继等。管理员可以使用 **undo dhcp enable** 命令禁用 DHCP 功能，执行该命令后，与 DHCP 相关的配置将会被删除。再次使用 **dhcp enable** 命令启用 DHCP 功能后，设备上与 DHCP 相关的配置将恢复为缺省配置。在缺省情况下，DHCP 功能处于禁用状态。

② **dhcp select global**：接口视图命令，使接口采用全局地址池的方式启用 DHCP 服务器功能。默认情况下，接口未启用 DHCP 服务器功能。在本实验中，我们需要在 AR1 的 G0/0/1 接口下配置这条命令。

例 12-9 展示了在 AR1 上配置 DHCP 服务器功能的命令。

例 12-9 配置 DHCP 服务器功能

```
[AR1]dhcp enable
Info: The operation may take a few seconds. Please wait for a moment.done.
[AR1]interface GigabitEthernet 0/0/1
[AR1-GigabitEthernet0/0/1]dhcp select global
```

步骤 8 配置 DHCP 客户端

为了使 PC1 和 PC2 能够获得 AR1 分配的 IP 地址，我们需要将它们配置为 DHCP 客户端，如图 12-3 所示。

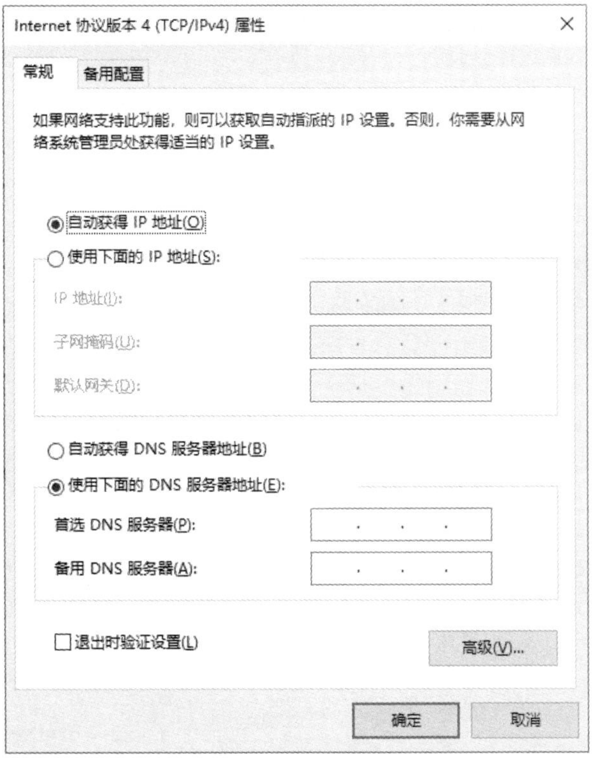

图 12-3 将 PC 配置为 DHCP 客户端

步骤 9 验证配置

首先我们在 PC1 和 PC2 上分别使用命令 **ipconfig** 确认它们获得的 IP 地址信息，具体见例 12-10 和例 12-11。以 PC1 为例，命令输出的内容中，除了显示 PC1 获得的 IP 地址为 192.168.10.10 外，还显示默认网关地址为 192.168.10.254，以及 DNS 服务器地址为 8.8.8.8。

例 12-10   查看 PC1 获得的 IP 地址信息

```
PC1>ipconfig

Link local IPv6 address..........: fe80::5689:98ff:fea1:53ba
IPv6 address.....................: :: / 128
IPv6 gateway.....................: ::
IPv4 address.....................: 192.168.10.10
Subnet mask......................: 255.255.255.0
Gateway..........................: 192.168.10.254
Physical address.................: 54-89-98-A1-53-BA
DNS server.......................: 8.8.8.8
```

例 12-11   查看 PC2 获得的 IP 地址信息

```
PC2>ipconfig

Link local IPv6 address..........: fe80::5689:98ff:fefa:be0
IPv6 address.....................: :: / 128
IPv6 gateway.....................: ::
IPv4 address.....................: 192.168.10.9
Subnet mask......................: 255.255.255.0
Gateway..........................: 192.168.10.254
Physical address.................: 54-89-98-FA-0B-E0
DNS server.......................: 8.8.8.8
```

然后我们可以在 AR1 上使用命令 **display ip pool name hcia-pool** 查看全局地址池 hcia-pool 的状态信息，具体见例 12-12。从命令输出中可以看出地址租期为 2 天，DNS 服务器地址为 8.8.8.8，网关地址为 192.168.10.254，已分配 2 个 IP 地址、空闲 8 个 IP 地址。

例 12-12   查看地址池 hcia-pool 的状态信息

```
[AR1]display ip pool name hcia-pool
  Pool-name          : hcia-pool
  Pool-No            : 1
  Lease              : 2 Days 0 Hours 0 Minutes
  Domain-name        : -
  DNS-server0        : 8.8.8.8
  NBNS-server0       : -
  Netbios-type       : -
  Position           : Local          Status            : Unlocked
  Gateway-0          : 192.168.10.254
  Mask               : 255.255.255.0
  VPN instance       : --
  ---------------------------------------------------------------------
       Start           End         Total  Used  Idle(Expired)  Conflict  Disable
  ---------------------------------------------------------------------
       192.168.10.1  192.168.10.254  253    2        8(0)          0       243
```

最后我们可以在 AR1 上测试一下它与两台 PC 的连通性，具体见例 12-13。

例 12-13   测试连通性

```
[AR1]ping 192.168.10.10
  PING 192.168.10.10: 56  data bytes, press CTRL_C to break
    Reply from 192.168.10.10: bytes=56 Sequence=1 ttl=128 time=160 ms
    Reply from 192.168.10.10: bytes=56 Sequence=2 ttl=128 time=30 ms
    Reply from 192.168.10.10: bytes=56 Sequence=3 ttl=128 time=100 ms
    Reply from 192.168.10.10: bytes=56 Sequence=4 ttl=128 time=50 ms
    Reply from 192.168.10.10: bytes=56 Sequence=5 ttl=128 time=120 ms

  --- 192.168.10.10 ping statistics ---
    5 packet(s) transmitted
    5 packet(s) received
    0.00% packet loss
    round-trip min/avg/max = 30/92/160 ms
[AR1]ping 192.168.10.9
  PING 192.168.10.9: 56  data bytes, press CTRL_C to break
    Reply from 192.168.10.9: bytes=56 Sequence=1 ttl=128 time=100 ms
    Reply from 192.168.10.9: bytes=56 Sequence=2 ttl=128 time=40 ms
    Reply from 192.168.10.9: bytes=56 Sequence=3 ttl=128 time=50 ms
    Reply from 192.168.10.9: bytes=56 Sequence=4 ttl=128 time=120 ms
    Reply from 192.168.10.9: bytes=56 Sequence=5 ttl=128 time=40 ms

  --- 192.168.10.9 ping statistics ---
    5 packet(s) transmitted
    5 packet(s) received
    0.00% packet loss
    round-trip min/avg/max = 40/70/120 ms
```

12.2.2 配置基于接口的 DHCP 服务器

在这个实验任务中,我们要以 AR1 为 DHCP 服务器,使用基于接口的方式向 AR2 分配 IP 地址。具体来说,我们需要将 AR2 的接口 G0/0/0 作为 DHCP 客户端,使它通过 DHCP 功能自动获取 IP 地址,配置要求为向 AR2 的 G0/0/0 接口分配 IP 地址 10.0.0.2,租期为 7 天。这种路由器接口自动获得 IP 地址的方法可以用于园区网与 ISP 之间,其中 DHCP 客户端是园区网的网关路由器(参考 AR2),DHCP 服务器是 ISP 的路由器设备(参考 AR1)。

基于接口的 DHCP 服务器配置如图 12-4 所示。

图 12-4 基于接口的 DHCP 服务器配置

步骤 1 完成基础配置

我们需要为 AR1 的 G0/0/0 接口配置 IP 地址。例 12-14 展示了相关配置。

例 12-14 AR1 上的基础配置

```
[AR1]interface GigabitEthernet 0/0/0
[AR1-GigabitEthernet0/0/0]ip address 10.0.0.1 24
```

步骤 2 配置 DHCP 服务器功能

为了使 AR1 向 AR2 分配 IP 地址,我们需要将 AR1 配置为 DHCP 服务器,并使用基于接口的方式分配 IP 地址。在使用接口地址池分配 IP 地址时,无须单独配置全局地址池,可分配的 IP 地址就是接口 IP 地址所处网段中的可用 IP 地址,并且只在这个接口下有效。我们需要使用以下命令。

① **dhcp enable**:系统视图命令,在配置 DHCP 服务器功能之前,管理员必须先在系统视图下使用这条命令启用 DHCP 功能。与 DHCP 功能相关的其他特性,在启用 DHCP 功能后才可以继续配置,比如 DHCP 客户端、DHCP 中继等。管理员可以使用 **undo dhcp enable** 命令禁用 DHCP 功能,执行该命令后,与 DHCP 相关的配置将会被删除。再次使用 **dhcp enable** 命令启用 DHCP 功能后,设备上与 DHCP 相关的配置将恢复为缺省配置。在缺省情况下,DHCP 功能处于禁用状态。

② **dhcp select interface**:接口视图命令,使接口采用基于接口地址池的 DHCP 服务器功能,为与该接口连接的 DHCP 客户端分配和该接口 IP 子网相同的 IP 地址。在缺省情况下,接口未启用基于接口地址池的 DHCP 服务器功能。

在接口上执行 **dhcp select interface** 命令后,设备会自动创建一个以接口名称命名的 DHCP 地址池,这个地址池中的 IP 地址与接口 IP 地址属于同一个网段,并且这个地址池中的 IP 地址只能分配给与该接口连接的 DHCP 客户端。

在 DHCP 地址池中有 3 个 IP 地址不会参与自动分配:网络地址,即主机号二进制全为 0 的地址;广播地址,即主机号二进制全为 1 的地址;已配置了客户端的网关 IP 地址。

例 12-15 展示了 AR1 上的 DHCP 服务器功能配置。

例 12-15 在 AR1 上配置 DHCP 服务器功能

```
[AR1]dhcp enable
Info: The operation may take a few seconds. Please wait for a moment.done.
[AR1]interface GigabitEthernet 0/0/0
[AR1-GigabitEthernet0/0/0]dhcp select interface
```

步骤 3 精确指定自动分配的 IP 地址范围

AR1 能够向 AR2 分配的 IP 地址范围很大,实际上,管理员能够精确指定要分配的 IP 地址。若想指定自动分配的 IP 地址范围,我们需要使用以下命令。

dhcp server excluded-ip-address *start-ip-address* [*end-ip-address*]:接口视图命令,用来指定接口地址池中不参与自动分配的 IP 地址。在缺省情况下,地址池中的所有 IP 地址都参与自动分配。需要注意的是,系统会从地址池中最后一个可用地址开始进行分配,比如在本实验中,会将 10.0.0.254 分配给 AR2。

本实验只需要分配一个 IP 地址,因此我们将 IP 地址范围 10.0.0.3~10.0.0.254 都排除在外,具体见例 12-16。

例 12-16 指定可分配的 IP 地址范围

```
[AR1]interface GigabitEthernet 0/0/0
[AR1-GigabitEthernet0/0/0]dhcp server excluded-ip-address 10.0.0.3 10.0.0.254
```

步骤 4 配置自动分配的 IP 地址租期

租期是 DHCP 服务器自动分配的 IP 地址中的一个参数,我们可以使用以下命令来配置租期。

dhcp server lease {**day** *day* [**hour** *hour* [**minute** *minute*]] | **unlimited**}:接口视图命令,用来配置 DHCP 服务器接口地址池中 IP 地址的租期。在缺省情况下,租期有效期为 1 天。

我们将 IP 地址有效期配置为 7 天,具体见例 12-17。

例 12-17 设置租期为 7 天

```
[AR1]interface GigabitEthernet 0/0/0
[AR1-GigabitEthernet0/0/0]dhcp server lease day 7
```

至此,DHCP 服务器的配置已经完成,例 12-18 展示了接口 G0/0/0 的完整配置信息。

例 12-18 查看 AR1 接口 G0/0/0 的配置信息

```
[AR1-GigabitEthernet0/0/0]display this
[V200R003C00]
#
interface GigabitEthernet0/0/0
 ip address 10.0.0.1 255.255.255.0
 dhcp select interface
 dhcp server excluded-ip-address 10.0.0.3 10.0.0.254
 dhcp server lease day 7 hour 0 minute 0
#
return
```

步骤 5 配置 DHCP 客户端

为了使 AR2 能够获得 AR1 分配的 IP 地址,我们需要将 AR2 配置为 DHCP 客户端。我们需要使用以下命令。

① **dhcp enable**:系统视图命令,在配置 DHCP 服务器功能之前,管理员必须先在系统视图下使用这条命令启用 DHCP 功能。

② **ip address dhcp-alloc**:接口视图命令,在接口上启用 DHCP 客户端功能。开启 DHCP 客户端功能的接口,可以通过 DHCP 服务器获取 IP 地址等配置参数。如果设备从

DHCP 服务器获得的 IP 地址与本地其他接口使用的 IP 地址属于同一网段，则设备不会使用 DHCP 分配的 IP 地址，并且不会向 DHCP 服务器重新申请 IP 地址。如果想要设备重新申请 IP 地址，需要先在接口上执行 **shutdown** 命令，然后再次执行 **undo shutdown** 命令；或者先执行 **undo ip address dhcp-alloc** 命令，然后再次执行 **ip address dhcp-alloc** 命令。

例 12-19 展示了 AR2 上的 DHCP 客户端配置。

例 12-19　在 AR2 上配置 DHCP 客户端功能

```
[AR2]dhcp enable
Info: The operation may take a few seconds. Please wait for a moment.done.
[AR2]interface GigabitEthernet 0/0/0
[AR2-GigabitEthernet0/0/0]ip address dhcp-alloc
```

步骤 6　验证配置

首先我们可以使用命令 **display ip pool interface GigabitEthernet0/0/0**，在 AR1 上查看接口 G0/0/0 下的 DHCP 地址池信息，具体见例 12-20。从命令输出内容中可以看出，AR1 接口 G0/0/0 下的 IP 地址租期为 7 天，已分配 1 个 IP 地址、禁用 252 个 IP 地址。

例 12-20　查看 AR1 上的接口 DHCP 地址池信息

```
[AR1]display ip pool interface GigabitEthernet0/0/0
  Pool-name        : GigabitEthernet0/0/0
  Pool-No          : 0
  Lease            : 7 Days 0 Hours 0 Minutes
  Domain-name      : -
  DNS-server0      : -
  NBNS-server0     : -
  Netbios-type     : -
  Position         : Interface       Status          : Unlocked
  Gateway-0        : 10.0.0.1
  Mask             : 255.255.255.0
  VPN instance     : --
  --------------------------------------------------------------------------
        Start           End          Total  Used  Idle(Expired)  Conflict  Disable
  --------------------------------------------------------------------------
        10.0.0.1        10.0.0.254   253    1     0(0)           0         252
  --------------------------------------------------------------------------
```

然后我们在 AR2 上使用命令 **display ip interface brief** 查看接口 G0/0/0 获得的 IP 地址，具体见例 12-21。从命令输出内容中可以看出，接口 G0/0/0 获得的 IP 地址是 10.0.0.2/24。

例 12-21　查看 AR2 接口 G0/0/0 获得的 IP 地址

```
[AR2]display ip interface brief
*down: administratively down
^down: standby
(l): loopback
(s): spoofing
The number of interface that is UP in Physical is 3
The number of interface that is DOWN in Physical is 1
The number of interface that is UP in Protocol is 3
The number of interface that is DOWN in Protocol is 1

Interface                    IP Address/Mask      Physical    Protocol
GigabitEthernet0/0/0         10.0.0.2/24          up          up
GigabitEthernet0/0/1         192.168.20.254/24    up          up
GigabitEthernet0/0/2         unassigned           down        down
NULL0                        unassigned           up          up(s)
```

接着验证 AR1 与 AR2 之间的连通性，具体见例 12-22。在 AR2 上使用命令 **ping 10.0.0.1** 来进行测试，结果显示连通性已建立。

例 12-22　验证 AR1 与 AR2 之间的连通性

```
[AR2]ping 10.0.0.1
  PING 10.0.0.1: 56  data bytes, press CTRL_C to break
    Reply from 10.0.0.1: bytes=56 Sequence=1 ttl=255 time=170 ms
    Reply from 10.0.0.1: bytes=56 Sequence=2 ttl=255 time=40 ms
    Reply from 10.0.0.1: bytes=56 Sequence=3 ttl=255 time=30 ms
    Reply from 10.0.0.1: bytes=56 Sequence=4 ttl=255 time=40 ms
    Reply from 10.0.0.1: bytes=56 Sequence=5 ttl=255 time=30 ms

  --- 10.0.0.1 ping statistics ---
    5 packet(s) transmitted
    5 packet(s) received
    0.00% packet loss
    round-trip min/avg/max = 30/62/170 ms
```

最后我们可以在 AR2 上使用命令 **display ip interface GigabitEthernet 0/0/0** 查看接口 IP 地址的详细信息，具体见例 12-23。从命令输出内容中可以看出，接口的 IP 地址是由 DHCP 获得的。另外，从输入数据包统计中可以看到接收到 5 个 Echo Reply 数据包。

例 12-23　查看 AR2 接口 G0/0/0 的详细 IP 信息

```
[AR2]display ip interface GigabitEthernet 0/0/0
GigabitEthernet0/0/0 current state : UP
Line protocol current state : UP
The Maximum Transmit Unit : 1500 bytes
input packets : 5, bytes : 420, multicasts : 0
output packets : 10, bytes : 2086, multicasts : 0
Directed-broadcast packets:
  received packets:            0, sent packets:          5
  forwarded packets:           0, dropped packets:       0
ARP packet input number:           1
  Request packet:                  0
  Reply packet:                    1
  Unknown packet:                  0
Internet Address is allocated by DHCP, 10.0.0.2/24
Broadcast address : 10.0.0.255
TTL being 1 packet number:         0
TTL invalid packet number:         0
ICMP packet input number:          5
  Echo reply:                      5
  Unreachable:                     0
  Source quench:                   0
  Routing redirect:                0
  Echo request:                    0
  Router advert:                   0
  Router solicit:                  0
  Time exceed:                     0
  IP header bad:                   0
  Timestamp request:               0
  Timestamp reply:                 0
  Information request:             0
  Information reply:               0
  Netmask request:                 0
  Netmask reply:                   0
  Unknown type:                    0
```

12.2.3　配置 DHCP 中继

在前两个实验任务中，DHCP 客户端都是直连在 DHCP 服务器本地接口的。当路由器需要作为 DHCP 服务器向其非直连的 DHCP 客户端分配 IP 地址时，就需要使用 DHCP 中继特性。在本实验任务中，我们需要将 AR1 配置为 DHCP 服务器，通过 DHCP 中继路由器 AR2，向 PC3 分配 IP 地址。具体要求为：IP 地址池名称为 pc3-pool，可分配的 IP 地址范围为 192.168.20.1～192.168.20.10，网关为 192.168.20.254，租期为 2 天，DNS 为 8.8.8.8。DHCP 中继配置如图 12-5 所示。

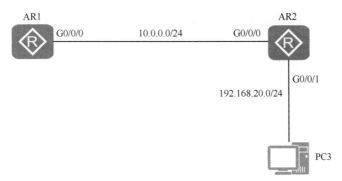

图 12-5 DHCP 中继配置

步骤 1 完成前置配置

我们需要在 AR1 的 G0/0/0 接口采用全局地址池的方式提供 IP 地址信息，因此需要清除该接口的原有配置，并且在 AR2 的 G0/0/0 接口上配置静态 IP 地址 10.0.0.2/24，具体见例 12-24 和例 12-25。

例 12-24 恢复 AR1 接口配置

```
[AR1]interface GigabitEthernet 0/0/0
[AR1-GigabitEthernet0/0/0]undo dhcp select interface
[AR1-GigabitEthernet0/0/0]display this
[V200R003C00]
#
interface GigabitEthernet0/0/0
 ip address 10.0.0.1 255.255.255.0
#
return
```

例 12-25 配置 AR2 接口 G0/0/0

```
[AR2]interface GigabitEthernet 0/0/0
[AR2-GigabitEthernet0/0/0]ip address 10.0.0.2 24
```

步骤 2 完成基础配置

我们需要完成的基础配置包括两个内容：在 AR1 上配置到达 DHCP 客户端所在 IP 网段（192.168.20.0/24）的路由；在 AR2 上配置接口 G0/0/1 的 IP 地址为 192.168.20.254/24，具体见例 12-26 和例 12-27。

例 12-26 完成 AR1 上的基础配置

```
[AR1]ip route-static 192.168.20.0 24 10.0.0.2
```

例 12-27 完成 AR2 上的基础配置

```
[AR2]interface GigabitEthernet 0/0/1
[AR2-GigabitEthernet0/0/1]ip address 192.168.20.254 24
```

步骤 3 创建全局地址池

配置命令已在 12.2.1 中进行过介绍，读者可以按照本实验任务的需求进行配置，具体见例 12-28。

例 12-28 创建全局地址池 pc3-pool

```
[AR1]ip pool pc3-pool
Info: It's successful to create an IP address pool.
[AR1-ip-pool-pc3-pool]network 192.168.20.0 mask 24
[AR1-ip-pool-pc3-pool]excluded-ip-address 192.168.20.11 192.168.20.253
[AR1-ip-pool-pc3-pool]gateway-list 192.168.20.254
[AR1-ip-pool-pc3-pool]lease day 2
[AR1-ip-pool-pc3-pool]dns-list 8.8.8.8
```

配置完成后,读者可以使用命令 **display ip pool name pc3-pool** 查看地址池状态信息,具体见例 12-29。

例 12-29　查看地址池 pc3-pool 的状态信息

```
[AR1]display ip pool name pc3-pool
  Pool-name          : pc3-pool
  Pool-No            : 0
  Lease              : 2 Days 0 Hours 0 Minutes
  Domain-name        : -
  DNS-server0        : 8.8.8.8
  NBNS-server0       : -
  Netbios-type       : -
  Position           : Local             Status           : Unlocked
  Gateway-0          : 192.168.20.254
  Mask               : 255.255.255.0
  VPN instance       : --
  ----------------------------------------------------------------------
        Start           End          Total  Used  Idle(Expired)  Conflict  Disable
  ----------------------------------------------------------------------
        192.168.20.1    192.168.20.254  253    0     10(0)          0         243
  ----------------------------------------------------------------------
```

步骤 4　采用全局地址池方式提供 DHCP 服务

我们需要在 AR1 的 G0/0/0 接口上启用基于全局地址池方式的 DHCP 服务,读者可自行完成配置,具体见例 12-30。如果读者是在 12.2.2 实验任务的基础上进行本实验任务,则无须再次配置例 12-30 中的第一条命令。

例 12-30　配置 DHCP 服务器功能

```
[AR1]dhcp enable
Info: The operation may take a few seconds. Please wait for a moment.done.
[AR1]interface GigabitEthernet 0/0/0
[AR1-GigabitEthernet0/0/0]dhcp select global
```

步骤 5　配置 DHCP 中继

充当 DHCP 中继的路由器需要启用 DHCP 功能,并在连接 DHCP 客户端的接口上启用 DHCP 代理功能,我们需要使用以下命令。

- **dhcp enable**:系统视图命令,在配置 DHCP 服务器功能之前,管理员必须先在系统视图下使用这条命令以启用 DHCP 功能。
- **dhcp select relay**:接口视图命令,在接口上启用 DHCP 中继功能,路由器从该接口接收到 DHCP 请求包后,自己作为 DHCP 中继设备,将 DHCP 请求包以单播形式转发给 DHCP 服务器。
- **dhcp relay server-ip** *ip-address*:接口视图命令,指明 DHCP 服务器的 IP 地址,本实验中就是 AR1 的 G0/0/0 接口地址 10.0.0.1。

例 12-31 展示了将 AR2 设为 DHCP 中继的配置命令。

例 12-31　配置 DHCP 中继

```
[AR2]dhcp enable
Info: The operation may take a few seconds. Please wait for a moment.done.
[AR2]interface GigabitEthernet 0/0/1
[AR2-GigabitEthernet0/0/1]dhcp select relay
[AR2-GigabitEthernet0/0/1]dhcp relay server-ip 10.0.0.1
```

步骤 6　配置 DHCP 客户端

为了使 PC3 能够通过 DHCP 自动获得 IP 地址信息,我们需要将其从手动模式配置为 DHCP 模式,如图 12-6 所示。

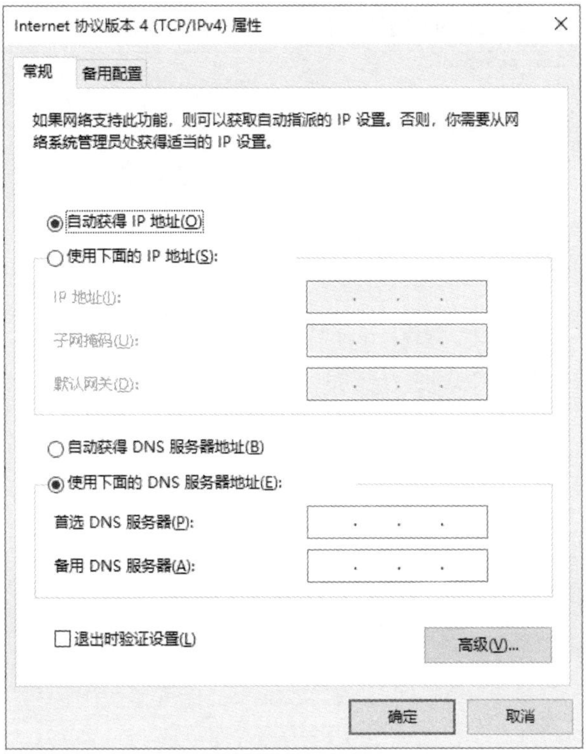

图 12-6 配置 DHCP 客户端

配置完成后，PC3 会发出 DHCP 请求，通过 AR2 的中继，最终由 AR1 为其分配 IP 地址，并提供相应的信息，包括默认网关地址、DNS 地址等。例 12-32 展示了 PC3 获得的 IP 地址信息，并且验证了它的网络连通性。

例 12-32 查看 PC3 的 IP 地址并验证网络连通性

```
PC3>ipconfig

Link Local IPv6 address...........: fe80::5689:98ff:fe50:23fd
IPv6 address......................: :: / 128
IPv6 gateway......................: ::
IPv4 address......................: 192.168.20.10
Subnet mask.......................: 255.255.255.0
Gateway...........................: 192.168.20.254
Physical address..................: 54-89-98-50-23-FD
DNS Server........................: 8.8.8.8

PC3>ping 10.0.0.1

ping 10.0.0.1: 32 data bytes, Press Ctrl_C to break
From 10.0.0.1: bytes=32 seq=1 ttl=254 time=31 ms
From 10.0.0.1: bytes=32 seq=2 ttl=254 time=31 ms
From 10.0.0.1: bytes=32 seq=3 ttl=254 time=16 ms
From 10.0.0.1: bytes=32 seq=4 ttl=254 time=32 ms
From 10.0.0.1: bytes=32 seq=5 ttl=254 time=16 ms

--- 10.0.0.1 ping statistics ---
  5 packet(s) transmitted
  5 packet(s) received
  0.00% packet loss
  round-trip min/avg/max = 16/25/32 ms

PC3>
```

我们可以使用命令查看 pc3-pool 的状态信息，具体见例 12-33。

例 12-33 查看 pc3-pool 的状态信息

```
[AR1]display ip pool name pc3-pool
  Pool-name        : pc3-pool
  Pool-No          : 0
  Lease            : 2 Days 0 Hours 0 Minutes
  Domain-name      : -
  DNS-server0      : 8.8.8.8
  NBNS-server0     : -
  Netbios-type     : -
  Position         : Local          Status          : Unlocked
  Gateway-0        : 192.168.20.254
  Mask             : 255.255.255.0
  VPN instance     : --
  ---------------------------------------------------------------------
       Start          End         Total  Used  Idle(Expired)  Conflict  Disable
  ---------------------------------------------------------------------
       192.168.20.1  192.168.20.254  253   1        9(0)          0       243
  ---------------------------------------------------------------------
```

第 13 章
构建基础 WLAN

本章主要内容

13.1 实验介绍

13.2 实验配置任务

对于没有接触过 AC（Access Controller，无线控制器）集中式管理的读者来说，配置 WLAN 的步骤和参数可能有些繁杂。本章将分步骤为读者介绍构建基础 WLAN 需要配置的参数，以及各个配置项之间的嵌套关系。

为了方便管理员对 WLAN 中多种功能进行管理和维护，华为针对不同的功能在配置中设计了不同类型的模板，这些模板之间有着相互引用的关系。只有了解它们之间的引用关系，才能厘清配置思路，从而顺利完成 WLAN 的部署。

图 13-1 所示为本实验使用的模板之间的相互关系，完整的模板及其之间的引用关系可以参考华为设备配置指南。在这里我们以使用 AP（Access Point，无线接入点）组的环境为例：从内向外进行介绍，SSID（Service Set Identifier，服务集标识）模板和安全模板在配置完成后，必须被应用在 VAP（Virtual Access Point，虚拟接入点）模板中才能生效；域管理模板和 VAP 模板在配置完成后，必须被应用在 AP 组中才可以生效；AP 组会进行最终的关联，将 WLAN 相关的配置与 AP 射频绑定，并由此将配置自动下发到 AP。

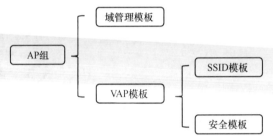

图 13-1 本实验使用的模板之间的相互关系

我们需要在上述模板中设定不同的参数，读者需要掌握的 WLAN 参数具体见表 13-1。

表 13-1 WLAN 参数

配置项	配置参数
AP 组	应用模板：域管理模板、VAP 模板
域管理模板	国家码
SSID 模板	SSID 名称
安全模板	安全策略、密码
VAP 模板	转发模式、业务 VLAN
	应用模板：SSID 模板、安全模板

13.1 实验介绍

13.1.1 关于本实验

本实验通过多个配置任务完成基础 WLAN 的构建。在本实验的 WLAN 中，AC 负责对 AP 进行统一管理，并且进行 WLAN 的配置，使无线客户端 STA（Station，站点）能够连接到 AP 并访问网络。本实验需要完成以下任务。

① 二层交换机作为 DHCP 服务器，分别为 AP 和 STA 分配 IP 地址。需要注意的是，AP 需要获得管理 VLAN 10 中的 IP 地址，STA 需要获得业务 VLAN 20 中的 IP 地址。

② AP 注册在 AC 上，并且由 AC 进行统一管理。

③ 在 AC 上进行 WLAN 配置，使 AP 广播 SSID，STA 通过 AP 连接到网络中。

在部署了 AC 的 WLAN 环境中，管理员的主要工作是配置 AC。读者可以按照以下步骤完成小型 WLAN 的配置。

步骤 1 实现网络设备之间的通信。本实验使用二层通信，即实现 AC、AP、交换机之间的二层通信。

步骤 2 配置 DHCP 服务器为 AP 和 STA 分配 IP 地址。在本环境中，我们既可以使用 AC 充当 DHCP 服务器，也可以使用交换机充当 DHCP 服务器。

步骤 3 配置 AP 上线，即 AP 能够注册在 AC 上，并接受 AC 的管理。

步骤 4 配置 WLAN 业务参数，即在 AC 中配置 STA 接入 WLAN 所需要的参数并将其下发给 AP，使 AP 根据相应的参数进行工作，STA 能够连接到网络中。

本章通过 3 个配置任务分别展示以上 4 个步骤的配置，其中步骤 1 和步骤 2 合并在配置任务 1 中。

由于 WLAN 的配置参数较多，表 13-2 中总结了 AC 上所使用的配置参数，每个模板的用途会在具体的配置中进行说明。

表 13-2 AC 所使用的配置参数

配置项	配置参数
AP 组	名称：hcia-group
	应用的模板：域管理模板 hcia-domain、VAP 模板 hcia-vap
域管理模板	名称：hcia-domain
	国家码：cn
SSID 模板	名称：hcia-ssid
	SSID 名称：hcia-wlan
安全模板	名称：hcia-sec
	安全策略：WPA-WPA2+PSK+AES
	密码：Huawei@123
VAP 模板	名称：hcia-vap
	转发模式：直接转发
	业务 VLAN：VLAN 20
	应用的模板：SSID 模板 hcia-ssid、安全模板 hcia-sec

13.1.2 实验目的

- 将 AP 注册在 AC 上。
- 使 AC 下发 WLAN 业务参数至 AP。

- 使 STA 通过 AP 连接到 WLAN。

13.1.3 实验组网介绍

基础 WLAN 实验拓扑如图 13-2 所示。

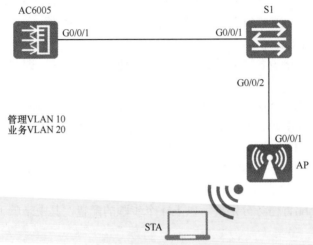

图 13-2 基础 WLAN 实验拓扑

表 13-3 中列出了本章使用的网络地址。

表 13-3 本章使用的网络地址

设备	接口	IP 地址	子网掩码	默认网关
AC6005	VLANIF 10	192.168.10.100	255.255.255.0	192.168.10.1
S1	VLANIF 10	192.168.10.1	255.255.255.0	—
	VLANIF 20	192.168.20.1	255.255.255.0	—
AP	G0/0/1	192.168.10.10	255.255.255.0	192.168.10.1
STA	E0/0/1	192.168.20.20	255.255.255.0	192.168.20.1

13.1.4 实验任务列表

配置任务 1：基本配置。
配置任务 2：配置 AP 上线。
配置任务 3：配置 WLAN 业务参数。

13.2 实验配置任务

13.2.1 基本配置

在这个实验任务中，读者需要根据图 13-2 完成连通性配置，并且设置 DHCP 服务器。

AC6005（在本实验中简称 AC）和 AP 是通过 S1 实现二层连接的，我们选择以 S1 作为 DHCP 服务器，为 AP 和 STA 分配 IP 地址，并且作为 AP 和 STA 的网关。

步骤 1　完成连通性

我们需要在 AC 上配置 VLAN 10，在 S1 上配置 VLAN 10 和 VLAN 20，将相应的端口配置为 Trunk 模式并放行相应 VLAN 的流量。

本环境使用的转发模式为直接转发，即终端流量直接由其所连接的 AP 转发到网络中，不通过 AC 进行中转。还有一种转发模式为隧道转发，即所有终端数据由 AC 进行中转。两种转发模式各有其适用的情景：隧道转发能够更好地实现集中管理和控制，直接转发能够在减少 AC 工作负载的同时，提高转发效率。

VLAN 10 作为管理 VLAN，负责传输 AP 与 AC 之间的 CAPWAP 消息，因此 AC 与 AP 之间的端口需要放行 VLAN 10 的流量。VLAN 20 作为业务 VLAN，负责传输 STA 的流量，因此 S1 的端口 G0/0/2 上既需要放行 VLAN 10 的流量，也需要放行 VLAN 20 的流量。

例 13-1 和例 13-2 分别展示了 AC 和 S1 上的 VLAN 配置。

例 13-1　AC 上的 VLAN 配置

```
[AC6005]vlan 10
Info: This operation may take a few seconds. Please wait for a moment...done.
[AC6005-vlan10]quit
[AC6005]interface GigabitEthernet 0/0/1
[AC6005-GigabitEthernet0/0/1]port link-type trunk
[AC6005-GigabitEthernet0/0/1]port trunk allow-pass vlan 10
```

例 13-2　S1 上的 VLAN 配置

```
[S1]vlan batch 10 20
Info: This operation may take a few seconds. Please wait for a moment....done.
[S1]interface GigabitEthernet 0/0/1
[S1-GigabitEthernet0/0/1]port link-type trunk
[S1-GigabitEthernet0/0/1]port trunk allow-pass vlan 10
[S1-GigabitEthernet0/0/1]quit
[S1]interface GigabitEthernet 0/0/2
[S1-GigabitEthernet0/0/2]port link-type trunk
[S1-GigabitEthernet0/0/2]port trunk pvid vlan 10
[S1-GigabitEthernet0/0/2]port trunk allow-pass vlan 10 20
```

在上述配置中，管理员使 AC 的端口 G0/0/1 放行了管理 VLAN 10 的相关流量，使 S1 的端口 G0/0/1 放行了管理 VLAN 10 的相关流量，端口 G0/0/2 放行了管理 VLAN 10 和业务 VLAN 20 的相关流量，且 G0/0/2 端口的 PVID 被设置为 VLAN 10。

除此之外，我们还需要在 AC 上配置默认路由，并将其指向 S1，具体见例 13-3。

例 13-3　在 AC 上配置默认路由

```
[AC6005]interface Vlanif 10
[AC6005-Vlanif10]ip address 192.168.10.100 24
[AC6005-Vlanif10]quit
[AC6005]ip route-static 0.0.0.0 0.0.0.0 192.168.10.1
```

步骤 2　S1 配置为 DHCP 服务器

S1 作为 DHCP 服务器，为 AP 和 STA 分配相应的 IP 地址。S1 通过接口 VLANIF 10 为 AP 提供 IP 地址，同时作为 AP 的网关。S1 通过接口 VLANIF 20 为 STA 提供 IP 地址，同时作为 STA 的网关。表 13-4 中列出了 DHCP 配置参数。

表 13-4 DHCP 配置参数

接口	接口/网关地址	DHCP 地址池范围
VLANIF 10	192.168.10.1/24	192.168.10.2～192.168.10.10
VLANIF 20	192.168.20.1/24	192.168.20.2～192.168.20.20

例 13-4 展示了 S1 上的 DHCP 配置。

例 13-4 S1 配置为 DHCP 服务器

```
[S1]dhcp enable
Info: The operation may take a few seconds. Please wait for a moment.done.
[S1]interface Vlanif 10
[S1-Vlanif10]ip address 192.168.10.1 24
[S1-Vlanif10]dhcp select interface
[S1-Vlanif10]dhcp server excluded-ip-address 192.168.10.11 192.168.10.254
[S1-Vlanif10]quit
[S1]interface Vlanif 20
[S1-Vlanif20]ip address 192.168.20.1 24
[S1-Vlanif20]dhcp select interface
[S1-Vlanif20]dhcp server excluded-ip-address 192.168.20.21 192.168.20.254
```

第 12 章已介绍过 DHCP 配置命令，首先通过系统命令 **dhcp enable** 在全局启用 DHCP 功能，接着在需要提供 DHCP 服务的接口上使用 **dhcp select interface** 命令，使设备根据接口地址池进行 IP 地址的分配，并且按照规划，排除相应的 IP 地址。

读者此时可以在 AC 上对 192.168.10.1 发起 ping 测试，以验证连通性，具体见例 13-5。

例 13-5 测试 AC 与 S1 之间的连通性

```
[AC6005]ping 192.168.10.1
  PING 192.168.10.1: 56  data bytes, press CTRL_C to break
    Reply from 192.168.10.1: bytes=56 Sequence=1 ttl=255 time=330 ms
    Reply from 192.168.10.1: bytes=56 Sequence=2 ttl=255 time=330 ms
    Reply from 192.168.10.1: bytes=56 Sequence=3 ttl=255 time=900 ms
    Reply from 192.168.10.1: bytes=56 Sequence=4 ttl=255 time=1210 ms
    Reply from 192.168.10.1: bytes=56 Sequence=5 ttl=255 time=480 ms

  --- 192.168.10.1 ping statistics ---
    5 packet(s) transmitted
    5 packet(s) received
    0.00% packet loss
    round-trip min/avg/max = 330/650/1210 ms
```

13.2.2 配置 AP 上线

在这个实验任务中，我们需要进入 AC 并在 AC 上完成相应的参数配置（如国家码），使 AP 能够成功上线。本实验任务中需要配置的参数如下。

① 配置 AP 组。
② 配置 AC 系统参数。
③ 选择 AC 的源接口。
④ 导入 AP。

步骤 1 配置 AP 组

使用 AC 对 AP 进行集中式管理的好处之一就是，管理员可以在 AC 上创建 AP 组，将所有使用相同配置的 AP 加入这个 AP 组中，这样就可以将相同的配置应用到多台 AP 上。在本实验中我们只部署一台 AP 作为示范。

我们可以使用以下命令创建 AP 组。

① **wlan**：系统视图命令，用来进入 WLAN 视图。在进行 WLAN 特性的相关配置

时，需要先使用 wlan 命令进入 WLAN 视图，与 WLAN 特性相关的所有配置命令均需要在 WLAN 视图或其子视图下进行配置。

② **ap-group name** *group-name*：WLAN 视图命令，用来创建 AP 组并进入 AP 组视图，若系统中已存在该 AP 组，则直接进入 AP 组视图。缺省情况下，系统中存在一个名为 default 的 AP 组。当管理员将多个需要使用相同参数的 AP 加入同一个 AP 组后，这些 AP 都会使用 AP 组中的配置，无须对每个 AP 进行单独配置，通过集中式管理简化了配置工作。

对于 AP 组有几点注意事项：对于一个 AP 来说，在 AP 组视图和 AP 视图下都配置了相应的参数，则优先使用 AP 视图下的参数；当 AP 组中存在 AP 时，无法删除 AP 组；无法删除缺省存在的 default AP 组。

例 13-6 中创建了名为 hcia-group 的 AP 组，从中可以看出当前管理员处于名为 hcia-group 的 AP 组视图中。

例 13-6　创建 AP 组

```
[AC6005]wlan
[AC6005-wlan-view]ap-group name hcia-group
Info: This operation may take a few seconds. Please wait for a moment.done.
[AC6005-wlan-ap-group-hcia-group]
```

步骤 2　配置 AC 系统参数

根据实验需求，管理员需要创建名为 hcia-domain 的域管理模板，并在其中设置国家码为 cn。我们可以使用以下命令进行配置。

① **regulatory-domain-profile** *profile-name*：WLAN 视图命令，用来创建域管理模板并进入域模板视图，若该模板已创建则直接进入模板视图。缺省情况下，系统中存在一个名为 default 的域管理模板，这个模板不可删除。在域管理模板中，管理员可以对国家码、调优信道和调优带宽等参数进行配置。创建域管理模板后，还需要在 AP 视图或 AP 组视图中应用该模板，当 AP 或 AP 组应用了域管理模板后，模板中的参数将会对相应的 AP 生效。已经被 AP 或 AP 组应用的域管理模板无法删除，需要先从 AP 视图或 AP 组视图中解除应用后才可以删除。

② **country-code** *country-code*：域管理模板视图命令，用来配置设备的国家码标识。缺省情况下，华为设备的国家码标识为 cn。当一个 AC 管理的 AP 分别位于不同国家或地区时，需要通过不同的国家码来满足不同国家或地区对于射频特性的要求（比如功率和信道规范）。域管理模板下的国家码配置发生变化后，应用了该模板的 AP 会自动重启。

例 13-7 中创建了名为 hcia-domain 的域管理模板，并将其国家码设置为 cn。从中可以看出，设备默认的国家码是 cn。

例 13-7　创建域管理模板

```
[AC6005]wlan
[AC6005-wlan-view]regulatory-domain-profile name hcia-domain
[AC6005-wlan-regulate-domain-hcia-domain]country-code cn
Info: The current country code is same with the input country code.
```

要想使创建的域管理模板生效，需要对其进行应用。根据本实验的环境，管理员需要在 AP 组 hcia-group 下进行应用，具体见例 13-8。在应用/更改 AP 组中的域管理模板时，设备会弹出警告信息并且要求管理员进行确认。管理员输入 y 后，更改生效。

例 13-8　应用域管理模板

```
[AC6005]wlan
[AC6005-wlan-view]ap-group name hcia-group
[AC6005-wlan-ap-group-hcia-group]regulatory-domain-profile hcia-domain
Warning: Modifying the country code will clear channel, power and antenna gain c
onfigurations of the radio and reset the AP. Continue?[Y/N]:y
[AC6005-wlan-ap-group-hcia-group]
```

步骤 3　选择 AC 的源接口

所谓 AC 的源接口，是指 AC 使用哪个接口与 AP 进行 CAPWAP 通信，我们可以使用以下命令进行配置。

capwap source interface {**loopback** *loopback-number* | **vlanif** *vlan-id* }：系统视图命令，用来指定 AC 与 AP 建立 CAPWAP 隧道的源接口，可以使用环回接口或 VLANIF 接口。缺省情况下未指定源接口。在指定源接口之前，设备上需要已经配置了相应的环回接口地址或 VLANIF 接口地址。

根据实验任务要求，我们使用 VLANIF 接口作为 AC 的源接口，具体见例 13-9。

例 13-9　选择 AC 的源接口

```
[AC6005]capwap source interface Vlanif 10
```

步骤 4　导入 AP

在导入 AP 时，为了保障网络的安全，我们可以对 AP 的身份进行认证。本实验使用 MAC 地址对 AP 进行认证，此外还可以使用设备序列号进行认证，或者不认证。本实验环境中 AP 的 MAC 地址为 00-E0-FC-63-7F-00，读者需要确认自己 AP 的 MAC 地址。我们可以使用以下命令以 MAC 地址对 AP 进行认证。

① **ap auth-mode** {**mac-auth** | **no-auth** | **sn-auth**}：WLAN 视图命令，用来指定 AP 的认证模式，必需关键字分别为使用 MAC 地址认证、不认证，以及使用设备序列号认证。

② **ap-id** *ap-id* {**ap-mac** *ap-mac* | **ap-sn** *ap-sn* | **ap-mac** *ap-mac* **ap-sn** *ap-sn* }：WLAN 视图命令，用来离线添加 AP 设备或进入 AP 视图。在添加 AP 时，必须输入 AP 的 MAC 地址、序列号，或者同时输入 MAC 地址和序列号。如果设置了 MAC 认证，则必须输入 MAC 地址；如果设置了序列号认证，则必须输入序列号。如需进入 AP 视图，只需要输入 AP ID 即可。

③ **ap-name** *ap-name*：AP 视图命令，用来为 AP 设置可识别的名称。一般来说，管理员可以根据 AP 所在位置来进行设置，本实验使用 hcia-ap 进行命名。如果没有为 AP 配置名称，则 AP 上线后会使用 MAC 地址作为名称。

④ **ap-group** *ap-group*：AP 视图命令，用来指定 AP 所属的 AP 组。每个 AP 都存在于一个 AP 组中，如果没有明确配置的话，AP 会自动加入到名为 default 的缺省 AP 组中。

例 13-10 中展示了 AC 上的上述配置。

例 13-10　导入 AP

```
[AC6005]wlan
[AC6005-wlan-view]ap auth-mode mac-auth
[AC6005-wlan-view]ap-id 0 ap-mac 00e0-fc63-7f00
[AC6005-wlan-ap-0]ap-name hcia-ap
[AC6005-wlan-ap-0]ap-group hcia-group
Warning: This operation may cause AP reset. If the country code changes, it will
 clear channel, power and antenna gain configurations of the radio, Whether to c
ontinue? [Y/N]:y
Info: This operation may take a few seconds. Please wait for a moment.. done.
[AC6005-wlan-ap-0]
```

配置完成后就可以启动 AP 进行测试了。在此之前,我们可以通过命令 **display ap all** 查看离线配置的 AP,具体见例 13-11。从中可以看出,这个 AP 已经写入配置中,但它当前的状态为 idle,说明它还未上线。

例 13-11 查看离线配置的 AP

```
[AC6005]display ap all
Info: This operation may take a few seconds. Please wait for a moment.done.
Total AP information:
idle : idle           [1]
--------------------------------------------------------------------------------
ID   MAC            Name       Group       IP Type              State STA Uptime
--------------------------------------------------------------------------------
0    00e0-fc63-7f00 hcia-ap hcia-group   -     -                idle  0    -
--------------------------------------------------------------------------------
Total: 1
```

现在可以启动 AP,AP 启动后会先获得 IP 地址,然后向 AC 进行注册。在此期间读者可以在 AP 与 S1 之间的链路上开启抓包,观察 DHCP 请求过程和 CAPWAP 隧道的建立过程。隧道建立后,在 AP 的 CLI 中可以看到例 13-12 中的信息提示,此时按下回车键可以发现 AP 的名称已经发生了变化。

例 13-12 在 AP 上发生的变化

```
<Huawei>
===== CAPWAP LINK IS UP!!! =====

<hcia-ap>
```

我们还可以在 AC 上使用命令 **display ap all** 查看 AP 的状态信息,具体见例 13-13。从中可以看出,添加的 AP 已经上线,它获得的 IP 地址为 192.168.10.10,属于 AP 组 hcia-group;状态为 nor 表示 AP 已成功上线且状态正常。

例 13-13 在 AC 上查看 AP 的状态信息

```
[AC6005]display ap all
Info: This operation may take a few seconds. Please wait for a moment.done.
Total AP information:
nor  : normal         [1]
--------------------------------------------------------------------------------
ID   MAC            Name       Group       IP            Type       State STA Uptime
--------------------------------------------------------------------------------
0    00e0-fc63-7f00 hcia-ap hcia-group 192.168.10.10 AP2050DN   nor   0   1M:46S
--------------------------------------------------------------------------------
Total: 1
```

13.2.3 配置 WLAN 业务参数

在这个实验任务中,我们需要在 AC 上配置与 WLAN 相关的参数,其中包括 SSID 模板、安全模板和 VAP 模板。具体如下:

① 创建 SSID 模板,并设置 SSID 参数;
② 创建安全模板,并设置安全策略;
③ 创建 VAP 模板,并设置数据转发模式、业务 VLAN、应用 SSID 模板和安全模板;
④ 在 AP 组中应用 VAP 模板。

步骤 1 创建 SSID 模板

不同的 SSID 指定了不同的无线网络,在 STA 对可接收范围内的无线网络进行搜索时,搜索到的无线网络名称就是 SSID。在华为 AC 中,我们需要在 SSID 模板中设置 SSID,同时还可以设置其他相关参数。在创建 SSID 模板后,需要在 VAP 模板中对 SSID 模板进行应用，SSID 模板中的设置才可以生效。

根据本实验任务的要求，我们需要创建名为 hcia-ssid 的 SSID 模板，并将 SSID 名称设置为 hcia-wlan，可以使用以下命令。

ssid-profile name *profile-name*：WLAN 视图命令，用来创建 SSID 模板并进入 SSID 模板视图，若该模板已存在则直接进入 SSID 模板视图。缺省情况下，系统中存在一个名为 default 的 SSID 模板，且该模板不能删除。在 SSID 模板中，管理员主要是为单个 SSID 配置与 STA 相关联的参数，比如 SSID 的名称、STA 的超时时间，以及 QoS 等。

SSID 模板有以下注意事项：在创建 SSID 模板后，需要在 VAP 模板中应用 SSID 模板，然后再在 AP 组视图中应用 VAP 模板，形成一个嵌套关系；已经在 VAP 模板中应用的 SSID 模板不能删除，如需删除的话，需要先从 VAP 模板中解除应用的 SSID 模板；当 VAP 模板已经被应用到 AP 组或单个 AP 时，对 SSID 模板的修改会导致业务中断。

ssid *ssid*：SSID 模板视图命令，用来指定当前 SSID 模板中的 SSID 名称。当 STA 搜索可接入的无线网络时，看到的网络名称就是 SSID。

例 13-14 展示了在 AC 上配置 SSID 模板的命令。

例 13-14　配置 SSID 模板

```
[AC6005]wlan
[AC6005-wlan-view]ssid-profile name hcia-ssid
[AC6005-wlan-ssid-prof-hcia-ssid]ssid hcia-wlan
Info: This operation may take a few seconds, please wait.done.
[AC6005-wlan-ssid-prof-hcia-ssid]
```

配置完成后，我们可以使用命令 **display ssid-profile all** 查看配置中的 SSID 模板，具体见例 13-15。从命令输出内容中我们可以看出，除了刚才创建的 SSID 模板之外，还有一个名为 default 的 SSID 模板，并且该模板已被应用。

例 13-15　查看设备中的 SSID 模板

```
[AC6005]display ssid-profile all
---------------------------------------------------------------
Profile name                          Reference
---------------------------------------------------------------
default                               1
hcia-ssid                             0
---------------------------------------------------------------
Total:2
```

我们还可以使用命令 **display ssid-profile name hcia-ssid** 查看 SSID 模板的详细信息，具体见例 13-16。

例 13-16　查看 SSID 模板的详细信息

```
[AC6005]display ssid-profile name hcia-ssid
---------------------------------------------------------------
Profile ID                    : 1
SSID                          : hcia-wlan
SSID hide                     : disable
Association timeout(min)      : 5
Max STA number                : 64
Reach max STA SSID hide       : enable
Legacy station                : enable
DTIM interval                 : 1
Beacon 2.4G rate(Mbps)        : 1
Beacon 5G rate(Mbps)          : 6
Deny-broadcast-probe          : disable
```

```
Probe-response-retry num        : 1
802.11r                         : disable
  802.11r authentication        : -
  Reassociation timeout (s)     : -
QOS CAR inbound CIR(kbit/s)     : -
QOS CAR inbound PIR(kbit/s)     : -
QOS CAR inbound CBS(byte)       : -
QOS CAR inbound PBS(byte)       : -
QOS CAR outbound CIR(kbit/s)    : -
QOS CAR outbound PIR(kbit/s)    : -
QOS CAR outbound CBS(byte)      : -
QOS CAR outbound PBS(byte)      : -
U-APSD                          : disable
Active dull client              : disable
MU-MIMO                         : disable
-------------------------------------------------------------
WMM EDCA client parameters:
-------------------------------------------------------------
       ECWmax  ECWmin  AIFSN  TXOPLimit
AC_VO  3       2       2      47
AC_VI  4       3       2      94
AC_BE  10      4       3      0
AC_BK  10      4       7      0
-------------------------------------------------------------
```

步骤 2 创建安全模板

在使用无线射频作为传输介质的环境中，攻击者可以对无线信道中传输的数据进行窃听和篡改。为了提高 WLAN 的安全性，我们可以在安全模板中设置安全策略机制，以此完善无线隧道建立过程中的认证方式、STA 上线时的用户认证方式、STA 数据业务传输时的数据加密方式。在创建安全模板后，需要在 VAP 模板中对安全模板进行应用，安全模板中的设置才可以生效。若没有配置任何安全策略，则设备会使用模板内的缺省安全策略，使 STA 用户在搜索到无线网络时，无须认证直接连接。这种缺省的安全策略存在很大的安全隐患，不建议在生产环境中部署。

根据本实验任务的要求，我们要创建一个名为 hcia-sec 的安全模板，将安全策略设置为 WPA-WPA2+PSK+AES，密码设置为 Huawei@123。我们可以使用以下命令。

① **security-profile name** *profile-name*：WLAN 视图命令，用来创建安全模板并进入安全模板视图。缺省情况下，系统中已有名为 default、default-wds 和 default-mesh 的安全模板。

② **security {wpa | wpa2 | wpa-wpa2} psk {pass-phrase | hex}** *key-value* **{aes | tkip | aes-tkip}**：安全视图命令，用来配置 WPA/WPA2 预共享密钥认证和加密。本实验任务需要选择 wpa-wpa2 混合方式，即 STA 使用 WPA 或 WPA2 进行认证；密码选择使用 pass-phrase 并将密码设置为 Huawei@123，这个密码在配置中会以密文形式显示；最后选择使用 aes 作为数据加密机制。

例 13-17 展示了 AC 上的安全模板配置命令。

例 13-17 安全模板配置命令

```
[AC6005]wlan
[AC6005-wlan-view]security-profile name hcia-sec
[AC6005-wlan-sec-prof-hcia-sec]security wpa-wpa2 psk pass-phrase Huawei@123 aes
```

配置完成后，我们可以使用命令 **display security-profile all** 查看设备中存在的安全模板，具体见例 13-18。从命令输出内容中可以看出，刚才创建的名为 hcia-sec 的安全模板还未被应用到 VAP 模板中。

例 13-18　查看设备中的安全模板

```
[AC6005]display security-profile all
-------------------------------------------------------------
Profile name                    Reference
-------------------------------------------------------------
default                         1
default-wds                     1
default-mesh                    1
hcia-sec                        0
-------------------------------------------------------------
Total: 4
```

我们还可以使用命令 **display security-profile name hcia-sec** 查看安全模板的详细信息，具体见例 13-19。从中可以看到，我们已按照实验任务要求设置了安全策略和加密策略。

例 13-19　查看安全模板的详细信息

```
[AC6005]display security-profile name hcia-sec
-------------------------------------------------------------
Security policy               : WPA-WPA2 PSK
Encryption                    : AES
-------------------------------------------------------------
WEP's configuration
Key 0                         : *****
Key 1                         : *****
Key 2                         : *****
Key 3                         : *****
Default key ID                : 0
-------------------------------------------------------------
WPA/WPA2's configuration
PTK update                    : disable
PTK update interval(s)        : 43200
-------------------------------------------------------------
WAPI's configuration
CA certificate filename       : -
ASU certificate filename      : -
AC certificate filename       : -
AC private key filename       : -
Authentication server IP      : -
WAI timeout(s)                : 60
BK update interval(s)         : 43200
BK lifetime threshold(%)      : 70
USK update method             : Time-based
USK update interval(s)        : 86400
MSK update method             : Time-based
MSK update interval(s)        : 86400
Cert auth retrans count       : 3
USK negotiate retrans count   : 3
MSK negotiate retrans count   : 3
-------------------------------------------------------------
```

为了展示密码在配置中是以加密形式显示的，可以在安全模板视图中使用命令 **display this** 查看安全模板中的配置，具体见例 13-20。从中可以看出，配置中没有显示出实际的密码 Huawei@123，而是显示出一串乱码。

例 13-20　查看配置中的密码

```
[AC6005-wlan-sec-prof-hcia-sec]display this
#
 security wpa-wpa2 psk pass-phrase %^%#JTbt%:2]%@vS({DZGPq,@b93E9Sr9/i"Wv(Kp4=-%^%# aes
#
return
[AC6005-wlan-sec-prof-hcia-sec]
```

步骤 3　创建 VAP 模板

管理员需要在 VAP 模板中应用相应的 SSID 模板和安全模板，并且还需要设置一些参数，比如数据转发模式、业务 VLAN 等。

根据本实验任务的要求，我们需要将转发模式设置为直接转发，并且设置业务 VLAN 为 VLAN 20，这些参数都是在 VAP 模板下进行设置的。我们可以使用以下命令设置上述参数，并且应用各种模板。

① **vap-profile name** *profile-name*：WLAN 视图命令，用来创建 VAP 模板并进入 VAP 模板视图，若该模板已存在则直接进入模板视图。缺省情况下，系统中存在一个名为 default 的 VAP 模板。管理员在创建 VAP 模板后，需要在 AP 组视图中应用这个 VAP 模板，模板中的设置才会生效。

② **forward-mode** {**direct-forward** | **tunnel**}：VAP 模板视图命令，用来配置数据转发方式，缺省情况下，VAP 模板中的数据转发方式为直接转发。

③ **service-vlan vlan-id** *vlan-id*：VAP 模板视图命令，用来配置 VAP 的业务 VLAN，缺省情况下，VAP 模板中的业务 VLAN 为 VLAN 1。需要注意的是，在配置已生效的情况下更改业务 VLAN 会导致 STA 业务中断。

④ **ssid-profile** *profile-name*：VAP 模板视图命令，用来应用 SSID 模板，缺省情况下，VAP 模板中应用的是名为 default 的 SSID 模板。只有在 VAP 模板中应用了 SSID 模板后，SSID 模板中的配置才会对使用了该 VAP 模板的所有 AP 生效。

⑤ **security-profile** *profile-name*：VAP 模板视图命令，用来应用安全模板，缺省情况下，VAP 模板中应用的是名为 default 的安全模板。

例 13-21 展示了 AC 中的 VAP 模板配置命令。

例 13-21 VAP 模板配置命令

```
[AC6005]wlan
[AC6005-wlan-view]vap-profile name hcia-vap
[AC6005-wlan-vap-prof-hcia-vap]forward-mode direct-forward
[AC6005-wlan-vap-prof-hcia-vap]service-vlan vlan-id 20
Info: This operation may take a few seconds, please wait.done.
[AC6005-wlan-vap-prof-hcia-vap]ssid-profile hcia-ssid
Info: This operation may take a few seconds, please wait.done.
[AC6005-wlan-vap-prof-hcia-vap]security-profile hcia-sec
Info: This operation may take a few seconds, please wait.done.
[AC6005-wlan-vap-prof-hcia-vap]
```

配置完成后，我们可以使用命令 **display vap-profile all** 查看设备中的 VAP 模板，具体见例 13-22。从命令输出内容中可以看出，刚才创建的名为 hcia-vap 的 VAP 模板还未被应用到任何 AP 或 AP 组中。

例 13-22 查看设备中的 VAP 模板

```
[AC6005]display vap-profile all
--------------------------------------------------------------------------------
Profile name                        Reference
--------------------------------------------------------------------------------
default                             0
hcia-vap                            0
--------------------------------------------------------------------------------
Total: 2
```

我们还可以使用命令 **display vap-profile name hcia-vap** 查看安全模板的详细信息，具体见例 13-23，从中可以看到管理员配置过的参数。

例 13-23　查看 VAP 模板的详细信息

```
[AC6005]display vap-profile name hcia-vap
--------------------------------------------------------------------------
 Profile ID                                       : 1
 Service mode                                     : enable
 Type                                             : service
 Forward mode                                     : direct-forward
 mDNS centralized-control                         : disable
 Offline management                               : disable
 Service VLAN ID                                  : 20
 Service VLAN Pool                                : -
 Permit VLAN ID                                   : -
 Auto off service switch                          : disable
 Auto off starttime                               : -
 Auto off endtime                                 : -
 Auto off time-range                              :
 STA access mode                                  : disable
 STA blacklist profile                            :
 STA whitelist profile                            :
 VLAN mobility group                              : 1
 Band steer                                       : enable
 Learn client address                             : enable
 Learn client DHCP strict                         : disable
 Learn client DHCP blacklist                      : disable
 Learn client DHCPv6 strict                       : disable
 Learn client DHCPv6 blacklist                    : disable
 IP source check                                  : disable
 ARP anti-attack check                            : disable
 DHCP option82 insert                             : disable
 DHCP option82 remote id format                   : Insert AP-MAC
 DHCP option82 circuit id format                  : Insert AP-MAC
 DHCP trust port                                  : disable
 ND trust port                                    : disable
 Zero roam                                        : disable
 Beacon multicast unicast                         : disable
 Anti-attack broadcast-flood                      : enable
   Anti-attack broadcast-flood sta-rate-threshold : 10
   Anti-attack broadcast-flood blacklist          : disable
 Anti-attack ARP flood                            : enable
   Anti-attack ARP flood sta-rate-threshold       : 5
   Anti-attack ARP flood blacklist                : disable
 Anti-attack ND flood                             : enable
   Anti-attack ND flood sta-rate-threshold        : 16
   Anti-attack ND flood blacklist                 : disable
 Anti-attack IGMP flood                           : enable
   Anti-attack IGMP flood sta-rate-threshold      : 4
   Anti-attack IGMP flood blacklist               : disable
 SSID profile                                     : hcia-ssid
 Security profile                                 : hcia-sec
 Traffic profile                                  : default
 Authentication profile                           :
 SAC profile                                      :
 Hotspot2.0 profile                               :
 User profile                                     :
 UCC profile                                      :
 Home agent                                       : ap
 Layer3 roam                                      : enable
--------------------------------------------------------------------------
```

步骤 4　在 AP 组中应用 VAP 模板

管理员需要在 AP 组视图（或 AP 等其他视图中）应用 VAP 模板，才能使 VAP 模板下的配置自动下发到指定的 AP 上。我们可以使用以下命令。

vap-profile *profile-name* **wlan** *wlan-id* **radio** {*radio-id* | **all**}：本实验需要在 AP 组视图下执行这条命令，这条命令会将 VAP 模板应用到具体的射频。这条命令中的 3 个参数分别是：profile-name，VAP 模板名称，即 hcia-vap；wlan-id，AC 中 VAP 的 ID，由管理员进行指定，本实验使用的 AP 支持的 ID 取值范围是 1～16，并且对于离线管理 VAP，

WLAN ID 的取值范围是[1～12, 15]，本实验使用 1；radio-id，射频 ID，射频 ID 0 是指 2.4 GHz 射频，射频 ID 1 是指 5 GHz 射频，根据 AP 类型的不同，有些 AP 支持三射频，具体信息请参考华为设备配置指南。本实验任务为两个射频应用相同的 VAP 模板，因此需要执行两次这条命令。

例 13-24 展示了在 AC 组中应用 VAP 模板的命令，我们将 VAP 模板 hcia-vap 与两个射频都进行了绑定。

例 13-24 在 AP 组中应用 VAP 模板

```
[AC6005]wlan
[AC6005-wlan-view]ap-group name hcia-group
[AC6005-wlan-ap-group-hcia-group]vap-profile hcia-vap wlan 1 radio 0
Info: This operation may take a few seconds, please wait...done.
[AC6005-wlan-ap-group-hcia-group]vap-profile hcia-vap wlan 1 radio 1
Info: This operation may take a few seconds, please wait...done.
[AC6005-wlan-ap-group-hcia-group]
```

我们可以在 AC 上使用命令 **display vap ssid hcia-wlan** 验证创建的 VAP，具体见例 13-25。从命令的输出内容中可以确认相应射频上已经成功创建了 VAP，WLAN 业务配置会由 AC 自动下发给 AP。AP 名称为 hcia-ap，状态显示为 ON，表示这个 AP 的相应射频上已经成功创建了 VAP。

例 13-25 验证创建的 VAP

```
[AC6005]display vap ssid hcia-wlan
Info: This operation may take a few seconds, please wait.
WID : WLAN ID
--------------------------------------------------------------------------------
AP ID AP name   RfID WID  BSSID          Status  Auth type      STA  SSID
--------------------------------------------------------------------------------
0     hcia-ap  0  1    00E0-FC63-7F00 ON      WPA/WPA2-PSK   0    hcia-wlan
0     hcia-ap  1  1    00E0-FC63-7F10 ON      WPA/WPA2-PSK   0    hcia-wlan
--------------------------------------------------------------------------------
Total: 2
```

我们可以将 STA 启动并连接 hcia-wlan（2.4 GHz），按照提示输入密码 Huawei@123。连接成功后，我们可以使用命令 **display station ssid hcia-wlan** 在 AC 中查看已连接的 STA，具体见例 13-26。从命令的输出内容中可以看出，STA 获得的 IP 地址是 VLAN 20 中的地址 192.168.20.20。

例 13-26 在 AC 上查看已连接的 STA

```
[AC6005]display station ssid hcia-wlan
Rf/WLAN: Radio ID/WLAN ID
Rx/Tx: link receive rate/link transmit rate(Mbps)
--------------------------------------------------------------------------------
STA MAC        AP ID Ap name   Rf/WLAN Band  Type  Rx/Tx   RSSI VLAN IP address
--------------------------------------------------------------------------------
5489-98c1-2a54 0     hcia-ap   0/1     2.4G  -     -/-     -    20   192.168.20.20
--------------------------------------------------------------------------------
Total: 1 2.4G: 1 5G: 0
```

至此，STA 已经正确连接到 WLAN，并且获得了业务 VLAN 中正确的 IP 地址。我们可以在 STA 的命令行中使用命令 **ipconfig** 查看 IP 地址信息，具体见例 13-27。

例 13-27 在 STA 中查看 IP 地址信息

```
STA>ipconfig

Link local IPv6 address...........: ::
IPv6 address......................: :: / 128
IPv6 gateway......................: ::
IPv4 address......................: 192.168.20.20
Subnet mask.......................: 255.255.255.0
Gateway...........................: 192.168.20.1
Physical address..................: 54-89-98-C1-2A-54
DNS server........................:
```

我们还可以在 STA 上对其网关发起 ping 测试，以验证 STA 的网络连通性，具体见例 13-28。

例 13-28 验证 STA 的连通性

```
STA>ping 192.168.20.1

Ping 192.168.20.1: 32 data bytes, Press Ctrl_C to break
From 192.168.20.1: bytes=32 seq=1 ttl=255 time=125 ms
From 192.168.20.1: bytes=32 seq=2 ttl=255 time=140 ms
From 192.168.20.1: bytes=32 seq=3 ttl=255 time=125 ms
From 192.168.20.1: bytes=32 seq=4 ttl=255 time=109 ms
From 192.168.20.1: bytes=32 seq=5 ttl=255 time=141 ms

--- 192.168.20.1 ping statistics ---
  5 packet(s) transmitted
  5 packet(s) received
  0.00% packet loss
  round-trip min/avg/max = 109/128/141 ms
```

第 14 章
构建简单 IPv6 网络

本章主要内容

14.1　实验介绍

14.2　实验配置任务

IPv6 是继 IPv4 之后广泛部署的 IP 版本，也称为下一代 IP。在 IPv4 耗尽之后，IPv6 的部署已越来越广泛，它的 IP 地址长度从 IPv4 使用的 32 位扩展为 128 位，满足了日益增长的联网设备数量需求。除了 IP 地址空间充足之外，IPv6 还具有简化的头部格式、层次化的地址结构、IP 地址的自动配置机制等优势。

掌握 IPv6 地址格式和书写方法是完成实验的先决条件。IPv6 地址的具体格式为 8 组十六进制数值，每组数值由 4 个十六进制字符构成，组之间以英文冒号间隔。十六进制字符包括数字 0~9 及字母 A~F。由于 IPv6 地址很长，为了简化，人们对于 IPv6 地址在书写方面提出了以下规则。

① 省略前导零：当 IPv6 地址中的每一组十六进制数值以 0 开头时，可以省略开头的一个或连续的多个 0。比如，0000 可以简单表示为 0，00DF 可以简单表示为 DF；完整示例比如 IPv6 地址 FC00:008A:0000:0000:0000:0000:0370:7300，可以缩写为 FC00:8A:0:0:0:0:370:7300。从这个示例可以看出，每组中开头的 0 可以省略，比如 008A 压缩为 8A，0370 压缩为 370；若 4 个字符全为 0，则保留一个 0，即 0000 压缩为 0；结尾的 0 不能被省略，比如 FC00 和 7300 无法被压缩。

② 双冒号：我们可以使用双冒号来指代 IPv6 地址中连续为 0 的多组十六进制数值，但一个 IPv6 地址中只可以使用一个双冒号。比如 IPv6 地址 FC00:8A:0:0:0:0:370:7300 中有连续 4 组十六进制数值都为 0，可以使用::代替这 4 组数值，即这个 IPv6 地址可以进一步被压缩为 FC00:8A::370:7300。

在配置 IPv6 地址时，除了手动配置之外，我们还可以让设备根据某些信息自动生成 IPv6 地址。具体来讲，在配置 IPv6 链路本地地址时，有以下两种方式。

① 手动配置：管理员手动配置具体的 IPv6 链路本地地址和掩码。

② 自动配置：以 FE80::开头，以 EUI-64 格式填充后 64 位。EUI-64 地址是通过 MAC 地址转换而来的。

在配置 IPv6 全球单播地址时，有以下两种方式。

① 手动配置：管理员手动配置具体的 IPv6 全球单播地址和掩码。

② 自动配置分为有状态方式和无状态方式。

a. 有状态方式

• 有状态 DHCPv6：通过 DHCPv6 获取 IPv6 地址和其他参数。

• 无状态 DHCPv6：通过 DHCPv6 获取其他参数，通过无状态方式获得 IPv6 地址。

b. 无状态方式

根据 RA（路由通告）报文中的前缀信息，以及 EUI-64 地址自动配置 IPv6 地址，我们称之为无状态方式，无状态方式也称为 SLAAC 方式。

14.1 实验介绍

14.1.1 关于本实验

在本实验中，读者需要根据表 14-1 配置接口的 IPv6 全球单播地址，令设备自动生成接口的 IPv6 链路本地地址。4 台路由器的配置略有不同，其中，AR1 与 AR2 通过手动配置的方式配置接口 IPv6 地址，AR3 通过 DHCPv6 从 AR2 获取 IPv6 地址，AR4 通过无状态配置从 AR2 获取 IPv6 地址。在完成直连通信后，再通过静态路由实现 AR1、AR3 与 AR4 之间的通信。

14.1.2 实验目的

- 掌握手动配置 IPv6 地址的方式。
- 掌握通过 DHCPv6 获得 IPv6 地址的方式。
- 掌握通过 SLAAC 获得 IPv6 地址的方式。
- 掌握 IPv6 静态路由的配置。

14.1.3 实验组网介绍

简单 IPv6 网络拓扑如图 14-1 所示。

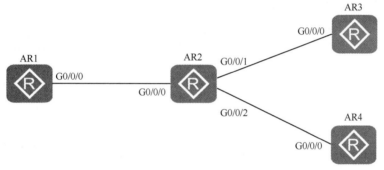

图 14-1 简单 IPv6 网络拓扑

表 14-1 中列出了本章使用的网络地址。

表 14-1 本章使用的网络地址

设备	接口	全球单播地址	子网掩码	默认网关
AR1	G0/0/0	2001:10:10:12::1	/64	—
AR2	G0/0/0	2001:10:10:12::2	/64	—
	G0/0/1	2001:10:10:23::2	/64	—
	G0/0/2	2001:10:10:24::2	/64	—
AR3	G0/0/0	DHCPv6 获取	DHCPv6 获取	AR2
AR4	G0/0/0	无状态方式	无状态方式	—

14.1.4 实验任务列表

配置任务 1：手动配置 IPv6 地址。
配置任务 2：使用 DHCPv6 获得 IPv6 地址。
配置任务 3：使用 SLAAC 获得 IPv6 地址。
配置任务 4：配置静态 IPv6 路由。

14.2 实验配置任务

14.2.1 手动配置 IPv6 地址

根据实验要求,我们需要手动配置 AR1 和 AR2 接口的 IPv6 全球单播地址,并且让设备自动生成 IPv6 链路本地地址,本实验手动配置 IPv6 地址如图 14-2 所示。

图 14-2 手动配置 IPv6 地址

在缺省情况下,华为路由器是没有启用 IPv6 路由功能的,管理员需要首先在系统视图中使用命令 **ipv6** 来启用 IPv6 路由。接着,在运行 IPv6 的接口视图中,我们还需要使用命令 **ipv6 enable** 在接口上启用 IPv6 功能。本部分使用命令如下。

① **ipv6**:系统视图命令,用来在设备全局启用 IPv6 路由转发功能,在缺省情况下,未启用 IPv6 路由转发功能。

② **ipv6 enable**:接口视图命令,用来在接口上启用 IPv6 功能。在缺省情况下,接口上未启用 IPv6 功能。

③ **ipv6 address auto link-local**:接口视图命令,用来在接口上自动生成链路本地地址,这也是缺省配置。读者也可以使用接口视图命令 **ipv6 address** *ipv6-address prefix-length* **link-local** 手动配置链路本地地址。本实验使用自动生成的设置,仅以 AR1 的接口 G0/0/0 为例展示该命令的配置,其他接口保持缺省设置。

④ **ipv6 address** *ipv6-address prefix-length*:接口视图命令,用来手动配置 IPv6 全球单播地址。

例 14-1 和例 14-2 中展示了 AR1 和 AR2 上的 IPv6 配置。为求统一,AR2 分别连接 AR3 和 AR4 的接口配置也在例中直接展示,后续任务中不再重复。

例 14-1 在 AR1 上配置 IPv6

```
[AR1]ipv6
[AR1]interface GigabitEthernet 0/0/0
[AR1-GigabitEthernet0/0/0]ipv6 enable
[AR1-GigabitEthernet0/0/0]ipv6 address auto link-local
[AR1-GigabitEthernet0/0/0]ipv6 address 2001:10:10:12::1 64
```

例 14-2 在 AR2 上配置 IPv6

```
[AR2]ipv6
[AR2]interface GigabitEthernet 0/0/0
[AR2-GigabitEthernet0/0/0]ipv6 enable
[AR2-GigabitEthernet0/0/0]ipv6 address 2001:10:10:12::2 64
[AR2-GigabitEthernet0/0/0]quit
[AR2]interface GigabitEthernet 0/0/1
[AR2-GigabitEthernet0/0/1]ipv6 enable
[AR2-GigabitEthernet0/0/1]ipv6 address 2001:10:10:23::2 64
[AR2-GigabitEthernet0/0/1]quit
[AR2]interface GigabitEthernet 0/0/2
[AR2-GigabitEthernet0/0/2]ipv6 enable
[AR2-GigabitEthernet0/0/2]ipv6 address 2001:10:10:24::2 64
```

配置完成后，读者可以使用 **display ipv6 interface brief** 来查看刚才配置的接口 IPv6 全球单播地址，以 AR2 为例，具体见例 14-3。从命令输出内容中，我们可以看到手动配置的 IPv6 全球单播地址。

例 14-3　在 AR2 上查看 IPv6 接口汇总信息

```
[AR2]display ipv6 interface brief
*down: administratively down
(l): loopback
(s): spoofing
Interface                         Physical              Protocol
GigabitEthernet0/0/0              up                    up
 [IPv6 Address] 2001:10:10:12::2
GigabitEthernet0/0/1              up                    up
 [IPv6 Address] 2001:10:10:23::2
GigabitEthernet0/0/2              up                    up
 [IPv6 Address] 2001:10:10:24::2
```

如果管理员没有在接口上手动指定 IPv6 链路本地地址，也没有设置自动生成链路本地地址的命令，但是接口配置了 IPv6 全球单播地址后，设备会根据 IEEE EUI-64 地址格式自动生成 IPv6 链路本地地址。以 AR2 的接口 G0/0/0 为例，读者可以使用命令 **display ipv6 interface G0/0/0** 来查看接口上的 IPv6 链路本地地址，具体见例 14-4。

例 14-4　查看 IPv6 接口的详细信息

```
[AR2]display ipv6 interface GigabitEthernet 0/0/0
GigabitEthernet0/0/0 current state : UP
IPv6 protocol current state : UP
IPv6 is enabled, link-local address is FE80::2E0:FCFF:FEF9:337B
  Global unicast address(es):
    2001:10:10:12::2, subnet is 2001:10:10:12::/64
  Joined group address(es):
    FF02::1:FF00:2
    FF02::2
    FF02::1
    FF02::1:FFF9:337B
  MTU is 1500 bytes
  ND DAD is enabled, number of DAD attempts: 1
  ND reachable time is 30000 milliseconds
  ND retransmit interval is 1000 milliseconds
  Hosts use stateless autoconfig for addresses
```

从命令的输出内容中我们可以看出，其与 IPv4 接口相同，命令输出内容的前两行分别显示了该接口的物理状态和协议状态。阴影部分展示了接口上自动生成的 IPv6 链路本地地址，以 FE80 开头，并使用接口的 MAC 地址进行了转换。我们先使用命令 **display interface G0/0/0** 向读者展示 G0/0/0 接口的 MAC 地址，然后通过两者的对比，使读者可以更直观地理解这种转换是如何实现的，详见例 14-5。

例 14-5　确认 AR2 G0/0/0 接口的 MAC 地址

```
[AR2]display interface GigabitEthernet 0/0/0
GigabitEthernet0/0/0 current state : UP
Line protocol current state : DOWN
Description:HUAWEI, AR Series, GigabitEthernet0/0/0 Interface
Route Port,The Maximum Transmit Unit is 1500
Internet protocol processing : disabled
IP Sending Frames' Format is PKTFMT_ETHNT_2, Hardware address is 00e0-fcf9-337b
Last physical up time   : 2021-07-17 12:08:40 UTC-08:00
Last physical down time : 2021-07-17 12:08:32 UTC-08:00
Current system time: 2021-07-17 12:12:31-08:00
Port Mode: FORCE COPPER
Speed : 1000,   Loopback: NONE
Duplex: FULL,   Negotiation: ENABLE
Mdi   : AUTO
Last 300 seconds input rate 0 bits/sec, 0 packets/sec
Last 300 seconds output rate 0 bits/sec, 0 packets/sec
Input peak rate 120 bits/sec,Record time: 2021-07-17 12:09:50
Output peak rate 248 bits/sec,Record time: 2021-07-17 12:10:30

Input:  2 packets, 156 bytes
  Unicast:                        0,  Multicast:                      2
  Broadcast:                      0,  Jumbo:                          0
  Discard:                        0,  Total Error:                    0
```

```
    CRC:                            0,   Giants:                    0
    Jabbers:                        0,   Throttles:                 0
    Runts:                          0,   Symbols:                   0
    Ignoreds:                       0,   Frames:                    0
Output:  2 packets, 156 bytes
    Unicast:                        0,   Multicast:                 2
    Broadcast:                      0,   Jumbo:                     0
    Discard:                        0,   Total Error:               0

    Collisions:                     0,   ExcessiveCollisions:       0
    Late Collisions:                0,   Deferreds:                 0

    Input bandwidth utilization  threshold : 100.00%
    Output bandwidth utilization threshold: 100.00%
    Input bandwidth utilization  :      0%
    Output bandwidth utilization :      0%
```

现在，读者可以观察 AR2 G0/0/0 接口的 MAC 地址（00e0-fcf9-337b）与 IPv6 链路本地地址（FE80::2E0:FCFF:FEF9:337B）的区别。接下来，我们来介绍一下 IPv6 链路本地地址是如何自动生成的。

首先，IPv6 的链路本地地址是定义在 FE80::/10 地址范围中的，因此自动生成的 IPv6 链路本地地址都是以 FE80 开头的，后面跟着的两段十六进制数值都为 0，因此前半部分 FE80:0:0:可以压缩为 FE80::。以上是 IPv6 链路本地地址的前 64 位，后 64 位是由接口 MAC 地址转换而来。MAC 地址转换为 EUI-64 格式的接口 ID 具体转换规则如图 14-3 所示。

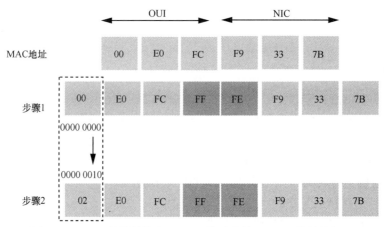

图 14-3　MAC 地址转换为 EUI-64 格式的接口 ID 具体转换规则

根据 MAC 地址自动生成的 64 位地址格式我们称为 EUI-64 地址，它是 IEEE 定义的地址格式，旨在通过 48 位 IEEE 802 地址（MAC 地址）自动生成 64 位地址，并将其作为 IPv6 地址的接口 ID 部分。它的转换方法可以总结为两个步骤。

① 在 MAC 地址的前 24 位和后 24 位的中间插入 16 位字符 FFFE，将 48 位 MAC 地址扩展为 64 位 EUI-64 标识符，也就是将 00-E0-FC-F9-33-7B 扩展为 00-E0-FC-FF-FE-F9-33-7B。

② 将 64 位 EUI-64 标识符的第 7 位反转，接口 ID 完成转换，也就是将 00-E0-FC-FF-FE-F9-33-7B 转换为了 02-E0-FC-FF-FE-F9-33-7B。

再回到例 14-4 命令输出内容，从 Joined group address(es)部分我们可以看到 4 个 IPv6 的地址，它们都是 IPv6 组播地址，也是当前该接口加入的组播组，具体说明如下。

① FF02::1:FF00:2：本地单播地址所对应的被请求节点组播地址，用来解析二层地址。

② FF02::2：表示所有路由器，也就是所有启用了 IPv6 的路由器都会加入这个组播组。

③ FF02::1：表示所有节点，也就是所有启用了 IPv6 的节点都会加入这个组播组。

④ FF02::1:FFF9:337B：链路本地地址所对应的被请求节点组播地址。

读者可以使用命令 **display ipv6 routing-table** 查看 IPv6 路由表，详见例 14-6。当前的 IPv6 路由表中全部是 AR2 的直连 IPv6 路由。

例 14-6　AR2 上的直连 IPv6 路由

```
[AR2]display ipv6 routing-table
Routing Table : Public
       Destinations : 8       Routes : 8

Destination  : ::1                              PrefixLength : 128
NextHop      : ::1                              Preference   : 0
Cost         : 0                                Protocol     : Direct
RelayNextHop : ::                               TunnelID     : 0x0
Interface    : InLoopBack0                      Flags        : D

Destination  : 2001:10:10:12::                  PrefixLength : 64
NextHop      : 2001:10:10:12::2                 Preference   : 0
Cost         : 0                                Protocol     : Direct
RelayNextHop : ::                               TunnelID     : 0x0
Interface    : GigabitEthernet0/0/0             Flags        : D

Destination  : 2001:10:10:12::2                 PrefixLength : 128
NextHop      : ::1                              Preference   : 0
Cost         : 0                                Protocol     : Direct
RelayNextHop : ::                               TunnelID     : 0x0
Interface    : GigabitEthernet0/0/0             Flags        : D

Destination  : 2001:10:10:23::                  PrefixLength : 64
NextHop      : 2001:10:10:23::2                 Preference   : 0
Cost         : 0                                Protocol     : Direct
RelayNextHop : ::                               TunnelID     : 0x0
Interface    : GigabitEthernet0/0/1             Flags        : D

Destination  : 2001:10:10:23::2                 PrefixLength : 128
NextHop      : ::1                              Preference   : 0
Cost         : 0                                Protocol     : Direct
RelayNextHop : ::                               TunnelID     : 0x0
Interface    : GigabitEthernet0/0/1             Flags        : D

Destination  : 2001:10:10:24::                  PrefixLength : 64
NextHop      : 2001:10:10:24::2                 Preference   : 0
Cost         : 0                                Protocol     : Direct
RelayNextHop : ::                               TunnelID     : 0x0
Interface    : GigabitEthernet0/0/2             Flags        : D

Destination  : 2001:10:10:24::2                 PrefixLength : 128
NextHop      : ::1                              Preference   : 0
Cost         : 0                                Protocol     : Direct
RelayNextHop : ::                               TunnelID     : 0x0
Interface    : GigabitEthernet0/0/2             Flags        : D

Destination  : FE80::                           PrefixLength : 10
NextHop      : ::                               Preference   : 0
Cost         : 0                                Protocol     : Direct
RelayNextHop : ::                               TunnelID     : 0x0
Interface    : NULL0                            Flags        : D
```

读者可以从 AR2 上测试直连链路的连通性，详见例 14-7，AR2 向 AR1 的接口 IPv6 地址 2001:10:10:12::1 发起了 ping 测试，结果表明已建立直连连通性。

例 14-7　测试直连链路的连通性

```
[AR2]ping ipv6 2001:10:10:12::1
  PING 2001:10:10:12::1 : 56  data bytes, press CTRL_C to break
    Reply from 2001:10:10:12::1
    bytes=56 Sequence=1 hop limit=64  time = 110 ms
    Reply from 2001:10:10:12::1
    bytes=56 Sequence=2 hop limit=64  time = 30 ms
    Reply from 2001:10:10:12::1
    bytes=56 Sequence=3 hop limit=64  time = 20 ms
    Reply from 2001:10:10:12::1
    bytes=56 Sequence=4 hop limit=64  time = 20 ms
    Reply from 2001:10:10:12::1
    bytes=56 Sequence=5 hop limit=64  time = 30 ms

  --- 2001:10:10:12::1 ping statistics ---
    5 packet(s) transmitted
    5 packet(s) received
    0.00% packet loss
    round-trip min/avg/max = 20/42/110 ms
```

读者还可以通过 AR1 与 AR2 接口上的链路本地地址来测试直连连通性。首先，读者可以根据 EUI-64 格式练习计算出 AR1 接口 G0/0/0 的 IPv6 链路本地地址，例 14-8 中展示了 AR1 接口 G0/0/0 的 MAC 地址 00e0-fc19-4e66。

例 14-8　查看 AR1 接口 G0/0/0 的 MAC 地址

```
[AR1]display interface GigabitEthernet 0/0/0
GigabitEthernet0/0/0 current state : UP
Line protocol current state : DOWN
Description:HUAWEI, AR Series, GigabitEthernet0/0/0 Interface
Route Port,The Maximum Transmit Unit is 1500
Internet protocol processing : disabled
IP Sending Frames' Format is PKTFMT_ETHNT_2,  Hardware address is 00e0-fc19-4e66
Last physical up time   : 2021-07-17 12:08:17 UTC-08:00
Last physical down time : 2021-07-17 12:05:44 UTC-08:00
Current system time: 2021-07-17 12:16:29-08:00
Port Mode: FORCE COPPER
Speed : 1000,  Loopback: NONE
Duplex: FULL,  Negotiation: ENABLE
Mdi   : AUTO
Last 300 seconds input rate 0 bits/sec, 0 packets/sec
Last 300 seconds output rate 0 bits/sec, 0 packets/sec
Input peak rate 248 bits/sec,Record time: 2021-07-17 12:09:47
Output peak rate 120 bits/sec,Record time: 2021-07-17 12:09:17

Input:  2 packets, 156 bytes
  Unicast:                 0, Multicast:               2
  Broadcast:               0, Jumbo:                   0
  Discard:                 0, Total Error:             0

  CRC:                     0, Giants:                  0
  Jabbers:                 0, Throttles:               0
  Runts:                   0, Symbols:                 0
  Ignoreds:                0, Frames:                  0

Output: 2 packets, 156 bytes
  Unicast:                 0, Multicast:               2
  Broadcast:               0, Jumbo:                   0
  Discard:                 0, Total Error:             0

  Collisions:              0, ExcessiveCollisions:     0
  Late Collisions:         0, Deferreds:               0

    Input bandwidth utilization threshold : 100.00%
    Output bandwidth utilization threshold: 100.00%
    Input bandwidth utilization  :    0%
    Output bandwidth utilization :    0%
```

然后，根据 MAC 地址 00e0-fc19-4e66，读者应该可以计算出该接口的 IPv6 链路本地地址为 FE80::02e0-fcff-fe19-4e66。我们可以通过命令来确认这一点，详见例 14-9。

例 14-9　确认 AR1 接口 G0/0/0 的 IPv6 链路本地地址

```
[AR1]display ipv6 interface GigabitEthernet 0/0/0
GigabitEthernet0/0/0 current state : UP
IPv6 protocol current state : UP
IPv6 is enabled, link-local address is FE80::2E0:FCFF:FE19:4E66
  Global unicast address(es):
    2001:10:10:12::1, subnet is 2001:10:10:12::/64
  Joined group address(es):
    FF02::1:FF00:1
    FF02::2
    FF02::1
    FF02::1:FF19:4E66
  MTU is 1500 bytes
  ND DAD is enabled, number of DAD attempts: 1
  ND reachable time is 30000 milliseconds
  ND retransmit interval is 1000 milliseconds
  Hosts use stateless autoconfig for addresses
```

最后，在针对 IPv6 链路本地地址进行 ping 测试时，我们需要指定所使用的源接口。读者可以使用的命令为 **ping ipv6** *ipv6-address* **-i** *interface-type interface-number*，我们仍以 AR2 对 AR1 发起 ping 测试为例，详见例 14-10。

例 14-10　测试 IPv6 链路本地地址的连通性

```
[AR2]ping ipv6 FE80::2E0:FCFF:FE19:4E66 -i GigabitEthernet 0/0/0
  PING FE80::2E0:FCFF:FE19:4E66 : 56  data bytes, press CTRL_C to break
    Reply from FE80::2E0:FCFF:FE19:4E66
    bytes=56 Sequence=1 hop limit=64  time = 170 ms
    Reply from FE80::2E0:FCFF:FE19:4E66
    bytes=56 Sequence=2 hop limit=64  time = 70 ms
    Reply from FE80::2E0:FCFF:FE19:4E66
    bytes=56 Sequence=3 hop limit=64  time = 30 ms
    Reply from FE80::2E0:FCFF:FE19:4E66
    bytes=56 Sequence=4 hop limit=64  time = 20 ms
    Reply from FE80::2E0:FCFF:FE19:4E66
    bytes=56 Sequence=5 hop limit=64  time = 30 ms

  --- FE80::2E0:FCFF:FE19:4E66 ping statistics ---
    5 packet(s) transmitted
    5 packet(s) received
    0.00% packet loss
    round-trip min/avg/max = 20/64/170 ms
```

14.2.2　使用 DHCPv6 获得 IPv6 地址

在本实验任务中，我们需要以 AR2 作为 DHCPv6 服务器，为 AR3 分配 IPv6 地址和默认网关。使用 DHCPv6 获取 IPv6 地址的实验拓扑如图 14-4 所示。

图 14-4　使用 DHCPv6 获取 IPv6 地址的实验拓扑

DHCPv6 服务器对 IPv6 地址自动分配时，分为有状态自动分配和无状态自动分配两种分配方式。我们可以用一句话概括"有状态"和"无状态"：它只是在说明 IPv6 地址的获取方式，"有状态"表示客户端使用 DHCPv6 服务器分配的 IPv6 地址，"无状态"表示客户端根据 RA 报文中的前缀自动生成 IPv6 地址。进一步说，DHCPv6 有状态方式不仅可以通过 DHCPv6 服务器自动分配 IPv6 地址，还可以分配其他所需参数（比如 DNS 等）。DHCPv6 无状态方式虽不能分配 IPv6 地址，但可以通过 DHCPv6 服务器来分配其他所需参数（比如 DNS 等），IPv6 地址信息会通过 RA 通告的方式令客户端进行自动配置。本章在下一个实验任务中会展示如何通过 RA 通告的方式自动配置 IPv6 地址。本实验使用 DHCP 有状态的方式分配 IPv6 地址和默认网关信息。

步骤 1 通过 DHCPv6 获取 IPv6 地址

要想使用 DHCPv6 分配 IPv6 地址，我们需要创建 IPv6 地址池，在其中定义能够用来分配的 IPv6 地址，并在作为 DHCPv6 服务器的接口上指定该地址池。这种方法与使用地址池进行 IPv4 地址分配的方式相同，命令也类似，具体命令释义如下。

① **dhcpv6 pool** *pool-name*：系统视图命令，用来创建 IPv6 地址池并进入 IPv6 地址池视图。在缺省情况下，设备上没有配置任何 IPv6 地址池。

② **address prefix** *ipv6-prefix/ipv6-prefix-length*：IPv6 地址池视图命令，用来设置通过 DHCPv6 进行自动分配的网络前缀。

③ **dhcp enable**：系统视图命令，在配置 DHCP 服务器功能之前，管理员必须先在系统视图下使用这条命令启用 DHCP 功能，与 DHCP 功能相关的其他特性也依赖于这条命令。我们需要启用 DHCP 功能后才可以继续配置其他 DHCP 特性，比如 DHCP 客户端、DHCP 中继等。管理员可以使用 **undo dhcp enable** 命令禁用 DHCP 功能，执行该命令后，与 DHCP 相关的配置将会删除；再次使用 **dhcp enable** 命令启用 DHCP 功能后，设备上与 DHCP 相关的配置将恢复为缺省配置。在缺省情况下，DHCP 功能处于禁用状态。

④ **dhcpv6 server** *pool-name*：接口视图命令，用来启用 DHCPv6 服务器功能，并指定自动分配 IPv6 地址所使用的 IPv6 地址池。在缺省情况下，接口下未启用 DHCPv6 服务器功能。

例 14-11 中展示了在 AR2 上进行 DHCPv6 服务器配置的方法。

例 14-11 将 AR2 配置为 DHCPv6 服务器

```
[AR2]dhcpv6 pool r3
[AR2-dhcpv6-pool-r3]address prefix 2001:10:10:23::/64
[AR2-dhcpv6-pool-r3]quit
[AR2]dhcp enable
Info: The operation may take a few seconds. Please wait for a moment.done.
[AR2]interface GigabitEthernet 0/0/1
[AR2-GigabitEthernet0/0/1]dhcpv6 server r3
```

配置完成后，读者可以使用 **display dhcpv6 server** 命令查看设备上 DHCPv6 服务器的配置，详见例 14-12。

例 14-12 查看 DHCPv6 服务器的配置

```
[AR2]display dhcpv6 server
 Interface                                         DHCPv6 pool
 GigabitEthernet0/0/1                              r3
```

接下来，我们需要配置 AR3。读者已经从前文的实验配置任务中使用过大部分命令，本实验任务就不再一一介绍了。首先，我们需要启用 IPv6 功能（在系统视图和接口视图中），然后，需要在系统视图中启用 DHCP 功能，最后，需要在接口下配置 DHCPv6 客户端（**ipv6 address auto dhcp**）。

需要注意的是，DHCPv6 不再像 DHCP 那样使用全 0 位 IP 地址进行 IP 地址请求，而是需要使用链路本地地址。我们曾经在前文实验中提到，在接口上配置了 IPv6 全球单播地址后，设备会自动生成链路本地地址。此时，设备需要通过 DHCPv6 来获取全球单播地址，因此，我们必须使用命令 **ipv6 address auto link-local** 先将接口生成链路本地地址。在此之后才可以将接口配置为 DHCPv6 客户端。

例 14-13 中展示了在 AR3 上进行 DHCPv6 客户端配置的方法。

例 14-13 将 AR3 配置为 DHCPv6 客户端

```
[AR3]ipv6
[AR3]dhcp enable
Info: The operation may take a few seconds. Please wait for a moment.done.
[AR3]interface GigabitEthernet 0/0/0
[AR3-GigabitEthernet0/0/0]ipv6 enable
[AR3-GigabitEthernet0/0/0]ipv6 address auto link-local
[AR3-GigabitEthernet0/0/0]ipv6 address auto dhcp
```

配置完成后，我们可以通过命令来查看 AR3 接口 G0/0/0 获取的 IPv6 地址，详见例 14-14。现在，我们可以看到 AR3 已经通过 DHCPv6 获得了 IPv6 全球单播地址 2001:10:10:23::1。

例 14-14 查看通过 DHCPv6 获得的地址

```
[AR3]display ipv6 interface brief
*down: administratively down
(l): loopback
(s): spoofing
Interface                    Physical            Protocol
GigabitEthernet0/0/0         up                  up
[IPv6 Address] 2001:10:10:23::1
```

步骤 2 通过 DHCPv6 获取缺省网关

在缺省情况下，DHCPv6 服务器不会自动为 DHCPv6 客户端分配 IPv6 网关地址，因此管理员需要通过额外的配置让 DHCPv6 服务器发送网关参数，以及让 DHCPv6 客户端请求网关参数。

我们应先确认 AR3 上当前并没有默认路由信息（::），具体见例 14-15。

例 14-15 查看当前 AR3 上没有默认路由信息

```
[AR3]display ipv6 routing-table
Routing Table : Public
        Destinations : 3        Routes : 3

Destination  : ::1                    PrefixLength : 128
NextHop      : ::1                    Preference   : 0
Cost         : 0                      Protocol     : Direct
RelayNextHop : ::                     TunnelID     : 0x0
Interface    : InLoopBack0            Flags        : D

Destination  : 2001:10:10:23::1       PrefixLength : 128
NextHop      : ::1                    Preference   : 0
Cost         : 0                      Protocol     : Direct
RelayNextHop : ::                     TunnelID     : 0x0
Interface    : GigabitEthernet0/0/0   Flags        : D

Destination  : FE80::                 PrefixLength : 10
NextHop      : ::                     Preference   : 0
Cost         : 0                      Protocol     : Direct
RelayNextHop : ::                     TunnelID     : 0x0
Interface    : NULL0                  Flags        : D
```

在 DHCPv6 服务器上，我们需要配置以下命令。

① **undo ipv6 nd ra halt**：接口视图命令，需要启用 DHCPv6 服务器的接口后进行配置，启用设备发送 RA 报文的功能。在缺省情况下，华为路由器不会发送 RA 报文，即：配置了 DHCPv6 功能后，在缺省情况下，DHCPv6 客户端只能通过 DHCPv6 获得 IPv6 地址信息，无法获得其他参数信息，因为其他参数信息是通过 RA 报文进行发送的。

② **ipv6 nd autoconfig managed-address-flag**：接口视图命令，需要启用 DHCPv6 服务器的接口后进行配置，用来设置 RA 报文中有状态自动配置地址的标志位（也就是将 managed-address-flag（简称为 M flag）设置为 1），令客户端使用服务器分配的 IPv6 地址。

在缺省情况下，RA 报文中的有状态自动配置地址的标志位为 0，也就是令客户端使用无状态配置方式，根据前缀信息自动生成 IPv6 地址。

③ **ipv6 nd autoconfig other-flag**：接口视图命令，需要启用 DHCPv6 服务器的接口后进行配置，用来设置 RA 报文中有状态自动配置其他信息的标志位（也就是将 other-flag（简称为 O flag）设置为 1）。设置了这个标志位后，DHCPv6 客户端可以通过有状态自动配置获得除了 IPv6 地址之外的其他配置信息，包括路由器生存时间、链路 MTU 等信息。在缺省情况下，RA 报文中的有状态自动配置其他信息的标志位为 0。

例 14-16 中展示了 AR2 上的相关配置命令。

例 14-16　AR2 上的相关配置命令

```
[AR2]interface GigabitEthernet 0/0/1
[AR2-GigabitEthernet0/0/1]undo ipv6 nd ra halt
[AR2-GigabitEthernet0/0/1]ipv6 nd autoconfig managed-address-flag
[AR2-GigabitEthernet0/0/1]ipv6 nd autoconfig other-flag
```

此时，AR2 接口 G0/0/1 上的完整配置见例 14-17。

例 14-17　AR2 接口 G0/0/1 上的完整配置

```
[AR2-GigabitEthernet0/0/1]display this
[V200R003C00]
#
interface GigabitEthernet0/0/1
 ipv6 enable
 ipv6 address 2001:10:10:23::2/64
 undo ipv6 nd ra halt
 ipv6 nd autoconfig managed-address-flag
 ipv6 nd autoconfig other-flag
 dhcpv6 server r3
#
return
```

在 DHCPv6 客户端上，我们需要配置以下命令。

ipv6 address auto global default：接口视图命令，需要在使用 DHCPv6 获取 IPv6 地址的接口上进行配置，使设备请求默认网关信息。

例 14-18 中展示了 AR3 上的相关配置命令。

例 14-18　AR3 上的相关配置命令

```
[AR3]interface GigabitEthernet 0/0/0
[AR3-GigabitEthernet0/0/0]ipv6 address auto global default
```

此时，AR3 接口 G0/0/0 上的完整配置见例 14-19。

例 14-19　AR3 接口 G0/0/0 上的完整配置

```
[AR3-GigabitEthernet0/0/0]display this
[V200R003C00]
#
interface GigabitEthernet0/0/0
 ipv6 enable
 ipv6 address auto link-local
 ipv6 address auto global default
 ipv6 address auto dhcp
#
return
```

配置完成后，读者可以在 AR3 接口 G0/0/0 上执行 **shutdown** 和 **undo shutdown** 命令，令 AR3 重新请求 IPv6 地址信息。在本次交互过程完成后，我们再次查看 AR3 的 IPv6 路由表，详见例 14-20。阴影部分为 AR3 上的默认路由，以 G0/0/0 接口为出接口，下一跳指向 AR2 G0/0/1 接口的链路本地地址。

例 14-20　查看 AR3 上的默认路由

```
[AR3]display ipv6 routing-table
Routing Table : Public
        Destinations : 4        Routes : 4

Destination     : ::                        PrefixLength : 0
NextHop         : FE80::2E0:FCFF:FEF9:337C  Preference   : 64
Cost            : 0                         Protocol     : Unr
RelayNextHop    : ::                        TunnelID     : 0x0
Interface       : GigabitEthernet0/0/0      Flags        : D

Destination     : ::1                       PrefixLength : 128
NextHop         : ::1                       Preference   : 0
Cost            : 0                         Protocol     : Direct
RelayNextHop    : ::                        TunnelID     : 0x0
Interface       : InLoopBack0               Flags        : D

Destination     : 2001:10:10:23::1          PrefixLength : 128
NextHop         : ::1                       Preference   : 0
Cost            : 0                         Protocol     : Direct
RelayNextHop    : ::                        TunnelID     : 0x0
Interface       : GigabitEthernet0/0/0      Flags        : D

Destination     : FE80::                    PrefixLength : 10
NextHop         : ::                        Preference   : 0
Cost            : 0                         Protocol     : Direct
RelayNextHop    : ::                        TunnelID     : 0x0
Interface       : NULL0                     Flags        : D
```

步骤 3　观察 DHCPv6 交互过程

我们在 AR2 与 AR3 相连的链路上开启抓包来观察 DHCPv6 的交互过程。

首先，我们通过抓包信息观察 DHCPv6 服务器 AR2 上的标志位设置，如图 14-5 所示。从图中的 Flags 部分可以看出，AR2 已经将 M flag 和 O flag 都置为 1。

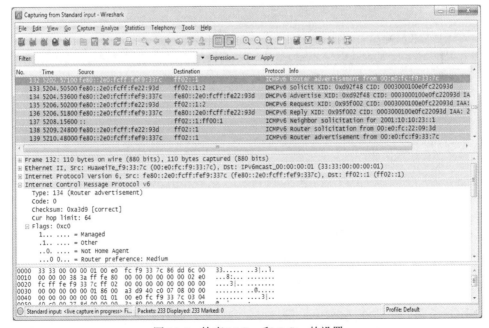

图 14-5　检查 M flag 和 O flag 的设置

DHCPv6 客户端会通过组播发送 DHCPv6 Solicit 消息来请求 IPv6 地址，如图 14-6 所示。

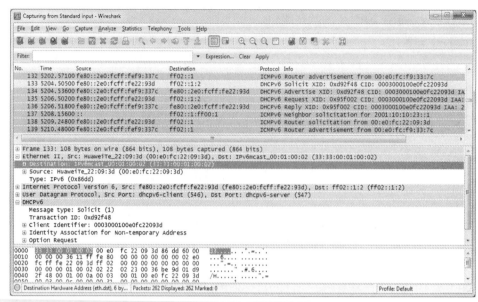

图 14-6　DHCPv6 客户端发送 DHCPv6 Solicit 消息

DHCPv6 服务器接收 Solicit 消息后，会以单播的方式回复 DHCPv6 Advertise 消息，并在其中提供详细的地址配置信息，如图 14-7 所示。

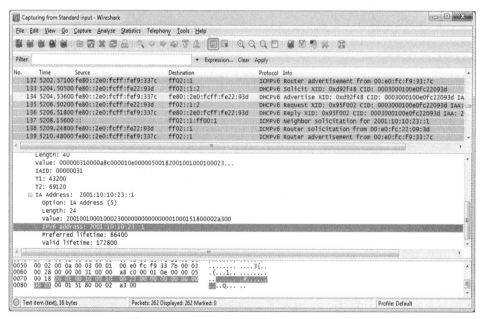

图 14-7　DHCPv6 服务器回复 Advertise 消息

DHCPv6 客户端接收 Advertise 消息后，会针对其提供的 IPv6 地址进行请求，请求使用这个地址作为自己的 IPv6 地址。图 14-8 中展示了 DHCPv6 客户端通过组播发送的 Request 消息。

最后，DHCPv6 服务器会以单播的方式发送 Reply 消息，确认 DHCPv6 客户端能够使用这个 IPv6 地址。这个 Reply 消息与 Advertise 消息相同，如图 14-9 所示。

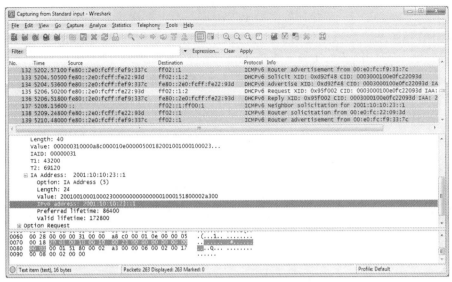

图 14-8　DHCPv6 客户端发出 Request 消息

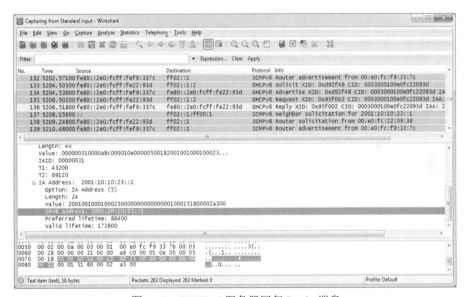

图 14-9　DHCPv6 服务器回复 Reply 消息

14.2.3　使用 SLAAC 获得 IPv6 地址

SLAAC（Stateless Address Autoconfiguration，无状态地址自动配置）是指 DHCPv6 客户端通过接收到的 RA 消息的前缀信息，自动生成 IPv6 地址信息。本实验要使 AR4 接口 G0/0/0 自动生成 IPv6 地址，实验拓扑具体如图 14-10 所示。

图 14-10　无状态地址自动配置实验拓扑

学习和理解了前一个实验任务中 DHCPv6 服务器上配置的各种参数后,读者应该可以判断出如何将 AR2 配置为使用无状态方式提供 IPv6 前缀信息,即:AR2 需要从 G0/0/2 接口发送 RA 消息,因此本实验需要在 AR2 上配置以下命令。

undo ipv6 nd ra halt:接口视图命令,我们需要在启用 DHCPv6 服务器的接口后进行配置,启用设备发送 RA 报文的功能。在缺省情况下,华为路由器不会发送 RA 报文。

例 14-21 中展示了 AR2 上的相关配置。

例 14-21　AR2 上的相关配置

```
[AR2]interface GigabitEthernet 0/0/2
[AR2-GigabitEthernet0/0/2]undo ipv6 nd ra halt
```

此时,AR2 接口 G0/0/2 上的完整配置见例 14-22。

例 14-22　AR2 接口 G0/0/2 上的完整配置

```
[AR2-GigabitEthernet0/0/2]display this
[V200R003C00]
#
interface GigabitEthernet0/0/2
 ipv6 enable
 ipv6 address 2001:10:10:24::2/64
 undo ipv6 nd ra halt
#
return
```

为了使 AR4 接口 G0/0/0 通过无状态方式生成 IPv6 地址,需要使用以下命令。

ipv6 address auto global:接口视图命令,用来无状态自动生成 IPv6 全球单播地址。

例 14-23 中展示了 AR4 上的相关配置。

例 14-23　AR4 上的相关配置

```
[AR4]ipv6
[AR4]interface GigabitEthernet 0/0/0
[AR4-GigabitEthernet0/0/0]ipv6 enable
[AR4-GigabitEthernet0/0/0]ipv6 address auto link-local
[AR4-GigabitEthernet0/0/0]ipv6 address auto global
```

现在,我们来检查 AR4 自动生成的 IPv6 全球单播地址,详见例 14-24。读者应该记得命令 **display this** 的用途,本例展示该命令的另一种用法:在 G0/0/0 接口视图下使用命令 **display this ipv6 interface** 获得的命令输出内容与在系统视图下使用命令 **display ipv6 interface G0/0/0** 获得的内容相同。例 14-24 的阴影部分突出显示了 AR4 G0/0/0 接口通过 SLAAC 生成的 IPv6 全球单播地址。

例 14-24　查看 AR4 的 IPv6 地址

```
[AR4-GigabitEthernet0/0/0]display this ipv6 interface
GigabitEthernet0/0/0 current state : UP
IPv6 protocol current state : UP
IPv6 is enabled, link-local address is FE80::2E0:FCFF:FE89:2897
  Global unicast address(es):
    2001:10:10:24:2E0:FCFF:FE89:2897,
    subnet is 2001:10:10:24::/64 [SLAAC 1970-01-01 03:23:14 2592000S]
  Joined group address(es):
    FF02::1:FF89:2897
    FF02::2
    FF02::1
  MTU is 1500 bytes
  ND DAD is enabled, number of DAD attempts: 1
  ND reachable time is 30000 milliseconds
  ND retransmit interval is 1000 milliseconds
  Hosts use stateless autoconfig for addresses
```

仔细观察例 14-24 中显示出的 IPv6 链路本地地址和 IPv6 全球单播地址。

① 链路本地地址:FE80::2E0:FCFF:FE89:2897。

② 全球单播地址:2001:10:10:24:2E0:FCFF:FE89:2897。

我们可以发现链路本地地址是以 FE80 开头，以接口 MAC 地址计算出的 EUI-64 格式地址结尾。全球单播地址是以链路上的前缀 2001:10:10:24 开头，以接口 MAC 地址计算出的 EUI-64 格式地址结尾。

无状态地址自动配置和有状态地址自动配置有其各自适用的场景。当管理员需要支持众多终端设备访问 IPv6 网络时，可以使用无状态地址自动配置，无须设置 DHCPv6 服务器。但如果需要 IPv6 地址的终端设备向外提供服务，则最好使用有状态地址进行自动配置，这种方法可使 IP 地址可控。

在有些场景中，根据具体的需求，管理员也可以将 SLAAC 与 DHCPv6 相结合，也就是通过 SLAAC 提供 IPv6 地址信息，依靠 DHCPv6 提供其他信息。

14.2.4 配置静态 IPv6 路由

在这个实验任务中，我们要通过静态路由来实现实验环境中所有 IPv6 地址的互通。本实验需要读者在 AR1 上使用一条静态汇总路由，通过它来访问 AR3 与 AR4。在 AR3 上配置一条静态默认路由，以及在 AR4 上配置两条静态路由。我们按照 AR4、AR3 和 AR1 的顺序分别进行配置。

IPv6 静态路由拓扑如图 14-11 所示。

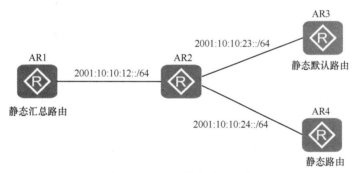

图 14-11　IPv6 静态路由拓扑

步骤 1　预先配置内容

为了顺利执行本实验任务的要求，我们对之前的配置略作修改，在 AR3 和 AR4 的 G0/0/0 接口上手动配置 IPv6 全球单播地址，并且使用表 14-2 中的 IPv6 地址。

表 14-2　需要手动配置的 IPv6 地址

路由器	接口	IPv6 全球单播地址
AR3	G0/0/0	2001:10:10:23::3/64
AR4	G0/0/0	2001:10:10:24::4/64

例 14-25 和例 14-26 中分别展示了 AR3 和 AR4 中的配置命令。

例 14-25　在 AR3 上手动配置 IPv6 地址

```
[AR3]interface GigabitEthernet 0/0/0
[AR3-GigabitEthernet0/0/0]undo ipv6 address auto global
[AR3-GigabitEthernet0/0/0]undo ipv6 address auto dhcp
[AR3-GigabitEthernet0/0/0]ipv6 address 2001:10:10:23::3 64
```

例 14-26　在 AR4 上手动配置 IPv6 地址

```
[AR4]interface GigabitEthernet 0/0/0
[AR4-GigabitEthernet0/0/0]undo ipv6 address auto global
[AR4-GigabitEthernet0/0/0]ipv6 address 2001:10:10:24::4 64
```

步骤 2　配置静态路由

在创建 IPv6 静态路由时，管理员可以将下一跳指定为出接口、IPv6 地址，或者同时指定出接口和 IPv6 地址，具体的选择与接口类型如下。

① 如果出接口是广播类型的接口，需要指定下一跳 IPv6 地址。下一跳地址可以是 IPv6 全球单播地址，也可以是 IPv6 链路本地地址。在使用 IPv6 链路本地地址作为下一跳时，还需要同时指定出接口。

② 如果出接口是点到点类型的接口，只需要指定出接口即可。

③ 如果出接口是 NBMA 类型的接口，需要指定下一跳 IPv6 地址。

本实验环境中，以太网接口属于广播类型的接口，因此我们选择使用 IPv6 全球单播地址作为下一跳进行配置。读者可以使用以下命令进行配置，完整的配置命令可参考华为设备配置指南。

ipv6 route-static *dest-ipv6-address ipv6-prefix-length nexthop-ipv6-address*：系统视图命令，在广播类型的环境中，我们选择配置下一跳 IPv6 地址作为下一跳参数，使用 IPv6 全球单播地址。

R4 上的 IPv6 静态路由配置详见例 14-27 所示。

例 14-27　在 AR4 上配置静态路由

```
[AR4]ipv6 route-static 2001:10:10:12:: 64 2001:10:10:24::2
[AR4]ipv6 route-static 2001:10:10:23:: 64 2001:10:10:24::2
```

配置完成后读者可以查看 IPv6 路由表，来确认静态路由的配置，详见例 14-28。阴影部分突出显示了两条静态路由，读者从中可以看出与 IPv4 静态路由相同，静态路由的优先级为 60。同时，路由器已经根据这个下一跳 IPv6 地址得到了出接口 G0/0/0。

例 14-28　查看 IPv6 路由表中的静态路由

```
[AR4]display ipv6 routing-table
Routing Table : Public
        Destinations : 6       Routes : 6

 Destination  : ::1                              PrefixLength : 128
 NextHop      : ::1                              Preference   : 0
 Cost         : 0                                Protocol     : Direct
 RelayNextHop : ::                               TunnelID     : 0x0
 Interface    : InLoopBack0                      Flags        : D

 Destination  : 2001:10:10:12::                  PrefixLength : 64
 NextHop      : 2001:10:10:24::2                 Preference   : 60
 Cost         : 0                                Protocol     : Static
 RelayNextHop : ::                               TunnelID     : 0x0
 Interface    : GigabitEthernet0/0/0             Flags        : RD

 Destination  : 2001:10:10:23::                  PrefixLength : 64
 NextHop      : 2001:10:10:24::2                 Preference   : 60
 Cost         : 0                                Protocol     : Static
 RelayNextHop : ::                               TunnelID     : 0x0
 Interface    : GigabitEthernet0/0/0             Flags        : RD

 Destination  : 2001:10:10:24::                  PrefixLength : 64
 NextHop      : 2001:10:10:24:2E0:FCFF:FE89:2897 Preference   : 0
 Cost         : 0                                Protocol     : Direct
 RelayNextHop : ::                               TunnelID     : 0x0
 Interface    : GigabitEthernet0/0/0             Flags        : D
```

```
Destination    : 2001:10:10:24:2E0:FCFF:FE89:2897  PrefixLength : 128
NextHop        : ::1                                Preference   : 0
Cost           : 0                                  Protocol     : Direct
RelayNextHop   : ::                                 TunnelID     : 0x0
Interface      : GigabitEthernet0/0/0               Flags        : D

Destination    : FE80::                             PrefixLength : 10
NextHop        : ::                                 Preference   : 0
Cost           : 0                                  Protocol     : Direct
RelayNextHop   : ::                                 TunnelID     : 0x0
Interface      : NULL0                              Flags        : D
```

步骤 3 配置静态默认路由

在创建静态默认路由时，使用的命令语法与静态路由相同，只不过此时的目的 IPv6 地址和掩码应该配置为:: 0。以 AR3 为例，例 14-29 中展示了具体的配置命令。

例 14-29 在 AR3 上配置静态默认路由

```
[AR3]ipv6 route-static :: 0 2001:10:10:23::2
```

配置完成后，读者可以查看 IPv6 路由表，来确认静态路由的配置，详见例 14-30。此时，读者可以回顾例 14-20 中根据 DHCPv6 学习到的默认路由，通过对比可以发现通过 DHCPv6 学习到的默认路由优先级为 64，低于手动配置的静态默认路由 60。

例 14-30 查看 IPv6 路由表中的静态默认路由

```
[AR3]display ipv6 routing-table
Routing Table : Public
        Destinations : 5      Routes : 5

Destination    : ::                                 PrefixLength : 0
NextHop        : 2001:10:10:23::2                   Preference   : 60
Cost           : 0                                  Protocol     : Static
RelayNextHop   : ::                                 TunnelID     : 0x0
Interface      : GigabitEthernet0/0/0               Flags        : RD

Destination    : ::1                                PrefixLength : 128
NextHop        : ::1                                Preference   : 0
Cost           : 0                                  Protocol     : Direct
RelayNextHop   : ::                                 TunnelID     : 0x0
Interface      : InLoopBack0                        Flags        : D

Destination    : 2001:10:10:23::                    PrefixLength : 64
NextHop        : 2001:10:10:23::3                   Preference   : 0
Cost           : 0                                  Protocol     : Direct
RelayNextHop   : ::                                 TunnelID     : 0x0
Interface      : GigabitEthernet0/0/0               Flags        : D

Destination    : 2001:10:10:23::3                   PrefixLength : 128
NextHop        : ::1                                Preference   : 0
Cost           : 0                                  Protocol     : Direct
RelayNextHop   : ::                                 TunnelID     : 0x0
Interface      : GigabitEthernet0/0/0               Flags        : D

Destination    : FE80::                             PrefixLength : 10
NextHop        : ::                                 Preference   : 0
Cost           : 0                                  Protocol     : Direct
RelayNextHop   : ::                                 TunnelID     : 0x0
Interface      : NULL0                              Flags        : D
```

步骤 4 配置静态汇总路由

在 AR1 上，我们需要使用一条静态路由汇总去往 AR3 和 AR4 的两条路由。对于本实验中需要汇总的两个网段，我们可以使用 2001:10:10:20::/61 进行汇总，如下。

① 2001:10:10:23::/64。

② 2001:10:10:24::/64。

我们通过如图 14-12 展示汇总路由的计算过程。

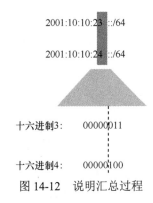

图 14-12 说明汇总过程

首先通过观察，我们发现这两个网段只有标记了阴影的部分有区别，因此我们需要将这一部分转换为二进制，以方便我们确认这两个网段中完全相同的字符位数。如上图所示，在将十六进制数值 3 和 4 转换为二进制后，我们可以看出前 5 位二进制是相同的，后面 3 位二进制有所区别。因此我们可以将汇总地址的前缀长度缩短至 3 位，使之成为 /61。因此得出汇总地址为 2001:10:10:20::/61。需要注意的是，本例的汇总方式会使汇总网段中涵盖的网段数量比实际上需要涵盖的网段数量多，因此，有可能会错误地涵盖不应该涵盖的网段。本书说明汇总网段的计算方法，读者在工作和练习中始终需要考虑汇总网段中是否涵盖了其他网段。

例 14-31 中展示了 AR1 上的静态汇总路由配置命令。

例 14-31　在 AR1 上配置静态汇总路由

```
[AR1]ipv6 route-static 2001:10:10:20:: 61 2001:10:10:12::2
```

配置完成后读者可以在 AR1 的 IPv6 路由表中查看这条路由，详见例 14-32。

例 14-32　查看 IPv6 路由表中的静态汇总路由

```
[AR1]display ipv6 routing-table
Routing Table : Public
        Destinations : 5      Routes : 5

 Destination  : ::1                         PrefixLength : 128
 NextHop      : ::1                         Preference   : 0
 Cost         : 0                           Protocol     : Direct
 RelayNextHop : ::                          TunnelID     : 0x0
 Interface    : InLoopBack0                 Flags        : D

 Destination  : 2001:10:10:12::             PrefixLength : 64
 NextHop      : 2001:10:10:12::1            Preference   : 0
 Cost         : 0                           Protocol     : Direct
 RelayNextHop : ::                          TunnelID     : 0x0
 Interface    : GigabitEthernet0/0/0        Flags        : D

 Destination  : 2001:10:10:12::1            PrefixLength : 128
 NextHop      : ::1                         Preference   : 0
 Cost         : 0                           Protocol     : Direct
 RelayNextHop : ::                          TunnelID     : 0x0
 Interface    : GigabitEthernet0/0/0        Flags        : D

 Destination  : 2001:10:10:20::             PrefixLength : 61
 NextHop      : 2001:10:10:12::2            Preference   : 60
 Cost         : 0                           Protocol     : Static
 RelayNextHop : ::                          TunnelID     : 0x0
 Interface    : GigabitEthernet0/0/0        Flags        : RD

 Destination  : FE80::                      PrefixLength : 10
 NextHop      : ::                          Preference   : 0
 Cost         : 0                           Protocol     : Direct
 RelayNextHop : ::                          TunnelID     : 0x0
 Interface    : NULL0                       Flags        : D
```

步骤 5 验证连通性

我们通过在 AR1 上对 AR3 和 AR4 发起 ping 测试，以及在 AR3 上对 AR4 发起 ping 测试，来验证全网连通性。

例 14-33 中展示了 AR1 上的 ping 测试，从命令输出可以看出 AR1 与 AR3 和 AR4 之间已建立连通性。

例 14-33 在 AR1 上进行验证

```
[AR1]ping ipv6 2001:10:10:23::3
  PING 2001:10:10:23::3 : 56  data bytes, press CTRL_C to break
    Reply from 2001:10:10:23::3
    bytes=56 Sequence=1 hop limit=63  time = 40 ms
    Reply from 2001:10:10:23::3
    bytes=56 Sequence=2 hop limit=63  time = 60 ms
    Reply from 2001:10:10:23::3
    bytes=56 Sequence=3 hop limit=63  time = 40 ms
    Reply from 2001:10:10:23::3
    bytes=56 Sequence=4 hop limit=63  time = 30 ms
    Reply from 2001:10:10:23::3
    bytes=56 Sequence=5 hop limit=63  time = 40 ms

  --- 2001:10:10:23::3 ping statistics ---
    5 packet(s) transmitted
    5 packet(s) received
    0.00% packet loss
    round-trip min/avg/max = 30/42/60 ms

[AR1]ping ipv6 2001:10:10:24::4
  PING 2001:10:10:24::4 : 56  data bytes, press CTRL_C to break
    Reply from 2001:10:10:24::4
    bytes=56 Sequence=1 hop limit=63  time = 30 ms
    Reply from 2001:10:10:24::4
    bytes=56 Sequence=2 hop limit=63  time = 40 ms
    Reply from 2001:10:10:24::4
    bytes=56 Sequence=3 hop limit=63  time = 50 ms
    Reply from 2001:10:10:24::4
    bytes=56 Sequence=4 hop limit=63  time = 40 ms
    Reply from 2001:10:10:24::4
    bytes=56 Sequence=5 hop limit=63  time = 30 ms

  --- 2001:10:10:24::4 ping statistics ---
    5 packet(s) transmitted
    5 packet(s) received
    0.00% packet loss
    round-trip min/avg/max = 30/38/50 ms
```

例 14-34 中展示了 AR3 上的 ping 测试，从命令输出可以看出 AR3 与 AR4 之间已建立连通性。

例 14-34 在 AR3 上进行验证

```
[AR3]ping ipv6 2001:10:10:24::4
  PING 2001:10:10:24::4 : 56  data bytes, press CTRL_C to break
    Reply from 2001:10:10:24::4
    bytes=56 Sequence=1 hop limit=63  time = 40 ms
    Reply from 2001:10:10:24::4
    bytes=56 Sequence=2 hop limit=63  time = 30 ms
    Reply from 2001:10:10:24::4
    bytes=56 Sequence=3 hop limit=63  time = 30 ms
    Reply from 2001:10:10:24::4
    bytes=56 Sequence=4 hop limit=63  time = 30 ms
    Reply from 2001:10:10:24::4
    bytes=56 Sequence=5 hop limit=63  time = 30 ms

  --- 2001:10:10:24::4 ping statistics ---
    5 packet(s) transmitted
    5 packet(s) received
    0.00% packet loss
    round-trip min/avg/max = 30/32/40 ms
```

第 15 章
网络编程与自动化基础

本章主要内容

15.1 实验介绍

15.2 实验配置任务

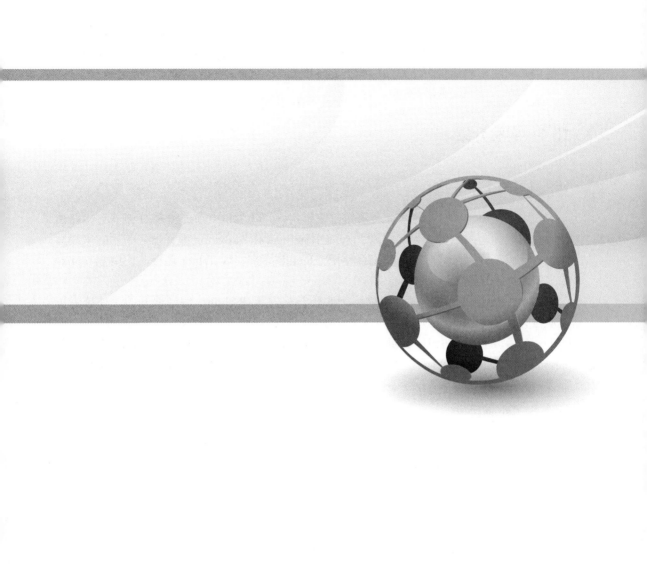

网络编程与自动化能够实现设备的自动巡检、配置、排错等工作，减少对人工的依赖，提高工作效率。网络工程师现在面临一个新的挑战，即掌握 Python 等计算机语言的代码编写能力，网络工程师将网络设备的配置文件编写为 Python 代码，并将其推送到设备上。

15.1 实验介绍

15.1.1 关于本实验

在本实验中，我们会为读者展示如何使用代码，以 Telnet 方式和 SSH 方式连接路由器，并读取和写入配置。本章将以 Jupyter Notebook 作为 Python 编译器，使用它来编写并执行代码。同时在代码中，我们会使用 telnetlib 库来实现 Telnet 连接，使用 paramiko 库来实现 SSH 连接。

本章将从头开始引导读者完成这个实验，从安装 Python、编译器开始，使读者可以从零开始搭建 Python 环境并练习网络编程与自动化技能。本章默认读者已掌握了 Python 编程的基础知识，因此不会对 Python 的语法进行介绍。对于 telnetlib 库和 paramiko 库的介绍也仅限于本实验所使用到的一些代码，更多内容读者可以在网上进行搜索并学习。

15.1.2 实验目的

- 搭建 Python 环境。
- 了解 telnetlib 库的用法。
- 了解 paramiko 库的用法。

15.1.3 实验组网介绍

本章实验拓扑如图 15-1 所示。

图 15-1　本章实验拓扑

在本章的实验拓扑中，我们仅连接一台路由器，读者可以按照本地实验环境的具体情况，实现本地 PC 与路由器之间的连通。图 15-1 中以云表示本地 PC。

表 15-1 中列出了本章使用的网络地址。

表 15-1　本章使用的网络地址

设备	接口	IP 地址	子网掩码	默认网关
AR1	G0/0/0	192.168.44.10	255.255.255.0	—
Cloud	—	192.168.44.1	255.255.255.0	—

15.1.4 实验任务列表

配置任务 1：搭建 Python 基础环境。
配置任务 2：通过 Telnet 获取路由器配置。
配置任务 3：通过 SSH 更改路由器配置。

15.2 实验配置任务

15.2.1 搭建 Python 基础环境

在这个实验中，我们需要在本地 PC 上安装 Python。首先我们逐步展示 Python 环境的搭建。在安装好 Python 后，我们还需要安装编译器，读者可以使用自己习惯使用的编译器。本章使用的编译器是 Jupyter Notebook。

步骤 1 搭建 Python 环境

在搭建 Python 环境时，我们可以在 Python 官方网站中选择适合本地 PC 操作系统的 Python 版本。以 Windows 系统为例，官方网站的 Windows 版本 Python 下载界面如图 15-2 所示。

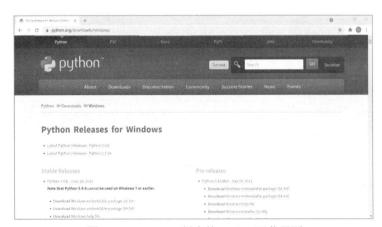

图 15-2 Windows 版本的 Python 下载界面

在确定 Python 版本时，读者首先需要考虑本地 PC 的操作系统版本，如 64 位或者 32 位。在 Python 版本中，x86-64 代表匹配的 Windows 版本为 64 位，x86 代表匹配的 Windows 版本为 32 位。

其次，Python 官方网站提供了以下 3 种格式的安装版本。

① executable installer：离线版，需要下载到本地后进行安装。对于刚入门的读者来说，建议下载离线版安装，因为软件在安装过程中会自动设置所需的环境变量。

② web-absed installer：在线安装版，在安装过程中需要保持联网状态。

③ embeddable zip file：便携版，解压后进行安装。

本章使用的是离线版，下载后的文件名为 python-3.9.0-amd64.exe。读者若选择其他版本的 Python，下载后的文件名会有所区别。下载完成后，双击可执行文件，读者可以看到如图 15-3 所示的安装界面。

图 15-3　安装 Python 界面（一）

在这个界面中，我们勾选底部的 Add Python 3.9 to PATH，让安装程序添加系统变量，然后点击 Customize installation（自定义安装）后，可以看到如图 15-4 所示界面。

图 15-4　安装 Python 界面（二）

将这个界面中的设置保持默认状态，单击 Next 按钮后，可以看到如图 15-5 所示界面。

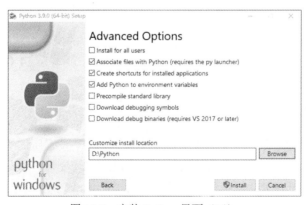

图 15-5　安装 Python 界面（三）

在这个界面中,读者可以更改 Python 的安装路径,本章将 Python 安装在 D 盘根目录中的 Python 文件夹内。另外,读者应先确认选项 Add Python to environment variables 已经勾选,然后单击 Install 按钮,开始自动安装,如图 15-6 所示。

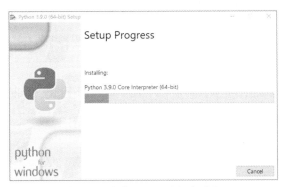

图 15-6　安装 Python 界面(四)

等待一段时间,进度条走完后 Python 就安装成功了,读者可以看到如图 15-7 所示界面。

图 15-7　安装 Python 界面(五)

安装完成后我们可以通过简单的命令来测试一下是否可以正常调用 Python。打开本地 PC 的命令提示符界面:读者可以按下 Win+R 组合键打开"运行"对话框,并在其中输入 cmd 后按下回车键。打开的命令提示符界面如图 15-8 所示,图中所示的路径可能与读者本地环境不同。

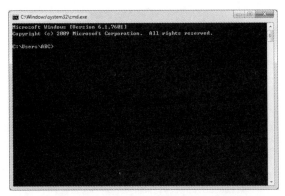

图 15-8　打开本地 PC 的命令提示符界面

在命令提示符界面中输入命令 **python**，若可以看到 Python 版本信息，则说明 Python 已安装成功，详见例 15-1。从命令输出内容中我们可以看到刚安装的 Python 版本为 3.9.0，最后一行阴影部分突出显示的符号（>>>）表示我们已经进入了 Python 虚拟环境。

例 15-1　调用 Python 虚拟环境

```
c:\Users\ABC>python
Python 3.9.0 (tags/v3.9.0:9cf6752, Oct  5 2020, 15:34:40) [MSC v.1927 64 bit (AMD64)] on win32
Type "help", "copyright", "credits" or "license" for more information.
>>>
```

在 Python 虚拟环境中，读者可以试着使用输入代码让 Python 进行回应，比如让 Python 显示出"Hello Huawei!"，详见例 15-2。

例 15-2　测试 Python 虚拟环境

```
c:\Users\ABC>python
Python 3.9.0 (tags/v3.9.0:9cf6752, Oct  5 2020, 15:34:40) [MSC v.1927 64 bit (AMD64)] on win32
Type "help", "copyright", "credits" or "license" for more information.
>>> print('Hello Huawei!')
Hello Huawei!
>>>
```

读者需要输入的命令是 **print('Hello Huawei!')**，print()函数的意思是让程序返回相应的信息，这条命令的意思是返回字符串"Hello Huawei!"。此时，我们可以退出 Python 虚拟环境，使用 Ctrl+Z 组合键后，屏幕上会出现"^Z"，此时按下回车键即可退出，详见例 15-3。

例 15-3　退出 Python 虚拟环境

```
c:\Users\ABC>python
Python 3.9.0 (tags/v3.9.0:9cf6752, Oct  5 2020, 15:34:40) [MSC v.1927 64 bit (AMD64)] on win32
Type "help", "copyright", "credits" or "license" for more information.
>>> print('Hello Huawei!')
Hello Huawei!
>>> ^Z

c:\Users\ABC>
```

退出后，我们需要测试 pip 是否与 Python 一起安装完成，很多 Python 的发行版中已经预装了 pip。pip 是用来安装第三方库的软件包管理系统，我们后续会使用它来安装 Jupyter Notebook 编译器和 paramiko 库。我们在本地 PC 的命令提示符中输入命令 **pip**，详见例 15-4。如果我们没有得到报错信息并且看到与 pip 相关的命令列表，说明 pip 已准备就绪，我们可以进行下一步的编译器安装。

例 15-4　检查 pip 工具

```
c:\Users\ABC>pip

Usage:
  pip <command> [options]

Commands:
  install                     Install packages.
  download                    Download packages.
  uninstall                   Uninstall packages.
  freeze                      Output installed packages in requirements format.
  list                        List installed packages.
  show                        Show information about installed packages.
  check                       Verify installed packages have compatible dependencies.
  config                      Manage local and global configuration.
  search                      Search PyPI for packages.
  cache                       Inspect and manage pip's wheel cache.
  wheel                       Build wheels from your requirements.
  hash                        Compute hashes of package archives.
```

```
    completion                  A helper command used for command completion.
    debug                       Show information useful for debugging.
    help                        Show help for commands.

General Options:
  -h, --help                    Show help.
  --isolated                    Run pip in an isolated mode, ignoring environment variables
and user configuration.
  -v, --verbose                 Give more output. Option is additive, and can be used up to 3
times.
  -V, --version                 Show version and exit.
  -q, --quiet                   Give less output. Option is additive, and can be used up to 3
times (corresponding to
                                WARNING, ERROR, and CRITICAL logging levels).
  --log <path>                  Path to a verbose appending log.
  --no-input                    Disable prompting for input.
  --proxy <proxy>               Specify a proxy in the form [user:passwd@]proxy.server:port.
  --retries <retries>           Maximum number of retries each connection should attempt
(default 5 times).
  --timeout <sec>               Set the socket timeout (default 15 seconds).
  --exists-action <action>      Default action when a path already exists: (s)witch,
(i)gnore, (w)ipe, (b)ackup,
                                (a)bort.
  --trusted-host <hostname>     Mark this host or host:port pair as trusted, even though it
does not have valid or any
                                HTTPS.
  --cert <path>                 Path to alternate CA bundle.
  --client-cert <path>          Path to SSL client certificate, a single file containing the
private key and the
                                certificate in PEM format.
  --cache-dir <dir>             Store the cache data in <dir>.
  --no-cache-dir                Disable the cache.
  --disable-pip-version-check
                                Don't periodically check PyPI to determine whether a new
version of pip is available for
                                download. Implied with --no-index.
  --no-color                    Suppress colored output
  --no-python-version-warning
                                Silence deprecation warnings for upcoming unsupported
Pythons.
  --use-feature <feature>       Enable new functionality, that may be backward incompatible.
  --use-deprecated <feature>    Enable deprecated functionality, that will be removed in the
future.

c:\Users\ABC>
```

步骤 2　安装编译器

下面将展示如何安装 Jupyter Notebook 编译器,读者也可以根据自己的习惯安装不同的编译器。以 Jupyter Notebook 为例,读者可以访问 Jupyter 官方网站查看安装和应用指南,图 15-9 中展示了使用 pip 进行安装时的安装指南。

图 15-9　Jupyter Notebook 安装指南

根据上述指南，我们在本地 PC 的命令提示符中输入命令 **pip install jupyterlab**，读者会看到例 15-5 所示的安装过程，阴影部分表示安装已完成。

例 15-5 通过 pip 安装 Jupyter Notebook

```
c:\Users\ABC>pip install jupyterlab
Collecting jupyterlab
  Downloading jupyterlab-3.0.16-py3-none-any.whl (8.2 MB)
     |████████████████████████████████| 8.2 MB 1.6 MB/s
Collecting jupyterlab-server~=2.3
  Downloading jupyterlab_server-2.6.1-py3-none-any.whl (56 kB)
     |████████████████████████████████| 56 kB 2.0 MB/s
Collecting ipython
  Downloading ipython-7.25.0-py3-none-any.whl (786 kB)
     |████████████████████████████████| 786 kB 6.4 MB/s
Collecting nbclassic~=0.2
  Downloading nbclassic-0.3.1-py3-none-any.whl (18 kB)
Collecting jupyter-server~=1.4
  Downloading jupyter_server-1.9.0-py3-none-any.whl (389 kB)
     |████████████████████████████████| 389 kB 6.4 MB/s
Collecting packaging
  Downloading packaging-21.0-py3-none-any.whl (40 kB)
     |████████████████████████████████| 40 kB 1.3 MB/s
Collecting jupyter-core
  Downloading jupyter_core-4.7.1-py3-none-any.whl (82 kB)
     |████████████████████████████████| 82 kB 368 kB/s
Collecting tornado>=6.1.0
  Downloading tornado-6.1-cp39-cp39-win_amd64.whl (422 kB)
     |████████████████████████████████| 422 kB ...
Collecting jinja2>=2.1
  Downloading Jinja2-3.0.1-py3-none-any.whl (133 kB)
     |████████████████████████████████| 133 kB 1.1 MB/s
Collecting json5
  Downloading json5-0.9.6-py2.py3-none-any.whl (18 kB)
Collecting babel
  Downloading Babel-2.9.1-py2.py3-none-any.whl (8.8 MB)
     |████████████████████████████████| 8.8 MB 1.2 MB/s
Collecting jsonschema>=3.0.1
  Downloading jsonschema-3.2.0-py2.py3-none-any.whl (56 kB)
     |████████████████████████████████| 56 kB 3.8 MB/s
Collecting requests
  Downloading requests-2.26.0-py2.py3-none-any.whl (62 kB)
     |████████████████████████████████| 62 kB 378 kB/s
Collecting decorator
  Downloading decorator-5.0.9-py3-none-any.whl (8.9 kB)
Collecting jedi>=0.16
  Downloading jedi-0.18.0-py2.py3-none-any.whl (1.4 MB)
     |████████████████████████████████| 1.4 MB 3.3 MB/s
Collecting prompt-toolkit!=3.0.0,!=3.0.1,<3.1.0,>=2.0.0
  Downloading prompt_toolkit-3.0.19-py3-none-any.whl (368 kB)
     |████████████████████████████████| 368 kB 3.2 MB/s
Collecting pickleshare
  Downloading pickleshare-0.7.5-py2.py3-none-any.whl (6.9 kB)
Collecting matplotlib-inline
  Downloading matplotlib_inline-0.1.2-py3-none-any.whl (8.2 kB)
Collecting pygments
  Downloading Pygments-2.9.0-py3-none-any.whl (1.0 MB)
     |████████████████████████████████| 1.0 MB 6.8 MB/s
Collecting traitlets>=4.2
  Downloading traitlets-5.0.5-py3-none-any.whl (100 kB)
     |████████████████████████████████| 100 kB 3.3 MB/s
Requirement already satisfied: setuptools>=18.5 in d:\python\lib\site-packages (from ipython->jupyterlab) (49.2.1)
Collecting backcall
  Downloading backcall-0.2.0-py2.py3-none-any.whl (11 kB)
Collecting colorama; sys_platform == "win32"
  Downloading colorama-0.4.4-py2.py3-none-any.whl (16 kB)
Collecting notebook<7
  Downloading notebook-6.4.0-py3-none-any.whl (9.5 MB)
     |████████████████████████████████| 9.5 MB 6.8 MB/s
```

```
Collecting jupyter-client>=6.1.1
  Downloading jupyter_client-6.1.12-py3-none-any.whl (112 kB)
     |████████████████████████████████| 112 kB 6.4 MB/s
Collecting requests-unixsocket
  Downloading requests_unixsocket-0.2.0-py2.py3-none-any.whl (11 kB)
Collecting terminado>=0.8.3
  Downloading terminado-0.10.1-py3-none-any.whl (14 kB)
Collecting prometheus-client
  Downloading prometheus_client-0.11.0-py2.py3-none-any.whl (56 kB)
     |████████████████████████████████| 56 kB 3.8 MB/s
Collecting websocket-client
  Downloading websocket_client-1.1.0-py2.py3-none-any.whl (68 kB)
     |████████████████████████████████| 68 kB 1.9 MB/s
Collecting anyio<4,>=3.1.0
  Downloading anyio-3.2.1-py3-none-any.whl (75 kB)
     |████████████████████████████████| 75 kB 605 kB/s
Collecting nbformat
  Downloading nbformat-5.1.3-py3-none-any.whl (178 kB)
     |████████████████████████████████| 178 kB 504 kB/s
Collecting argon2-cffi
  Downloading argon2_cffi-20.1.0-cp39-cp39-win_amd64.whl (42 kB)
     |████████████████████████████████| 42 kB 1.1 MB/s
Collecting pyzmq>=17
  Downloading pyzmq-22.1.0-cp39-cp39-win_amd64.whl (1.0 MB)
     |████████████████████████████████| 1.0 MB 156 kB/s
Collecting nbconvert
  Downloading nbconvert-6.1.0-py3-none-any.whl (551 kB)
     |████████████████████████████████| 551 kB 19 kB/s
Collecting ipython-genutils
  Downloading ipython_genutils-0.2.0-py2.py3-none-any.whl (26 kB)
Collecting Send2Trash
  Downloading Send2Trash-1.7.1-py3-none-any.whl (17 kB)
Collecting pyparsing>=2.0.2
  Downloading pyparsing-2.4.7-py2.py3-none-any.whl (67 kB)
     |████████████████████████████████| 67 kB 193 kB/s
Collecting pywin32>=1.0; sys_platform == "win32"
  Downloading pywin32-301-cp39-cp39-win_amd64.whl (9.3 MB)
     |████████████████████████████████| 9.3 MB 2.2 MB/s
Collecting MarkupSafe>=2.0
  Downloading MarkupSafe-2.0.1-cp39-cp39-win_amd64.whl (14 kB)
Collecting pytz>=2015.7
  Downloading pytz-2021.1-py2.py3-none-any.whl (510 kB)
     |████████████████████████████████| 510 kB 2.2 MB/s
Collecting pyrsistent>=0.14.0
  Downloading pyrsistent-0.18.0-cp39-cp39-win_amd64.whl (62 kB)
     |████████████████████████████████| 62 kB 511 kB/s
Collecting attrs>=17.4.0
  Downloading attrs-21.2.0-py2.py3-none-any.whl (53 kB)
     |████████████████████████████████| 53 kB 1.9 MB/s
Collecting six>=1.11.0
  Downloading six-1.16.0-py2.py3-none-any.whl (11 kB)
Collecting idna<4,>=2.5; python_version >= "3"
  Downloading idna-3.2-py3-none-any.whl (59 kB)
     |████████████████████████████████| 59 kB 1.7 MB/s
Collecting charset-normalizer~=2.0.0; python_version >= "3"
  Downloading charset_normalizer-2.0.3-py3-none-any.whl (35 kB)
Collecting certifi>=2017.4.17
  Downloading certifi-2021.5.30-py2.py3-none-any.whl (145 kB)
     |████████████████████████████████| 145 kB 2.2 MB/s
Collecting urllib3<1.27,>=1.21.1
  Downloading urllib3-1.26.6-py2.py3-none-any.whl (138 kB)
     |████████████████████████████████| 138 kB 819 kB/s
Collecting parso<0.9.0,>=0.8.0
  Downloading parso-0.8.2-py2.py3-none-any.whl (94 kB)
     |████████████████████████████████| 94 kB 1.7 MB/s
Collecting wcwidth
  Downloading wcwidth-0.2.5-py2.py3-none-any.whl (30 kB)
Collecting ipykernel
  Downloading ipykernel-6.0.3-py3-none-any.whl (122 kB)
     |████████████████████████████████| 122 kB 123 kB/s
Collecting python-dateutil>=2.1
  Downloading python_dateutil-2.8.2-py2.py3-none-any.whl (247 kB)
     |████████████████████████████████| 247 kB 175 kB/s
```

```
Collecting pywinpty>=1.1.0; os_name == "nt"
  Downloading pywinpty-1.1.3-cp39-none-win_amd64.whl (1.3 MB)
     |████████████████████████████████| 1.3 MB 327 kB/s
Collecting sniffio>=1.1
  Downloading sniffio-1.2.0-py3-none-any.whl (10 kB)
Collecting cffi>=1.0.0
  Downloading cffi-1.14.6-cp39-cp39-win_amd64.whl (180 kB)
     |████████████████████████████████| 180 kB 182 kB/s
Collecting mistune<2,>=0.8.1
  Downloading mistune-0.8.4-py2.py3-none-any.whl (16 kB)
Collecting entrypoints>=0.2.2
  Downloading entrypoints-0.3-py2.py3-none-any.whl (11 kB)
Collecting bleach
  Downloading bleach-3.3.1-py2.py3-none-any.whl (146 kB)
     |████████████████████████████████| 146 kB 364 kB/s
Collecting pandocfilters>=1.4.1
  Downloading pandocfilters-1.4.3.tar.gz (16 kB)
Collecting testpath
  Downloading testpath-0.5.0-py3-none-any.whl (84 kB)
     |████████████████████████████████| 84 kB 471 kB/s
Collecting defusedxml
  Downloading defusedxml-0.7.1-py2.py3-none-any.whl (25 kB)
Collecting nbclient<0.6.0,>=0.5.0
  Downloading nbclient-0.5.3-py3-none-any.whl (82 kB)
     |████████████████████████████████| 82 kB 254 kB/s
Collecting jupyterlab-pygments
  Downloading jupyterlab_pygments-0.1.2-py2.py3-none-any.whl (4.6 kB)
Collecting debugpy<2.0,>=1.0.0
  Downloading debugpy-1.4.0-cp39-cp39-win_amd64.whl (4.4 MB)
     |████████████████████████████████| 4.4 MB 1.1 MB/s
Collecting pycparser
  Downloading pycparser-2.20-py2.py3-none-any.whl (112 kB)
     |████████████████████████████████| 112 kB 6.8 MB/s
Collecting webencodings
  Downloading webencodings-0.5.1-py2.py3-none-any.whl (11 kB)
Collecting nest-asyncio
  Downloading nest_asyncio-1.5.1-py3-none-any.whl (5.0 kB)
Collecting async-generator
  Downloading async_generator-1.10-py3-none-any.whl (18 kB)
Using legacy 'setup.py install' for pandocfilters, since package 'wheel' is not installed.
Installing collected packages: six, python-dateutil, pywin32, ipython-genutils, traitlets,
jupyter-core, tornado, pyzmq, jupyter-client, idna, charset-normalizer, certifi, urllib3,
requests, requests-unixsocket, pywinpty, terminado, prometheus-client, websocket-client,
sniffio, anyio, pyrsistent, attrs, jsonschema, nbformat, pycparser, cffi, argon2-cffi,
mistune, entrypoints, webencodings, pyparsing, packaging, bleach, pygments, pandocfilters,
testpath, defusedxml, nest-asyncio, async-generator, nbclient, jupyterlab-pygments,
MarkupSafe, jinja2, nbconvert, Send2Trash, jupyter-server, json5, pytz, babel, jupyterlab-
server, decorator, parso, jedi, wcwidth, prompt-toolkit, pickleshare, matplotlib-inline,
backcall, colorama, ipython, debugpy, ipykernel, notebook, nbclassic, jupyterlab
    Running setup.py install for pandocfilters ... done
Successfully installed MarkupSafe-2.0.1 Send2Trash-1.7.1 anyio-3.2.1 argon2-cffi-20.1.0
async-generator-1.10 attrs-21.2.0 babel-2.9.1 backcall-0.2.0 bleach-3.3.1 certifi-2021.5.30
cffi-1.14.6 charset-normalizer-2.0.3 colorama-0.4.4 debugpy-1.4.0 decorator-5.0.9
defusedxml-0.7.1 entrypoints-0.3 idna-3.2 ipykernel-6.0.3 ipython-7.25.0 ipython-genutils-
0.2.0 jedi-0.18.0 jinja2-3.0.1 json5-0.9.6 jsonschema-3.2.0 jupyter-client-6.1.12 jupyter-
core-4.7.1 jupyter-server-1.9.0 jupyterlab-3.0.16 jupyterlab-pygments-0.1.2 jupyterlab-
server-2.6.1 matplotlib-inline-0.1.2 mistune-0.8.4 nbclassic-0.3.1 nbclient-0.5.3
nbconvert-6.1.0 nbformat-5.1.3 nest-asyncio-1.5.1 notebook-6.4.0 packaging-21.0
pandocfilters-1.4.3 parso-0.8.2 pickleshare-0.7.5 prometheus-client-0.11.0 prompt-toolkit-
3.0.19 pycparser-2.20 pygments-2.9.0 pyparsing-2.4.7 pyrsistent-0.18.0 python-dateutil-
2.8.2 pytz-2021.1 pywin32-301 pywinpty-1.1.3 pyzmq-22.1.0 requests-2.26.0 requests-
unixsocket-0.2.0 six-1.16.0 sniffio-1.2.0 terminado-0.10.1 testpath-0.5.0 tornado-6.1
traitlets-5.0.5 urllib3-1.26.6 wcwidth-0.2.5 webencodings-0.5.1 websocket-client-1.1.0

c:\Users\ABC>
```

接着，我们按照使用指南中的命令（**jupyter-lab**）运行 Jupyter Notebook，详见例 15-6。

例 15-6　运行 Jupyter Notebook

```
c:\Users\ABC>jupyter-lab
[I 2021-07-21 16:34:44.513 ServerApp] jupyterlab | extension was successfully linked.
[I 2021-07-21 16:34:44.527 ServerApp] Writing Jupyter server cookie secret to C:\Users\ABC
\AppData\Roaming\jupyter\runtime\jupyter_cookie_secret
[W 2021-07-21 16:34:44.549 ServerApp] The 'min_open_files_limit' trait of a ServerApp
instance expected an int, not the NoneType None.
[I 2021-07-21 16:34:45.020 ServerApp] nbclassic | extension was successfully loaded.
[I 2021-07-21 16:34:45.021 LabApp] Jupyterlab extension loaded from d:\python\lib\site-
packages\jupyterlab
[I 2021-07-21 16:34:45.022 LabApp] Jupyterlab application directory is D:\Python\share\
jupyter\lab
[I 2021-07-21 16:34:45.025 ServerApp] jupyterlab | extension was successfully loaded.
[I 2021-07-21 16:34:45.026 ServerApp] Serving notebooks from local directory: c:\
[I 2021-07-21 16:34:45.026 ServerApp] Jupyter Server 1.9.0 is running at:
[I 2021-07-21 16:34:45.026 ServerApp] http://localhost:8888/lab?token = 68e6ab812ba1cfb705cb
38652ac696723155c4d245221ebc
[I 2021-07-21 16:34:45.026 ServerApp] or http://127.0.0.1:8888/lab?token = 68e6ab812ba1cfb
705cb38652ac696723155c4d245221ebc
[I 2021-07-21 16:34:45.026 ServerApp] Use Control-C to stop this server and shut down all
kernels (twice to skip confirmation).
[C 2021-07-21 16:34:45.069 ServerApp]

    To access the server, open this file in a browser:
        file:///C:/Users/ABC/AppData/Roaming/jupyter/runtime/jpserver-5072-open.html
    Or copy and paste one of these URLs:
        http://localhost:8888/lab?token=68e6ab812ba1cfb705cb38652ac696723155c4d245221ebc
     or http://127.0.0.1:8888/lab?token=68e6ab812ba1cfb705cb38652ac696723155c4d245221ebc
[W 2021-07-21 16:35:02.893 LabApp] Could not determine jupyterlab build status without nodejs
```

上述运行过程结束后，本地 PC 上会自动弹出默认浏览器并显示 Jupyter Notebook 的操作界面，如图 15-10 所示。

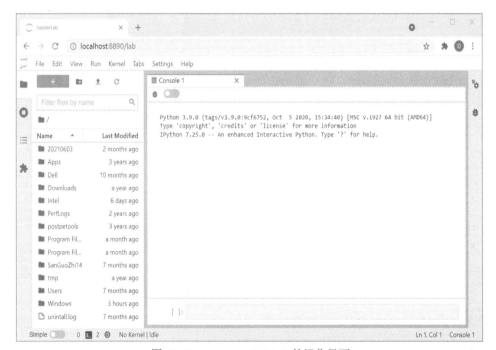

图 15-10　Jupyter Notebook 的操作界面

在这个操作界面中，我们需要在右侧的输入框中输入代码，使用组合键 Shift+Enter 来运行代码。再次以 **print('Hello Huawei!')** 为例，我们在代码输入框中输入代码，如图 15-11 所示。

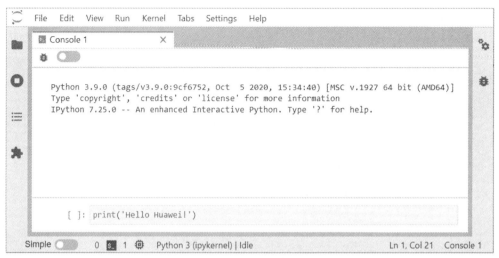

图 15-11　编写简单代码

此时按下 Shift+Enter 组合键后，编译器就会执行这行代码，并显示如图 15-12 所示结果。

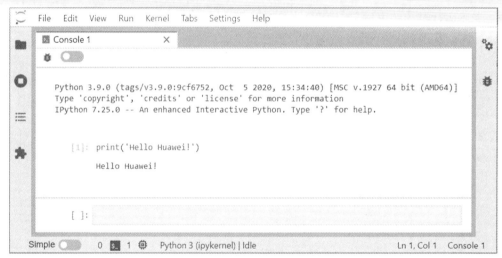

图 15-12　执行简单代码

至此，Python 基础环境已经搭建完成，读者需要根据本地网络环境，实现本地 PC 与路由器之间的连通。读者通过学习并练习前文的各种实验，相信已经能够驾轻就熟地实现本地 PC 与路由器之间的连接，因此基础网络通信的部分我们将不再展示。

15.2.2　通过 Telnet 获取路由器配置

在这个实验任务中，我们需要使用代码，以 Telnet 的方式登录路由器，并从中获取一些配置信息。读者使用密码 Huawei@123 进行 Telnet 登录，从路由器中获取接口的汇总信息。

本实验需要使用另一种华为数通设备所支持的登录设置方法，即不需用户名，仅凭密码登录 Telnet。我们可以使用以下命令进行配置。

① **user-interface vty 0 4**：系统视图命令，进入 VTY 用户界面。

② **authentication-mode password**：VTY 视图命令，将认证模式设置为使用密码进行认证。输入命令后系统会自动弹出提示，让管理员输入密码。

③ **protocol inbound telnet**：VTY 视图命令，允许接入用户的接入类型为 Telnet。

④ **user privilege level 3**：VTY 视图命令，为接入用户赋予用户级别 3。

例 15-7 中展示了路由器上启用 Telnet 功能的配置命令。

例 15-7 启用 Telnet 功能

```
[Huawei]user-interface vty 0 4
[Huawei-ui-vty0-4]authentication-mode password
Please configure the login password (maximum length 16):Huawei@123
[Huawei-ui-vty0-4]protocol inbound telnet
[Huawei-ui-vty0-4]user privilege level 3
```

我们先使用本地 PC 中的 PuTTY 终端程序验证 Telnet 功能是否已启用，与验证基础连通性相似，这一步测试通过后，可以尝试通过网络编程的方式连接路由器。与此同时，这一操作还有一个目的，就是观察网络设备回复的具体字符，在 Telnet 程序代码中会用到。

在第 1 章中，我们曾经使用 PuTTY 连接过路由器的 Console 口，这次我们要使用 Telnet 功能。图 15-13 中展示了 PuTTY 中的设置，需要填写 IP 地址 192.168.44.10，连接类型选择 Other，并在下拉菜单中选择 Telnet，选中后端口会自动变为 23。

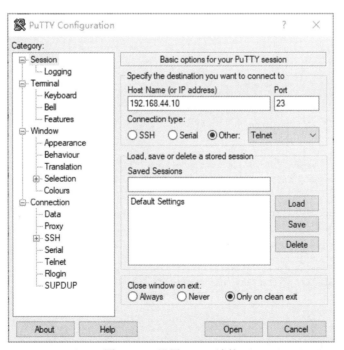

图 15-13　设置 Telnet 连接

设置完成后单击 Open 按钮会启动 Telnet 连接请求，读者需要在收到输入密码的提示后输入 Huawei@123，就可以连接路由器，如图 15-14 所示。请读者注意观察，路由器在返回"Password:"后，便等待用户输入密码登录。

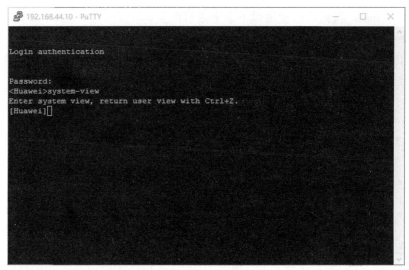

图 15-14 通过 PuTTY 测试 Telnet 设置

接下来，我们可以在 Jupyter Notebook 中创建代码，实现通过 Telnet 登录后，获取路由器接口的汇总信息。

例 15-8 中列出了本实验所使用的代码及其注释。

例 15-8 Telnet 代码及其注释

```
#导入telnetlib库和time库，telnetlib库提供了Telnet功能，time库负责处理与时间相关的操作
import telnetlib
import time

#创建变量
ip = '192.168.44.10'
pw = 'Huawei@123'

#调用telnetlib类下的Telnet()方法，并将其赋值给hcia，使后续代码可以看起来比较简明
hcia = telnetlib.Telnet(ip)

#读取到Password:后再继续执行后续代码（通过PuTTY测试获得了设备返回的精确字符），b表示将Python3中默认的
unicode编码转换为bytes，这是telnetlib库所要求的必需参数
hcia.read_until(b'Password:')

#提供登录密码，pw.encode('ascii')表示将pw所代表的字符串转换为ASCII编码,'\n'表示回车键
hcia.write(pw.encode('ascii') + b'\n')

#向设备发送所要执行的华为CLI命令，读者可以将命令更改为其他命令，也可以编写多条命令让设备执行
hcia.write(b'display ip interface brief \n')

#等待2s后将继续执行后续代码，这是为了让设备能够有时间将命令的输出内容显示完整
time.sleep(2)

#使用print()函数显示命令的输出内容，hcia.read_very_eager()表示读取尽可能多的数据，decode('ascii')表
示将读取到的数据解码为ASCII
print(hcia.read_very_eager().decode('ascii'))

#使用close()方法退出Telnet
hcia.close()
```

在 Jupyter Notebook 中，读者可以按照如图 15-15 所示输入代码。

图 15-15　编写 Telnet 代码

我们按下 Shift+Enter 后，编译器会执行该代码，以 Telnet 的方式连接路由器，在路由器中输入命令 **display ip interface brief**，然后显示出路由器返回的信息，如图 15-16 所示。

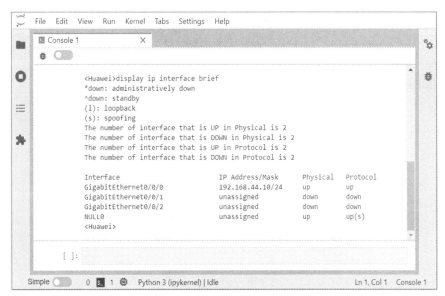

图 15-16　执行 Telnet 代码

至此，配置任务顺利完成。telnetlib 库和 time 库中还有很多未使用到的函数，对此部分内容感兴趣的读者可以自行查阅相关文档。

15.2.3　通过 SSH 更改路由器配置

在这个实验任务中，我们需要使用代码以 SSH 的方式登录路由器，并在路由器上实施一些配置。具体要求为使用用户名 python 和密码 Huawei@123 以 SSH 方式登录路由器，为路由器配置主机名为 AR1。

首先，我们可以使用以下命令来配置 SSH 功能。

① **stelnet server enable**：系统视图命令，启用 SSH 服务器功能。

② **user-interface vty 0 4**：系统视图命令，进入 VTY 用户界面。

③ **authentication-mode aaa**：VTY 视图命令，将认证模式设置为本地认证。

④ **protocol inbound ssh**：VTY 视图命令，允许接入用户的接入类型为 SSH。

⑤ **local-user** *user-name* **password cipher** *password*：AAA 视图命令，创建用户并设置密码。

⑥ **local-user** *user-name* **service-type ssh**：AAA 视图命令，将用户的接入类型设置为 SSH。

⑦ **local-user** *user-name* **privilege level** *level*：AAA 视图命令，设置用户的用户级别。

例 15-9 中展示了路由器上的 SSH 配置命令。

例 15-9　启用 SSH 功能

```
[Huawei]stelnet server enable
Info: Succeeded in starting the STELNET server.
[Huawei]user-interface vty 0 4
[Huawei-ui-vty0-4]authentication-mode aaa
[Huawei-ui-vty0-4]protocol inbound ssh
[Huawei]aaa
[Huawei-aaa]local-user python password cipher Huawei@123
Info: Add a new user.
[Huawei-aaa]local-user python service-type ssh
[Huawei-aaa]local-user python privilege level 3
```

配置完成后，我们仍通过 PuTTY 测试 SSH 的功能是否已启用。读者可以按照如图 15-17 所示设置 SSH 连接，在 IP 地址部分输入 192.168.44.10，保留默认的 SSH 设置并单击 Open 按钮。

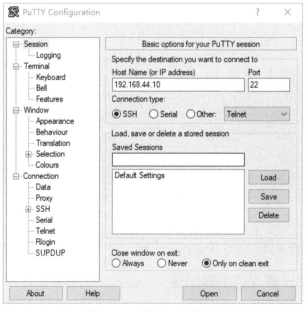

图 15-17　设置 SSH 连接

单击 Open 按钮后，读者需要按照要求输入用户名和密码，登录界面如图 15-18 所示。

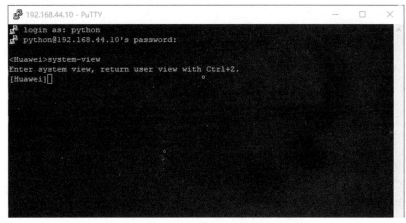

图 15-18　通过 PuTTY 测试 SSH 设置

接下来，我们通过 paramiko 库完成这个实验任务。paramiko 库的主要功能是实现 SSHv2 的连接，它同时提供了 SSH 客户端和 SSH 服务器功能。与 telnetlib 库不同的是，paramiko 库需要安装。我们应先通过 pip 来安装 paramiko，详见例 15-10。

例 15-10　安装 paramiko 库

```
c:\Users\ABC>pip3 install paramiko
Collecting paramiko
  Downloading paramiko-2.7.2-py2.py3-none-any.whl (206 kB)
     |                                | 206 kB 1.3 MB/s
Collecting pynacl>=1.0.1
  Downloading PyNaCl-1.4.0-cp35-abi3-win_amd64.whl (206 kB)
     |                                | 206 kB 726 kB/s
Collecting bcrypt>=3.1.3
  Downloading bcrypt-3.2.0-cp36-abi3-win_amd64.whl (28 kB)
Collecting cryptography>=2.5
  Downloading cryptography-3.4.7-cp36-abi3-win_amd64.whl (1.6 MB)
     |                                | 1.6 MB 819 kB/s
Requirement already satisfied: cffi>=1.4.1 in d:\python\lib\site-packages (from pynacl>=1.0.1->paramiko)(1.14.6)
Requirement already satisfied: six in d:\python\lib\site-packages (from pynacl>=1.0.1->paramiko)(1.16.0)
Requirement already satisfied: pycparser in d:\python\lib\site-packages (from cffi>=1.4.1->pynacl>=1.0.1->paramiko)(2.20)
Installing collected packages: pynacl, bcrypt, cryptography, paramiko
Successfully installed bcrypt-3.2.0 cryptography-3.4.7 paramiko-2.7.2 pynacl-1.4.0

c:\Users\ABC>
```

安装完成后，读者可以尝试在 Python 中导入该库，如果没有出现报错提醒，就说明 paramiko 已经安装完成并可使用，详见例 15-11。

例 15-11　导入 paramiko 进行测试

```
c:\Users\ABC>python
Python 3.9.0 (tags/v3.9.0:9cf6752, Oct  5 2020, 15:34:40) [MSC v.1927 64 bit (AMD64)] on win32
Type "help", "copyright", "credits" or "license" for more information.
>>> import paramiko
>>>
```

接下来，我们可以在 Jupyter Notebook 中创建代码，实现通过 SSH 登录，将路由器的主机名更改为 AR1。

例 15-12 中列出了本实验所使用的代码及其注释。

例 15-12　SSH 代码及其注释

```
#导入 paramiko 库和 time 库，paramiko 库提供了 SSH 功能，time 库负责处理与时间相关的操作
import paramiko
import time

#创建变量
ip = "192.168.44.10"
user = "python"
pw = "Huawei@123"

#调用 paramiko 类下的 SSHClient()方法，并将其赋值给 hcia，使后续代码看起来比较简明
hcia = paramiko.SSHClient()

#允许连接不在 know_hosts 文件中的主机
hcia.set_missing_host_key_policy(paramiko.AutoAddPolicy())

#提供远程主机信息：IP 地址、用户名和密码
hcia.connect(hostname=ip, username=user, password=pw)

#登录成功后显示消息提示和设备 IP 地址
print("You are on router Huawei." , ip)

#调用 CLI 命令行界面并将 invoke_shell()方法赋值给 cmd，使后续代码看起来比较简明
cmd = hcia.invoke_shell()

#向设备发送需要执行的命令
cmd.send("system-view \n")
cmd.send("sysname AR1 \n")
cmd.send("quit \n")

#等待 2s 后继续执行后续代码
time.sleep(2)

#显示所有命令内容，并设置最多显示 1000 个字符
output = cmd.recv(1000)
print(output.decode("ascii"))

#使用 close()方法退出 SSH
hcia.close()
```

在 Jupyter Notebook 中，读者可以按照如图 15-19 所示输入代码。

图 15-19　编写 SSH 代码

我们按下 Shift+Enter 后，编译器会执行该代码，以 SSH 的方式连接路由器，在路由器中执行指定命令并返回相应信息，如图 15-20 所示。

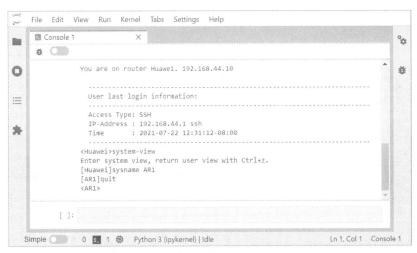

图 15-20　执行 SSH 代码

至此，配置任务顺利完成。paramiko 库中还有很多未使用到的函数，对此部分内容感兴趣的读者可以自行查阅相关文档。

第 16 章
园区网络项目实战

本章主要内容

16.1 实验介绍

16.2 实验配置任务

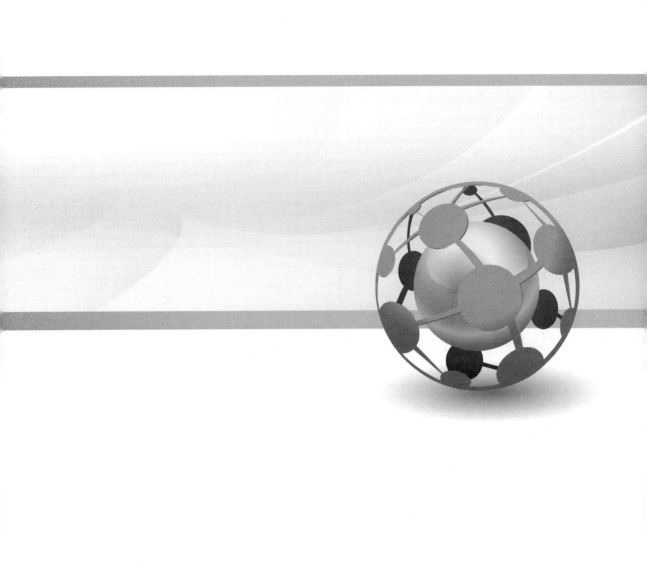

读者通过本书前文的实验练习更好地理解了每种协议和功能,并且熟悉了华为数通设备上的相关配置后,可以尝试从设计开始,按照一些需求构建园区网络。

本实验中涉及的技术都在前面的章节中进行过介绍,读者可以在此验证自己的学习成果。同时,本章会着重讲述设计部分,并不会针对每个设计结果给出相应的配置。读者如果需要配置参考,本章最后一部分会给出基本的配置。需要注意的是,本章给出的配置仅仅作为参考,并不是标准答案,更不是唯一答案。如果读者希望先按照本章的思路完成实验,可以在本章提供的基本内容之上进行扩展,考虑如何进行优化。

16.1 实验介绍

16.1.1 关于本实验

本实验要求读者按照一个中小型企业的网络和业务要求设计并部署一个中小型网络。实验中会涉及读者所学习到的各个知识点,在配置之余,本章将着重于网络设计。本实验将按照网络技术的类别,将完整的需求分解为多个实验任务。针对每个实验任务,读者需要分析需求并提出设计方案,进而在设计方案的基础上进行部署和配置。

读者需要为应用系统集成商 ABC 公司设计并部署有线及无线网络。该公司旨在提供行业信息化解决方案,共有大约 500 人,组织结构中包括管理部、大客户部、集成业务一部、集成业务二部、人力资源部、财务部,具体信息见表 16-1。

表 16-1 部门及人数

部门	人数/人
管理部	10
大客户部	120
集成业务一部	150
集成业务二部	150
人力资源部	20
财务部	10

ABC 集成商的办公环境采用开放式办公区,分为两层楼,共 4 个办公室。一层有两个开放式办公室,分别为集成业务一部和集成业务二部的办公区域。二层也有两个开放式办公室,一间办公室是大客户部的办公区域,另一间办公室是其他部门(管理部、人力资源部、财务部)的办公区域。

在 ABC 公司网络的设计中,要求读者对安全性、冗余性和可扩展性做充分考量。每个实验任务首先会站在 ABC 公司的角度上提出需求,进行分析后,设计人员再从网络工作人员的角度将企业的需求转化为网络部署设计。在实际工作中,负责提出需求的客户(比如 ABC 公司)往往不是网络专家,因此有时客户提出的需求比较笼统和模糊,需要经验丰富的网络专家对客户的需求进行解读,同时最好还能够根据客户的业务特点,提出更多具有针对性的建议。

本章暂不涉及设备选型及与此相关的成本和可扩展性考量。在实际工作中，设备选型也是项目中的重点，对此，不仅要考虑设备所提供的功能是否满足企业网络的要求，还要考虑可扩展性与成本。这需要读者不仅精通网络技术，还要熟悉产品。本章仅在本书前文所覆盖的技术范围内提出适当的需求，为读者提供设计思路，并展示如何将企业（客户）的实际情况转化为网络设计。

16.1.2 实验目的

- 了解园区网经典设计架构。
- 了解网络设计中的关键考量因素。
- 了解网络园区的生命周期。
- 熟悉网络规划的设计思路。

16.1.3 实验任务列表

实验任务 1：组网方案设计。
实验任务 2：二层网络设计。
实验任务 3：IP 地址规划。
实验任务 4：三层网络设计。
实验任务 5：WLAN 设计。
实验任务 6：网络设备调试。

16.2 实验配置任务

16.2.1 组网方案设计

目前 ABC 公司拥有不到 500 名员工，具有很好的发展前景，并且计划在未来的几年中进行扩展，比如增加员工，或者在其他省市开设分支机构，其组网方案设计步骤如下。

步骤 1 网络架构规划

组网方案是构建网络的基础，需要具有灵活性和可扩展性。对于一个中小型企业网络来说，典型的组网方案是三层模型：核心层、汇聚层和接入层。尽管 ABC 公司当前的规模并不算很大，但考虑到它未来的发展，我们根据当前的规模，可以按图 16-1 来进行规划。

图 16-1 所示的组网方案展示出了网络的三层结构。考虑到 ABC 公司当前的规模，每层的设计如下所示。

① **接入层**：每个办公室中部署一台接入层交换机，以支持该办公室中的有线终端设备联网。在实际工作中，考虑到集成公司的业务特点，多数工程师并不会在办公室中进行办公，而是会往返于各个客户之间；同时工程师基本上都会使用笔记本电脑，因此集成公司为工程师保留的有线终端接入点往往不需要很多，因此我们可以预留少部分有线端口（具体数量可以让客户提供，本章仅以 24 端口交换机作为实验设备），并提供无线接入。

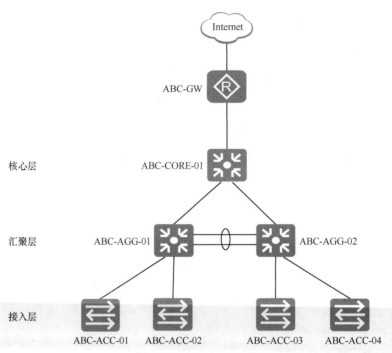

图 16-1　组网方案设计

② **汇聚层**：每个楼层部署一台汇聚层交换机，负责汇聚其下两台接入层交换机的流量。在实际工作中，读者可以考虑在接入层交换机与汇聚层交换机之间构建冗余链路，即每台接入层交换机同时连接两台汇聚层交换机。这样汇聚层不仅负责汇聚终端流量，还可以灵活地分担流量并实现相互备份。

③ **核心层**：以一台交换机作为核心层设备。在中小型企业中，互联网出口往往可以与核心层功能集成在同一台设备中。如果需要提供出口冗余或备份，可以通过两条链路连接 ISP 以实现链路级冗余。在大型企业网络中，核心层与互联网出口的角色分别由不同的设备承担（本章也将核心层与互联网出口相分离），同时双核心（两台核心层设备）的部署规划也很常见。一般来说，核心层会部署交换机以利用其强大的交换能力，互联网出口会部署路由器。而且在很多情况下，互联网出口与核心层之间还会被部署防火墙设备，以便为企业网络提供更多的安全保护。

网络架构规划图还包含设备名称，设备命名的原则首先是要有统一的格式，其次要采纳并汇总一些重要的信息。图 16-1 中网络设备的命名规则如下所示：以公司名称缩写（ABC）为前缀，核心层设备为 CORE，汇聚层设备为 AGG（英文 Aggregation 的缩写），接入层设备为 ACC（英文 Access 的缩写）。目前除了核心层之外，每一层都有多台设备，因此在设备命名中也添加了编号；出于未来扩展的考量，也为核心层设备添加了编号。在实际工作中，读者也可以考虑在此命名规则的基础上添加设备类型（比如 R 表示路由器、S 表示交换机等）和设备型号（比如 AR2200）。在为设备进行命名时，读者也可以根据实际需求考虑在名称中添加设备的物理位置信息（比如楼层和房间），以及设备所属的网络区域信息（比如数据中心）。

步骤 2 端口互联规划

在设计设备之间的互联端口时,读者需要考虑企业业务对于流量的需求,并按照需求选择适当类型和速率的端口。读者可以将本实验的端口按照图 16-2 进行规划。

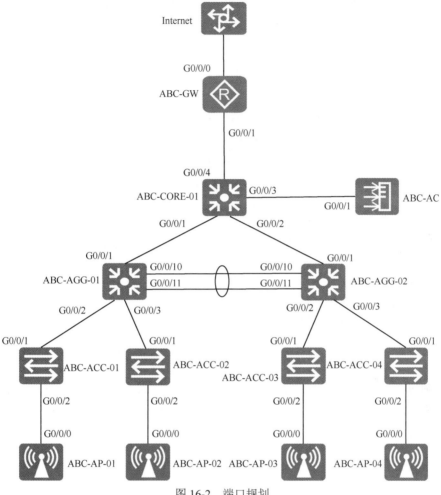

图 16-2 端口规划

对于网络端口规划,拓扑图的形式并不是最好的记录方式,因此在对这个规划进行归档时,我们可以按照表 16-2 中的格式对其进行记录。

表 16-2 设备端口规划

设备名称	端口	对端设备名称	对端端口	端口描述
ABC-GW	G0/0/0	ISP	—	TO ISP
	G0/0/1	ABC-CORE-01	G0/0/4	TO ABC-CORE-01 G0/0/4
ABC-CORE-01	G0/0/1	ABC-AGG-01	G0/0/1	TO ABC-AGG-01 G0/0/1
	G0/0/2	ABC-AGG-02	G0/0/1	TO ABC-AGG-02 G0/0/1
	G0/0/3	ABC-AC	G0/0/1	TO ABC-AC
	G0/0/4	ABC-GW	G0/0/1	TO ABC-GW G0/0/1

表 16-2 设备端口规划（续）

设备名称	端口	对端设备名称	对端端口	端口描述
ABC-AGG-01	G0/0/1	ABC-CORE-01	G0/0/1	TO ABC-CORE-01 G0/0/1
	G0/0/2	ABC-ACC-01	G0/0/1	TO ABC-ACC-01 G0/0/1
	G0/0/3	ABC-ACC-02	G0/0/1	TO ABC-ACC-02 G0/0/1
	G0/0/10	ABC-AGG-02	G0/0/10	TO ABC-AGG-02 G0/0/10
	G0/0/11	ABC-AGG-02	G0/0/11	TO ABC-AGG-02 G0/0/11
ABC-AGG-02	G0/0/1	ABC-CORE-01	G0/0/2	TO ABC-CORE-01 G0/0/2
	G0/0/2	ABC-ACC-03	G0/0/1	TO ABC-ACC-03 G0/0/1
	G0/0/3	ABC-ACC-04	G0/0/1	TO ABC-ACC-04 G0/0/1
	G0/0/10	ABC-AGG-01	G0/0/10	TO ABC-AGG-01 G0/0/10
	G0/0/11	ABC-AGG-01	G0/0/11	TO ABC-AGG-01 G0/0/11
ABC-ACC-01	G0/0/1	ABC-AGG-01	G0/0/2	TO ABC-AGG-01 G0/0/2
	G0/0/2	ABC-AP-01	G0/0/0	TO ABC-AP-01 G0/0/0
ABC-ACC-02	G0/0/1	ABC-AGG-01	G0/0/3	TO ABC-AGG-01 G0/0/3
	G0/0/2	ABC-AP-02	G0/0/0	TO ABC-AP-02 G0/0/0
ABC-ACC-03	G0/0/1	ABC-AGG-02	G0/0/2	TO ABC-AGG-02 G0/0/2
	G0/0/2	ABC-AP-03	G0/0/0	TO ABC-AP-03 G0/0/0
ABC-ACC-04	G0/0/1	ABC-AGG-02	G0/0/3	TO ABC-AGG-02 G0/0/3
	G0/0/2	ABC-AP-04	G0/0/0	TO ABC-AP-04 G0/0/0
ABC-AC	G0/0/1	ABC-CORE-01	G0/0/3	TO ABC-CORE-01 G0/0/3
ABC-AP-01	G0/0/0	ABC-ACC-01	G0/0/2	TO ABC-ACC-01 G0/0/2
ABC-AP-02	G0/0/0	ABC-ACC-02	G0/0/2	TO ABC-ACC-02 G0/0/2
ABC-AP-03	G0/0/0	ABC-ACC-03	G0/0/2	TO ABC-ACC-03 G0/0/2
ABC-AP-04	G0/0/0	ABC-ACC-04	G0/0/2	TO ABC-ACC-04 G0/0/2

除了表 16-2 中记录的内容之外，读者还可以在其中添加与设备端口相关的其他信息，比如设备所在的机柜和位置信息、端口类型信息（比如光口、电口）。这些信息对于实施阶段的设备上架和布线至关重要，清晰的表格说明可以让工程师更快速、准确地完成布线工作。

16.2.2 二层网络设计

在这部分，读者需要针对多种第二层特性进行规划。具体说来，在本实验中，读者需要考虑 VLAN、STP 及链路聚合的规划，二层网络设计的具体步骤如下。

步骤 1 VLAN 规划

VLAN 规划包括 VLAN 设计和 VLAN 划分方式。在 VLAN 设计中，读者可以根据部门进行规划（本章将使用这种方式），也可以根据流量类型进行规划，或者使用其他原则进行规划。在 VLAN 划分方式中，读者可以根据端口（本章将使用这种方式）、MAC 地址等信息进行划分。

若根据部门进行 VLAN 划分,由于 ABC 公司拥有管理部、大客户部、集成业务一部、集成业务二部、人力资源部和财务部,且需要部署无线环境,因此还需要考虑划分无线所使用的两个 VLAN:管理 VLAN 和业务 VLAN。

VLAN 划分信息见表 16-3。

表 16-3 VLAN 划分表

VLAN 编号	说明
2	ABC-GW 与 ABC-CORE-01 之间互联 VLAN
10	管理部
11	大客户部
12	集成业务一部
13	集成业务二部
14	人力资源部
15	财务部
16	AP 管理 VLAN
17	ABC-AP-01 业务 VLAN
19	ABC-AP-02 业务 VLAN
21	ABC-AP-03 业务 VLAN
23	ABC-AP-04 业务 VLAN

在本实验设计的网络环境中,并不是所有交换机都需要部署所有的 VLAN,读者可以以设备为单位,或者以 VLAN 为单位,将传输该 VLAN 流量的设备与 VLAN ID 对应起来。

在接入层设备的端口与 VLAN 的对应关系上,本实验的设计比较简单,其中ABC-ACC-01、ABC-ACC-02 和 ABC-ACC-03 都分别只连接了一台 VLAN 的主机,ABC-ACC-04 分别连接属于 3 台不同 VLAN 的主机。

本章将使用 24 端口的接入层交换机来进行模拟实验,暂时抛开端口是否充足的问题(设备选型的问题),只针对相关的要点内容及其配置进行练习。

读者可以参考表 16-4~表 16-10,列出每台设备的二层端口情况。

表 16-4 核心层交换机 ABC-CORE-01 的二层端口情况

端口 ID	端口类型	VLAN	说明
G0/0/1	Trunk	ALL	连接汇聚层交换机 ABC-AGG-01
G0/0/2	Trunk	ALL	连接汇聚层交换机 ABC-AGG-02
G0/0/3	Access	16	连接 ABC-AC
G0/0/4	Access	2	连接网关路由器 ABC-GW

表 16-5 汇聚层交换机 ABC-AGG-01 的二层端口情况

端口 ID	端口类型	VLAN	说明
G0/0/1	Trunk	ALL	连接核心层交换机 ABC-CORE-01

表 16-5 汇聚层交换机 ABC-AGG-01 的二层端口情况（续）

端口 ID	端口类型	VLAN	说明
G0/0/2	Trunk	12、16、17	连接接入层交换机 ABC-ACC-01
G0/0/3	Trunk	13、16、19	连接接入层交换机 ABC-ACC-02
Eth-Trunk1	Eth-Trunk	ALL	端口 ID 为 G0/0/10 和 G0/0/11 连接汇聚层交换机 ABC-AGG-02

表 16-6 汇聚层交换机 ABC-AGG-02 的二层端口情况

端口 ID	端口类型	VLAN	说明
G0/0/1	Trunk	ALL	连接核心层交换机 ABC-CORE-01
G0/0/2	Trunk	11、16、21	连接接入层交换机 ABC-ACC-03
G0/0/3	Trunk	10、14~16、23	连接接入层交换机 ABC-ACC-04
Eth-Trunk1	Eth-Trunk	ALL	端口 ID 为 G0/0/10 和 G0/0/11 连接汇聚层交换机 ABC-AGG-01

表 16-7 接入层交换机 ABC-ACC-01 的二层端口情况

端口 ID	端口类型	VLAN	说明
G0/0/1	Trunk	12、16、17	连接汇聚层交换机 ABC-AGG-01
G0/0/2	Trunk	16、17	连接 ABC-AP-01
G0/0/3~G0/0/24	Access	12	连接 VLAN 12 主机

表 16-8 接入层交换机 ABC-ACC-02 的二层端口情况

端口 ID	端口类型	VLAN	说明
G0/0/1	Trunk	13、16、19	连接汇聚层交换机 ABC-AGG-01
G0/0/2	Trunk	16、19	连接 ABC-AP-02
G0/0/3~G0/0/24	Access	13	连接 VLAN 13 主机

表 16-9 接入层交换机 ABC-ACC-03 的二层端口情况

端口 ID	端口类型	VLAN	说明
G0/0/1	Trunk	11、16、21	连接汇聚层交换机 ABC-AGG-02
G0/0/2	Trunk	16、21	连接 ABC-AP-03
G0/0/3~G0/0/24	Access	11	连接 VLAN 11 主机

表 16-10 接入层交换机 ABC-ACC-04 的二层端口情况

端口 ID	端口类型	VLAN	说明
G0/0/1	Trunk	10、14~16、23	连接汇聚层交换机 ABC-AGG-02
G0/0/2	Trunk	16、23	连接 ABC-AP-04
G0/0/3~G0/0/8	Access	10	连接 VLAN 10 主机
G0/0/9~G0/0/16	Access	14	连接 VLAN 14 主机
G0/0/17~G0/0/24	Access	15	连接 VLAN 15 主机

步骤 2　STP 设计

在本实验的拓扑设计中,核心层设备(ABC-CORE-01)与接入层设备(ABC-AGG-01 和 ABC-AGG-02)之间形成了物理环路。交换机会默认运行 STP 并且打断环路,一般来说,在规模越大的网络环境中,管理员越需要对 STP 打断环路的选择进行干预,以防止阻塞不当的端口,导致次优的流量路径。同时,管理员也可以根据网络的其他需求,对 STP 的参数进行调整。

在本实验中,我们必须配置的参数是将核心层设备(ABC-CORE-01)设置为 STP 根交换机。要想进一步对网络流量进行优化,读者可以从 VLAN 分布的角度进行考虑,并使用 MSTP 进行优化。

步骤 3　链路聚合

在链路聚合的规划中,读者首先需要考虑的是带宽问题,也就是说,需要有多少条链路捆绑在一起才能够满足业务需求。在本实验中,为了实现不同 VLAN(部门)之间的互访,我们在汇聚层交换机之间建立了以太网聚合链路,将两条千兆以太网端口捆绑在一起。

考虑到未来的业务扩展,有可能会根据需要添加更多链路。为了获得更高的灵活性,建议读者使用 LACP 模式来配置链路聚合。

16.2.3　IP 地址规划

对于私有网络来说,需要使用 RFC 1918 中定义的私有 IP 网段来规划 IP 地址:

① 10.0.0.0～10.255.255.255;
② 172.16.0.0～172.31.255.255;
③ 192.168.0.0～192.168.255.255。

这些地址是无法在互联网中进行路由的,需要进行网络地址转换。与公网 IP 地址相比,在使用这些地址时无须进行申请,因为这些是免费使用的局域网 IP 地址,ID 地址规划步骤如下。

步骤 1　IP 地址规划

在大型网络中,为了有效地利用 IP 地址,建议合理规划子网掩码并连续分配 IP 地址段。在本章中,我们选用 192.168.0.0/16 这个私有 IP 地址范围进行 IP 地址划分。将 IP 地址的第 3 位十进制数设置为与相应的 VLAN ID 相同,以此对应每个 VLAN,所有 IP 地址段都选用 24 位掩码。同时,读者还需要规划三层端口之间的互联 IP 地址段,由于这种链路上只需要两个 IP 地址,因此可以使用 30 位掩码来节省 IP 地址空间。

读者可以选用其他网段并按照其他原则进行规划,并且需要注意子网掩码及其能够支持的主机数量。规划完成后,读者可以将 IP 地址的规划记录在表 16-11 所示的表格中。

表 16-11　IP 地址规划

IP 地址段	VLAN ID	说明
192.168.10.0/24	10	管理部,网关位于 ABC-AGG-02
192.168.11.0/24	11	大客户部,网关位于 ABC-AGG-02
192.168.12.0/24	12	集成业务一部,网关位于 ABC-AGG-01
192.168.13.0/24	13	集成业务二部,网关位于 ABC-AGG-01
192.168.14.0/24	14	人力资源部,网关位于 ABC-AGG-02

表 16-11　IP 地址规划（续）

IP 地址段	VLAN ID	说明
192.168.15.0/24	15	财务部，网关位于 ABC-AGG-02
192.168.16.0/24	16	ABC-AP-01 无线管理网段
192.168.17.0/24	17	ABC-AP-01 无线终端网段，网关位于 ABC-AGG-01
192.168.19.0/24	19	ABC-AP-02 无线终端网段，网关位于 ABC-AGG-01
192.168.21.0/24	21	ABC-AP-03 无线终端网段，网关位于 ABC-AGG-02
192.168.23.0/24	23	ABC-AP-04 无线终端网段，网关位于 ABC-AGG-02
192.168.0.0/30	2	ABC-GW 与 ABC-CORE-01 之间直联链路
IP:103.31.200.2/30	—	ABC-GW G0/0/0 端口从 ISP 获得的公网 IP 地址

步骤 2　IP 地址分配方式规划

在对 IP 地址进行配置和分配时，我们可以在设备上手动配置 IP 地址，也可以使用动态 IP 地址分配。对于路由器和交换机之类的网络设备来说，一般我们会使用静态配置的方式；有时 WAN 出口的地址可能需要使用 DHCP 向 ISP 进行申请。服务器及特殊终端设备（比如打印机）一般也采用静态的方式分配 IP 地址。对于用户终端，无论是有线终端还是无线终端，都可以通过 DHCP 的方式自动进行分配。

IP 地址分配方式的选择比较灵活，具体见表 16-12。如果为了安全性，选择为每个终端设备分配静态的 IP 地址，则配置工作量会比较大，后期维护会比较困难。读者可以根据企业的实际情况进行合理选择。

表 16-12　IP 地址分配方式

IP 地址段	分配方式	说明
192.168.10.0/24	DHCP	由网关 ABC-AGG-02 分配
192.168.11.0/24	DHCP	由网关 ABC-AGG-02 分配
192.168.12.0/24	DHCP	由网关 ABC-AGG-01 分配
192.168.13.0/24	DHCP	由网关 ABC-AGG-01 分配
192.168.14.0/24	DHCP	由网关 ABC-AGG-02 分配
192.168.15.0/24	DHCP	由网关 ABC-AGG-02 分配
192.168.16.0/24	DHCP	ABC-AC 地址静态配置，AP 地址由 ABC-CORE-01 分配
192.168.17.0/24	DHCP	由网关 ABC-AGG-01 分配
192.168.19.0/24	DHCP	由网关 ABC-AGG-01 分配
192.168.21.0/24	DHCP	由网关 ABC-AGG-02 分配
192.168.23.0/24	DHCP	由网关 ABC-AGG-02 分配
192.168.0.1/30	静态	ABC-CORE-01 与 ABC-GW 之间直联链路
103.31.200.2/30	静态	ABC-CORE-01 G0/0/0 端口从 ISP 获得的公网 IP 地址

16.2.4　三层网络设计

在企业网的三层路由设计中，我们需要考虑企业内部路由，以及企业外部互联路由

（包括与互联网 ISP 之间的路由，以及与合作伙伴之间的连接）。

在本实验环境中，我们需要考虑以下路由信息。

① ABC-CORE-01 与 ISP 之间的路由：在 ABC-CORE-01 上配置静态缺省路由，指向 ISP。

② 不同 VLAN 之间的路由：由于本实验拓扑的特殊性（每台接入层交换机上支持的 VLAN ID 不同），VLAN 的网关分别部署在两台汇聚层设备上，因此读者需要通过配置来实现全内网的路由互通。本章使用 OSPF 路由协议，并且在网关路由器 ABC-GW 上发布缺省路由。

③ AP 设备与 ABC-AC 之间的路由：AP 设备通过 DHCP 获得 IP 地址后，会自动生成缺省路由。

为了更清晰地展示本章的路由设计，读者可以参考表 16-13。

表 16-13 IP 路由规划

IP 地址段	路由配置	网关设备
192.168.10.0/24	通过网关设备在 OSPF 中宣告	ABC-AGG-02
192.168.11.0/24	通过网关设备在 OSPF 中宣告	ABC-AGG-02
192.168.12.0/24	通过网关设备在 OSPF 中宣告	ABC-AGG-01
192.168.13.0/24	通过网关设备在 OSPF 中宣告	ABC-AGG-01
192.168.14.0/24	通过网关设备在 OSPF 中宣告	ABC-AGG-02
192.168.15.0/24	通过网关设备在 OSPF 中宣告	ABC-AGG-02
192.168.16.0/24	通过网关设备在 OSPF 中宣告	ABC-CORE-01
192.168.17.0/24	通过网关设备在 OSPF 中宣告	ABC-AGG-01
192.168.19.0/24	通过网关设备在 OSPF 中宣告	ABC-AGG-01
192.168.21.0/24	通过网关设备在 OSPF 中宣告	ABC-AGG-02
192.168.23.0/24	通过网关设备在 OSPF 中宣告	ABC-AGG-02
192.168.0.0/30	启用 OSPF 并建立邻居关系	—

具体来说，在网关路由器 ABC-GW 上需要参与 OSPF 路由的是以下端口：
G0/0/1，IP 地址为 192.168.0.1。

核心层交换机 ABC-CORE-01 上需要参与 OSPF 路由的是以下端口：

① Vlanif 2，IP 地址为 192.168.0.2；

② Vlanif 10，IP 地址为 192.168.10.253；

③ Vlanif 11，IP 地址为 192.168.11.253；

④ Vlanif 12，IP 地址为 192.168.12.253；

⑤ Vlanif 13，IP 地址为 192.168.13.253；

⑥ Vlanif 14，IP 地址为 192.168.14.253；

⑦ Vlanif 15，IP 地址为 192.168.15.253；

⑧ Vlanif 17，IP 地址为 192.168.17.253；

⑨ Vlanif 19，IP 地址为 192.168.19.253；

⑩ Vlanif 21，IP 地址为 192.168.21.253；

⑪ Vlanif 23，IP 地址为 192.168.23.253。

汇聚层交换机 ABC-AGG-01 上需要参与 OSPF 路由的是以下端口：
① Vlanif 12，IP 地址为 192.168.12.254；
② Vlanif 13，IP 地址为 192.168.13.254；
③ Vlanif 17，IP 地址为 192.168.17.254；
④ Vlanif 19，IP 地址为 192.168.19.254。

汇聚层交换机 ABC-AGG-02 上需要参与 OSPF 路由的是以下端口：
① Vlanif 10，IP 地址为 192.168.10.254；
② Vlanif 11，IP 地址为 192.168.11.254；
③ Vlanif 14，IP 地址为 192.168.14.254；
④ Vlanif 15，IP 地址为 192.168.15.254；
⑤ Vlanif 21，IP 地址为 192.168.21.254；
⑥ Vlanif 23，IP 地址为 192.168.23.254。

由于网关路由器 ABC-GW 会通过 OSPF 协议发布缺省路由，因此读者可以在其他 3 台参与 OSPF 进程的交换机路由表中看到一条通过 OSPF 学习到的缺省路由。以 ABC-CORE-01 为例，例 16-1 展示了 IP 路由表中的 OSPF 缺省路由。

例 16-1　OSPF 缺省路由

```
[CO1]display ip routing-table protocol ospf
Route Flags: R - relay, D - download to fib
------------------------------------------------------------------------------
Public routing table : OSPF
        Destinations : 1       Routes : 1

OSPF routing table status : <Active>
        Destinations : 1       Routes : 1

Destination/Mask    Proto   Pre  Cost       Flags NextHop         Interface
      0.0.0.0/0    O_ASE    150  1            D   192.168.0.1     Vlanif2

OSPF routing table status : <Inactive>
        Destinations : 0       Routes : 0
```

16.2.5　WLAN 设计

在本实验环境中，4 台 AP 设备分别部署在 4 间办公室中，它们都由一台 AC 进行管理和控制。本章中 4 台 AP 需要使用的参数见表 16-14~表 16-17，读者也可以自行进行设计并优化。

表 16-14　ABC-AP-01 所需参数

配置项	配置参数
SSID 模板	名称：ABC-F1R1
	SSID 名称：ABC-F1R1
安全模板	名称：F1R1
	安全策略：WPA-WPA2+PSK+AES
	密码：ABC-F1R1
VAP 模板	名称：ABC-F1R1
	业务 VLAN：VLAN 17

表 16-15 ABC-AP-02 所需参数

配置项	配置参数
SSID 模板	名称：ABC-F1R2
	SSID 名称：ABC-F1R2
安全模板	名称：F1R2
	安全策略：WPA-WPA2+PSK+AES
	密码：ABC-F1R2
VAP 模板	名称：ABC-F1R2
	业务 VLAN：VLAN 19

表 16-16 ABC-AP-03 所需参数

配置项	配置参数
SSID 模板	名称：ABC-F2R1
	SSID 名称：ABC-F2R1
安全模板	名称：F2R1
	安全策略：WPA-WPA2+PSK+AES
	密码：ABC-F2R1
VAP 模板	名称：ABC-F2R1
	业务 VLAN：VLAN 21

表 16-17 ABC-AP-04 所需参数

配置项	配置参数
SSID 模板	名称：ABC-F2R2
	SSID 名称：ABC-F2R2
安全模板	名称：F2R2
	安全策略：WPA-WPA2+PSK+AES
	密码：ABC-F2R2
VAP 模板	名称：ABC-F2R2
	业务 VLAN：VLAN 23

16.2.6 网络设备调试

在一个网络项目的实施过程中，当我们制订了总体设计方案和详细配置方案后，在设备上架并调试之前，往往已经对配置进行了实验室验证。因此在网络设备上架并连线后，我们在这一步需要做的是将配置导入设备并测试（有可能在上架前已经进行了设备导入并测试）。

如果读者希望按照本章提供的设计思路进行实验练习，可以参考以下设备的配置。本配置提供了最基本的配置命令，其中有大量可供优化的部分，比如 STP 的设计，当前的配置会使 Eth-Trunk 链路被阻塞，无法实现最优路径。

核心层交换机 ABC-CORE-01 的配置见例 16-2。

例 16-2 核心层交换机 ABC-CORE-01 的配置参考

```
[CO1]display current-configuration
#
sysname CO1
#
vlan batch 2 10 to 17 19 21 23
#
stp instance 0 root primary
#
cluster enable
ntdp enable
ndp enable
#
drop illegal-mac alarm
#
dhcp enable
#
diffserv domain default
#
drop-profile default
#
ip pool ap
 gateway-list 192.168.16.254
 network 192.168.16.0 mask 255.255.255.0
 excluded-ip-address 192.168.16.100
#
aaa
 authentication-scheme default
 authorization-scheme default
 accounting-scheme default
 domain default
 domain default_admin
 local-user admin password simple admin
 local-user admin service-type http
#
interface Vlanif1
#
interface Vlanif2
 ip address 192.168.0.2 255.255.255.252
#
interface Vlanif10
 ip address 192.168.10.253 255.255.255.0
#
interface Vlanif11
 ip address 192.168.11.253 255.255.255.0
#
interface Vlanif12
 ip address 192.168.12.253 255.255.255.0
#
interface Vlanif13
 ip address 192.168.13.253 255.255.255.0
#
interface Vlanif14
 ip address 192.168.14.253 255.255.255.0
#
interface Vlanif15
 ip address 192.168.15.253 255.255.255.0
#
interface Vlanif16
 ip address 192.168.16.254 255.255.255.0
 dhcp select global
#
interface Vlanif17
 ip address 192.168.17.253 255.255.255.0
#
```

```
interface Vlanif19
 ip address 192.168.19.253 255.255.255.0
#
interface Vlanif21
 ip address 192.168.21.253 255.255.255.0
#
interface Vlanif23
 ip address 192.168.23.253 255.255.255.0
#
interface MEth0/0/1
#
interface GigabitEthernet0/0/1
 port link-type trunk
 port trunk allow-pass vlan 2 to 4094
#
interface GigabitEthernet0/0/2
 port link-type trunk
 port trunk allow-pass vlan 2 to 4094
#
interface GigabitEthernet0/0/3
 port link-type access
 port default vlan 16
#
interface GigabitEthernet0/0/4
 port link-type access
 port default vlan 2
#
interface GigabitEthernet0/0/5
#
interface GigabitEthernet0/0/6
#
interface GigabitEthernet0/0/7
#
interface GigabitEthernet0/0/8
#
interface GigabitEthernet0/0/9
#
interface GigabitEthernet0/0/10
#
interface GigabitEthernet0/0/11
#
interface GigabitEthernet0/0/12
#
interface GigabitEthernet0/0/13
#
interface GigabitEthernet0/0/14
#
interface GigabitEthernet0/0/15
#
interface GigabitEthernet0/0/16
#
interface GigabitEthernet0/0/17
#
interface GigabitEthernet0/0/18
#
interface GigabitEthernet0/0/19
#
interface GigabitEthernet0/0/20
#
interface GigabitEthernet0/0/21
#
interface GigabitEthernet0/0/22
#
interface GigabitEthernet0/0/23
#
interface GigabitEthernet0/0/24
#
interface NULL0
#
ospf 10
 area 0.0.0.0
  network 192.168.0.2 0.0.0.0
  network 192.168.10.253 0.0.0.0
  network 192.168.11.253 0.0.0.0
  network 192.168.12.253 0.0.0.0
```

```
  network 192.168.13.253 0.0.0.0
  network 192.168.14.253 0.0.0.0
  network 192.168.15.253 0.0.0.0
  network 192.168.16.253 0.0.0.0
  network 192.168.17.253 0.0.0.0
  network 192.168.19.253 0.0.0.0
  network 192.168.21.253 0.0.0.0
  network 192.168.23.253 0.0.0.0
#
user-interface con 0
user-interface vty 0 4
#
return
```

汇聚层交换机 ABC-AGG-01 的配置见例 16-3。

例 16-3 汇聚层交换机 ABC-AGG-01 的配置参考

```
[ABC-AGG-01]display current-configuration
#
sysname ABC-AGG-01
#
vlan batch 10 to 17 19 21 23
#
cluster enable
ntdp enable
ndp enable
#
drop illegal-mac alarm
#
dhcp enable
#
diffserv domain default
#
drop-profile default
#
ip pool acc-01
 gateway-list 192.168.12.254
 network 192.168.12.0 mask 255.255.255.0
 excluded-ip-address 192.168.12.253
#
ip pool acc-02
 gateway-list 192.168.13.254
 network 192.168.13.0 mask 255.255.255.0
 excluded-ip-address 192.168.13.253
#
ip pool wlan-01
 gateway-list 192.168.17.254
 network 192.168.17.0 mask 255.255.255.0
 excluded-ip-address 192.168.17.253
#
ip pool wlan-02
 gateway-list 192.168.19.254
 network 192.168.19.0 mask 255.255.255.0
 excluded-ip-address 192.168.19.253
#
aaa
 authentication-scheme default
 authorization-scheme default
 accounting-scheme default
 domain default
 domain default_admin
 local-user admin password simple admin
 local-user admin service-type http
#
interface Vlanif1
#
interface Vlanif12
 ip address 192.168.12.254 255.255.255.0
 dhcp select global
#
interface Vlanif13
 ip address 192.168.13.254 255.255.255.0
 dhcp select global
#
```

```
interface Vlanif17
 ip address 192.168.17.254 255.255.255.0
 dhcp select global
#
interface Vlanif19
 ip address 192.168.19.254 255.255.255.0
 dhcp select global
#
interface MEth0/0/1
#
interface Eth-Trunk1
 mode lacp-static
#
interface GigabitEthernet0/0/1
 port link-type trunk
 port trunk allow-pass vlan 2 to 4094
#
interface GigabitEthernet0/0/2
 port link-type trunk
 port trunk pvid vlan 12
 port trunk allow-pass vlan 12 16 to 17
#
interface GigabitEthernet0/0/3
 port link-type trunk
 port trunk pvid vlan 13
 port trunk allow-pass vlan 13 16 19
#
interface GigabitEthernet0/0/4
#
interface GigabitEthernet0/0/5
#
interface GigabitEthernet0/0/6
#
interface GigabitEthernet0/0/7
#
interface GigabitEthernet0/0/8
#
interface GigabitEthernet0/0/9
#
interface GigabitEthernet0/0/10
 eth-trunk 1
#
interface GigabitEthernet0/0/11
 eth-trunk 1
#
interface GigabitEthernet0/0/12
#
interface GigabitEthernet0/0/13
#
interface GigabitEthernet0/0/14
#
interface GigabitEthernet0/0/15
#
interface GigabitEthernet0/0/16
#
interface GigabitEthernet0/0/17
#
interface GigabitEthernet0/0/18
#
interface GigabitEthernet0/0/19
#
interface GigabitEthernet0/0/20
#
interface GigabitEthernet0/0/21
#
interface GigabitEthernet0/0/22
#
interface GigabitEthernet0/0/23
#
interface GigabitEthernet0/0/24
#
interface NULL0
#
```

```
 ospf 10
  area 0.0.0.0
   network 192.168.12.254 0.0.0.0
   network 192.168.13.254 0.0.0.0
   network 192.168.17.254 0.0.0.0
   network 192.168.19.254 0.0.0.0
 #
 user-interface con 0
 user-interface vty 0 4
 #
 return
```

汇聚层交换机 ABC-AGG-02 的配置见例 16-4。

例 16-4　汇聚层交换机 ABC-AGG-02 的配置参考

```
[ABC-AGG-02]display current-configuration
#
sysname ABC-AGG-02
#
vlan batch 10 to 17 19 21 23
#
cluster enable
ntdp enable
ndp enable
#
drop illegal-mac alarm
#
dhcp enable
#
diffserv domain default
#
drop-profile default
#
ip pool acc-03
 gateway-list 192.168.11.254
 network 192.168.11.0 mask 255.255.255.0
 excluded-ip-address 192.168.11.253
#
ip pool acc-04-vlan10
 gateway-list 192.168.10.254
 network 192.168.10.0 mask 255.255.255.0
 excluded-ip-address 192.168.10.253
#
ip pool acc-04-vlan14
 gateway-list 192.168.14.254
 network 192.168.14.0 mask 255.255.255.0
 excluded-ip-address 192.168.14.253
#
ip pool acc-04-vlan15
 gateway-list 192.168.15.254
 network 192.168.15.0 mask 255.255.255.0
 excluded-ip-address 192.168.15.253
#
ip pool wlan-03
 gateway-list 192.168.21.254
 network 192.168.21.0 mask 255.255.255.0
 excluded-ip-address 192.168.21.253
#
ip pool wlan-04
 gateway-list 192.168.23.254
 network 192.168.23.0 mask 255.255.255.0
 excluded-ip-address 192.168.23.253
#
aaa
 authentication-scheme default
 authorization-scheme default
 accounting-scheme default
 domain default
 domain default_admin
 local-user admin password simple admin
 local-user admin service-type http
#
interface Vlanif1
#
```

```
interface Vlanif10
 ip address 192.168.10.254 255.255.255.0
 dhcp select global
#
interface Vlanif11
 ip address 192.168.11.254 255.255.255.0
 dhcp select global
#
interface Vlanif14
 ip address 192.168.14.254 255.255.255.0
 dhcp select global
#
interface Vlanif15
 ip address 192.168.15.254 255.255.255.0
 dhcp select global
#
interface Vlanif21
 ip address 192.168.21.254 255.255.255.0
 dhcp select global
#
interface Vlanif23
 ip address 192.168.23.254 255.255.255.0
 dhcp select global
#
interface MEth0/0/1
#
interface Eth-Trunk1
 mode lacp-static
#
interface GigabitEthernet0/0/1
 port link-type trunk
 port trunk allow-pass vlan 2 to 4094
#
interface GigabitEthernet0/0/2
 port link-type trunk
 port trunk pvid vlan 11
 port trunk allow-pass vlan 11 16 21
#
interface GigabitEthernet0/0/3
 port link-type trunk
 port trunk pvid vlan 10
 port trunk allow-pass vlan 10 14 to 16 23
#
interface GigabitEthernet0/0/4
#
interface GigabitEthernet0/0/5
#
interface GigabitEthernet0/0/6
#
interface GigabitEthernet0/0/7
#
interface GigabitEthernet0/0/8
#
interface GigabitEthernet0/0/9
#
interface GigabitEthernet0/0/10
 eth-trunk 1
#
interface GigabitEthernet0/0/11
 eth-trunk 1
#
interface GigabitEthernet0/0/12
#
interface GigabitEthernet0/0/13
#
interface GigabitEthernet0/0/14
#
interface GigabitEthernet0/0/15
#
interface GigabitEthernet0/0/16
#
interface GigabitEthernet0/0/17
#
interface GigabitEthernet0/0/18
#
interface GigabitEthernet0/0/19
#
```

```
interface GigabitEthernet0/0/20
#
interface GigabitEthernet0/0/21
#
interface GigabitEthernet0/0/22
#
interface GigabitEthernet0/0/23
#
interface GigabitEthernet0/0/24
#
interface NULL0
#
ospf 10
 area 0.0.0.0
  network 192.168.10.254 0.0.0.0
  network 192.168.11.254 0.0.0.0
  network 192.168.14.254 0.0.0.0
  network 192.168.15.254 0.0.0.0
  network 192.168.21.254 0.0.0.0
  network 192.168.23.254 0.0.0.0
#
user-interface con 0
user-interface vty 0 4
#
return
```

接入层交换机 ABC-ACC-01 的配置见例 16-5。

例 16-5 接入层交换机 ABC-ACC-01 的配置参考

```
<ABC-ACC-01>display current-configuration
#
sysname ABC-ACC-01
#
vlan batch 12 16 to 17
#
cluster enable
ntdp enable
ndp enable
#
drop illegal-mac alarm
#
diffserv domain default
#
drop-profile default
#
aaa
 authentication-scheme default
 authorization-scheme default
 accounting-scheme default
 domain default
 domain default_admin
 local-user admin password simple admin
 local-user admin service-type http
#
interface Vlanif1
#
interface MEth0/0/1
#
interface GigabitEthernet0/0/1
 port link-type trunk
 port trunk pvid vlan 12
 port trunk allow-pass vlan 12 16 to 17
#
interface GigabitEthernet0/0/2
 port link-type trunk
 port trunk pvid vlan 16
 port trunk allow-pass vlan 16 to 17
#
interface GigabitEthernet0/0/3
 port link-type access
 port default vlan 12
#
interface GigabitEthernet0/0/4
 port link-type access
```

```
 port default vlan 12
#
interface GigabitEthernet0/0/5
 port link-type access
 port default vlan 12
#
interface GigabitEthernet0/0/6
 port link-type access
 port default vlan 12
#
interface GigabitEthernet0/0/7
 port link-type access
 port default vlan 12
#
interface GigabitEthernet0/0/8
 port link-type access
 port default vlan 12
#
interface GigabitEthernet0/0/9
 port link-type access
 port default vlan 12
#
interface GigabitEthernet0/0/10
 port link-type access
 port default vlan 12
#
interface GigabitEthernet0/0/11
 port link-type access
 port default vlan 12
#
interface GigabitEthernet0/0/12
 port link-type access
 port default vlan 12
#
interface GigabitEthernet0/0/13
 port link-type access
 port default vlan 12
#
interface GigabitEthernet0/0/14
 port link-type access
 port default vlan 12
#
interface GigabitEthernet0/0/15
 port link-type access
 port default vlan 12
#
interface GigabitEthernet0/0/16
 port link-type access
 port default vlan 12
#
interface GigabitEthernet0/0/17
 port link-type access
 port default vlan 12
#
interface GigabitEthernet0/0/18
 port link-type access
 port default vlan 12
#
interface GigabitEthernet0/0/19
 port link-type access
 port default vlan 12
#
interface GigabitEthernet0/0/20
 port link-type access
 port default vlan 12
#
interface GigabitEthernet0/0/21
 port link-type access
 port default vlan 12
#
interface GigabitEthernet0/0/22
 port link-type access
 port default vlan 12
#
interface GigabitEthernet0/0/23
 port link-type access
 port default vlan 12
```

```
#
interface GigabitEthernet0/0/24
 port link-type access
 port default vlan 12
#
interface NULL0
#
user-interface con 0
user-interface vty 0 4
#
return
```

接入层交换机 ABC-ACC-02 的配置见例 16-6。

例 16-6　接入层交换机 ABC-ACC-02 的配置参考

```
<ABC-ACC-02>display current-configuration
#
sysname ABC-ACC-02
#
vlan batch 13 16 19
#
cluster enable
ntdp enable
ndp enable
#
drop illegal-mac alarm
#
diffserv domain default
#
drop-profile default
#
aaa
 authentication-scheme default
 authorization-scheme default
 accounting-scheme default
 domain default
 domain default_admin
 local-user admin password simple admin
 local-user admin service-type http
#
interface Vlanif1
#
interface MEth0/0/1
#
interface GigabitEthernet0/0/1
 port link-type trunk
 port trunk pvid vlan 13
 port trunk allow-pass vlan 13 16 19
#
interface GigabitEthernet0/0/2
 port link-type trunk
 port trunk pvid vlan 16
 port trunk allow-pass vlan 16 19
#
interface GigabitEthernet0/0/3
 port link-type access
 port default vlan 13
#
interface GigabitEthernet0/0/4
 port link-type access
 port default vlan 13
#
interface GigabitEthernet0/0/5
 port link-type access
 port default vlan 13
#
interface GigabitEthernet0/0/6
 port link-type access
 port default vlan 13
#
interface GigabitEthernet0/0/7
 port link-type access
 port default vlan 13
#
interface GigabitEthernet0/0/8
 port link-type access
 port default vlan 13
```

```
#
interface GigabitEthernet0/0/9
 port link-type access
 port default vlan 13
#
interface GigabitEthernet0/0/10
 port link-type access
 port default vlan 13
#
interface GigabitEthernet0/0/11
 port link-type access
 port default vlan 13
#
interface GigabitEthernet0/0/12
 port link-type access
 port default vlan 13
#
interface GigabitEthernet0/0/13
 port link-type access
 port default vlan 13
#
interface GigabitEthernet0/0/14
 port link-type access
 port default vlan 13
#
interface GigabitEthernet0/0/15
 port link-type access
 port default vlan 13
#
interface GigabitEthernet0/0/16
 port link-type access
 port default vlan 13
#
interface GigabitEthernet0/0/17
 port link-type access
 port default vlan 13
#
interface GigabitEthernet0/0/18
 port link-type access
 port default vlan 13
#
interface GigabitEthernet0/0/19
 port link-type access
 port default vlan 13
#
interface GigabitEthernet0/0/20
 port link-type access
 port default vlan 13
#
interface GigabitEthernet0/0/21
 port link-type access
 port default vlan 13
#
interface GigabitEthernet0/0/22
 port link-type access
 port default vlan 13
#
interface GigabitEthernet0/0/23
 port link-type access
 port default vlan 13
#
interface GigabitEthernet0/0/24
 port link-type access
 port default vlan 13
#
interface NULL0
#
user-interface con 0
user-interface vty 0 4
#
return
```

接入层交换机 ABC-ACC-03 的配置见例 16-7。

例 16-7 接入层交换机 ABC-ACC-03 的配置参考

```
<ABC-ACC-03>display current-configuration
#
sysname ABC-ACC-03
#
vlan batch 11 16 21
#
cluster enable
ntdp enable
ndp enable
#
drop illegal-mac alarm
#
diffserv domain default
#
drop-profile default
#
aaa
 authentication-scheme default
 authorization-scheme default
 accounting-scheme default
 domain default
 domain default_admin
 local-user admin password simple admin
 local-user admin service-type http
#
interface Vlanif1
#
interface MEth0/0/1
#
interface GigabitEthernet0/0/1
 port link-type trunk
 port trunk pvid vlan 11
 port trunk allow-pass vlan 11 16 21
#
interface GigabitEthernet0/0/2
 port link-type trunk
 port trunk pvid vlan 16
 port trunk allow-pass vlan 16 21
#
interface GigabitEthernet0/0/3
 port link-type access
 port default vlan 11
#
interface GigabitEthernet0/0/4
 port link-type access
 port default vlan 11
#
interface GigabitEthernet0/0/5
 port link-type access
 port default vlan 11
#
interface GigabitEthernet0/0/6
 port link-type access
 port default vlan 11
#
interface GigabitEthernet0/0/7
 port link-type access
 port default vlan 11
#
interface GigabitEthernet0/0/8
 port link-type access
 port default vlan 11
#
interface GigabitEthernet0/0/9
 port link-type access
 port default vlan 11
#
interface GigabitEthernet0/0/10
 port link-type access
 port default vlan 11
```

```
#
interface GigabitEthernet0/0/11
 port link-type access
 port default vlan 11
#
interface GigabitEthernet0/0/12
 port link-type access
 port default vlan 11
#
interface GigabitEthernet0/0/13
 port link-type access
 port default vlan 11
#
interface GigabitEthernet0/0/14
 port link-type access
 port default vlan 11
#
interface GigabitEthernet0/0/15
 port link-type access
 port default vlan 11
#
interface GigabitEthernet0/0/16
 port link-type access
 port default vlan 11
#
interface GigabitEthernet0/0/17
 port link-type access
 port default vlan 11
#
interface GigabitEthernet0/0/18
 port link-type access
 port default vlan 11
#
interface GigabitEthernet0/0/19
 port link-type access
 port default vlan 11
#
interface GigabitEthernet0/0/20
 port link-type access
 port default vlan 11
#
interface GigabitEthernet0/0/21
 port link-type access
 port default vlan 11
#
interface GigabitEthernet0/0/22
 port link-type access
 port default vlan 11
#
interface GigabitEthernet0/0/23
 port link-type access
 port default vlan 11
#
interface GigabitEthernet0/0/24
 port link-type access
 port default vlan 11
#
interface NULL0
#
user-interface con 0
user-interface vty 0 4
#
return
```

接入层交换机 ABC-ACC-04 的配置见例 16-8。

例 16-8　接入层交换机 ABC-ACC-04 的配置参考

```
<ABC-ACC-04>display current-configuration
#
sysname ABC-ACC-04
#
vlan batch 10 14 to 16 23
#
cluster enable
ntdp enable
ndp enable
#
drop illegal-mac alarm
#
diffserv domain default
#
drop-profile default
#
aaa
 authentication-scheme default
 authorization-scheme default
 accounting-scheme default
 domain default
 domain default_admin
 local-user admin password simple admin
 local-user admin service-type http
#
interface Vlanif1
#
interface MEth0/0/1
#
interface GigabitEthernet0/0/1
 port link-type trunk
 port trunk allow-pass vlan 10 14 to 16 23
#
interface GigabitEthernet0/0/2
 port link-type trunk
 port trunk allow-pass vlan 16 23
#
interface GigabitEthernet0/0/3
 port link-type access
 port default vlan 10
#
interface GigabitEthernet0/0/4
 port link-type access
 port default vlan 10
#
interface GigabitEthernet0/0/5
 port link-type access
 port default vlan 10
#
interface GigabitEthernet0/0/6
 port link-type access
 port default vlan 10
#
interface GigabitEthernet0/0/7
 port link-type access
 port default vlan 10
#
interface GigabitEthernet0/0/8
 port link-type access
 port default vlan 10
#
interface GigabitEthernet0/0/9
 port link-type access
 port default vlan 14
#
interface GigabitEthernet0/0/10
 port link-type access
 port default vlan 14
#
interface GigabitEthernet0/0/11
 port link-type access
 port default vlan 14
#
```

```
interface GigabitEthernet0/0/12
 port link-type access
 port default vlan 14
#
interface GigabitEthernet0/0/13
 port link-type access
 port default vlan 14
#
interface GigabitEthernet0/0/14
 port link-type access
 port default vlan 14
#
interface GigabitEthernet0/0/15
 port link-type access
 port default vlan 14
#
interface GigabitEthernet0/0/16
 port link-type access
 port default vlan 14
#
interface GigabitEthernet0/0/17
 port link-type access
 port default vlan 15
#
interface GigabitEthernet0/0/18
 port link-type access
 port default vlan 15
#
interface GigabitEthernet0/0/19
 port link-type access
 port default vlan 15
#
interface GigabitEthernet0/0/20
 port link-type access
 port default vlan 15
#
interface GigabitEthernet0/0/21
 port link-type access
 port default vlan 15
#
interface GigabitEthernet0/0/22
 port link-type access
 port default vlan 15
#
interface GigabitEthernet0/0/23
 port link-type access
 port default vlan 15
#
interface GigabitEthernet0/0/24
 port link-type access
 port default vlan 15
#
interface NULL0
#
user-interface con 0
user-interface vty 0 4
#
return
```

网关路由器 ABC-GW 的配置见例 16-9。

例 16-9　网关路由器 ABC-GW 的配置参考

```
[ABC-GW]display current-configuration
[V200R003C00]
#
 sysname ABC-GW
#
 snmp-agent local-engineid 800007DB03000000000000
 snmp-agent
#
 clock timezone China-Standard-Time minus 08:00:00
#
portal local-server load flash:/portalpage.zip
#
```

```
 drop illegal-mac alarm
#
 wlan ac-global carrier id other ac id 0
#
 set cpu-usage threshold 80 restore 75
#
acl number 2000
 rule 5 permit source 192.168.8.0 0.0.7.255
 rule 10 permit source 192.168.17.0 0.0.0.255
 rule 15 permit source 192.168.19.0 0.0.0.255
 rule 20 permit source 192.168.21.0 0.0.0.255
 rule 25 permit source 192.168.23.0 0.0.0.255
#
aaa
 authentication-scheme default
 authorization-scheme default
 accounting-scheme default
 domain default
 domain default_admin
 local-user admin password cipher %$%$K8m.Nt84DZ}e#<0`8bmE3Uw}%$%$
 local-user admin service-type http
#
firewall zone Local
 priority 15
#
interface GigabitEthernet0/0/0
 ip address 103.31.200.2 255.255.255.252
 nat outbound 2000
#
interface GigabitEthernet0/0/1
 ip address 192.168.0.1 255.255.255.252
#
interface GigabitEthernet0/0/2
#
interface NULL0
#
ospf 10
 default-route-advertise always
 area 0.0.0.0
  network 192.168.0.1 0.0.0.0
#
ip route-static 0.0.0.0 0.0.0.0 GigabitEthernet0/0/0 103.31.200.1
ip route-static 192.168.0.0 255.255.0.0 GigabitEthernet0/0/1 192.168.0.2
#
user-interface con 0
 authentication-mode password
user-interface vty 0 4
user-interface vty 16 20
#
wlan ac
#
return
```

ABC-AC 控制器的配置见例 16-10。

例 16-10　ABC-AC 控制器的配置参考

```
[ABC-AC]display current-configuration
#
 sysname ABC-AC
#
 set memory-usage threshold 0
#
ssl renegotiation-rate 1
#
vlan batch 16
#
authentication-profile name default_authen_profile
authentication-profile name dot1x_authen_profile
authentication-profile name mac_authen_profile
authentication-profile name portal_authen_profile
authentication-profile name macportal_authen_profile
#
diffserv domain default
#
radius-server template default
#
```

```
pki realm default
 rsa local-key-pair default
 enrollment self-signed
#
ike proposal default
 encryption-algorithm aes-256
 dh group14
 authentication-algorithm sha2-256
 authentication-method pre-share
 integrity-algorithm hmac-sha2-256
 prf hmac-sha2-256
#
free-rule-template name default_free_rule
#
portal-access-profile name portal_access_profile
#
aaa
 authentication-scheme default
 authentication-scheme radius
  authentication-mode radius
 authorization-scheme default
 accounting-scheme default
 domain default
  authentication-scheme radius
  radius-server default
 domain default_admin
  authentication-scheme default
 local-user admin password irreversible-cipher $1a$@D5uP]jG/1$RW/V44&&e"it1-R1,h
D/bt-r(OlQaAy/RlN*yi{H$
 local-user admin privilege level 15
 local-user admin service-type http
#
interface Vlanif16
 ip address 192.168.16.100 255.255.255.0
#
interface GigabitEthernet0/0/1
 port link-type access
 port default vlan 16
#
interface GigabitEthernet0/0/2
#
interface GigabitEthernet0/0/3
#
interface GigabitEthernet0/0/4
#
interface GigabitEthernet0/0/5
#
interface GigabitEthernet0/0/6
#
interface GigabitEthernet0/0/7
 undo negotiation auto
 duplex half
#
interface GigabitEthernet0/0/8
 undo negotiation auto
 duplex half
#
interface NULL0
#
 snmp-agent local-engineid 800007DB03000000000000
 snmp-agent
#
ssh server secure-algorithms cipher aes256_ctr aes128_ctr
ssh server key-exchange dh_group14_sha1
ssh client secure-algorithms cipher aes256_ctr aes128_ctr
ssh client secure-algorithms hmac sha2_256
ssh client key-exchange dh_group14_sha1
#
ip route-static 0.0.0.0 0.0.0.0 192.168.16.254
#
capwap source interface vlanif16
#
user-interface con 0
 authentication-mode password
user-interface vty 0 4
 protocol inbound all
user-interface vty 16 20
```

```
  protocol inbound all
 #
 wlan
  traffic-profile name default
  security-profile name F1R1
   security wpa-wpa2 psk pass-phrase %^%#OqM79QR!~&`yPP7`2@i<zT}iJ`s^;Y-P+06G^D"Z
%^%# aes
  security-profile name F1R2
   security wpa-wpa2 psk pass-phrase %^%###3[#kkVu"-cr0X`.gt~IFRQ0cwhG=!cn<-rO_X&
%^%# aes
  security-profile name F2R1
   security wpa-wpa2 psk pass-phrase %^%#NCP.VFsqB#EP3f>uu('$SxVQP5Dha$]@X{2;w;yO
%^%# aes
  security-profile name F2R2
   security wpa-wpa2 psk pass-phrase %^%#LZt)%GrGA~3}5.U,#ey)]q;47^G:bLBljlAux(O2
%^%# aes
  security-profile name default
  security-profile name default-wds
  security-profile name default-mesh
  ssid-profile name default
  ssid-profile name ABC-F1R1
   ssid ABC-F1R1
  ssid-profile name ABC-F1R2
   ssid F1R2
  ssid-profile name ABC-F2R1
   ssid ABC-F2R1
  ssid-profile name ABC-F2R2
   ssid ABC-F2R2
  vap-profile name default
  vap-profile name ABC-F1R1
   service-vlan vlan-id 17
   ssid-profile ABC-F1R1
   security-profile F1R1
  vap-profile name ABC-F1R2
   service-vlan vlan-id 19
   ssid-profile ABC-F1R2
   security-profile F1R2
  vap-profile name ABC-F2R1
   service-vlan vlan-id 21
   ssid-profile ABC-F2R1
   security-profile F2R1
  vap-profile name ABC-F2R2
   service-vlan vlan-id 23
   ssid-profile ABC-F2R2
   security-profile F2R2
  wds-profile name default
  mesh-handover-profile name default
  mesh-profile name default
  regulatory-domain-profile name default
  air-scan-profile name default
  rrm-profile name default
  radio-2g-profile name default
  radio-5g-profile name default
  wids-spoof-profile name default
  wids-profile name default
  wireless-access-specification
  ap-system-profile name default
  port-link-profile name default
  wired-port-profile name default
  serial-profile name preset-enjoyor-toeap
  ap-group name F1R1
   radio 0
    vap-profile ABC-F1R1 wlan 1
   radio 1
    vap-profile ABC-F1R1 wlan 1
  ap-group name F1R2
   radio 0
    vap-profile ABC-F1R2 wlan 2
   radio 1
    vap-profile ABC-F1R2 wlan 2
  ap-group name F2R1
   radio 0
    vap-profile ABC-F2R1 wlan 3
   radio 1
    vap-profile ABC-F2R1 wlan 3
  ap-group name F2R2
   radio 0
```

```
[ABC-AC]display current-configuration
#
 sysname ABC-AC
#
 set memory-usage threshold 0
#
ssl renegotiation-rate 1
#
vlan batch 16
#
authentication-profile name default_authen_profile
authentication-profile name dot1x_authen_profile
authentication-profile name mac_authen_profile
authentication-profile name portal_authen_profile
authentication-profile name macportal_authen_profile
#
diffserv domain default
#
radius-server template default
#
```

附录
命令索引

实验 1　基础配置命令

命令	命令解析
system-view	从用户视图进入系统视图
quit	从任意配置视图退到上一层配置视图
return	从任意配置视图直接退到用户视图。"Ctrl+Z"组合键也可以实现相同目的
help [**all** | *command-name*]	获取有关命令的在线帮助信息。 参数解析如下所示： ① **all** 可选关键词用来显示视图下的所有命令； ② *command-name* 参数用来显示指定命令的格式、参数及其含义
display history-command	查看当前用户输入且成功执行的历史命令记录。在缺省情况下，最多可以显示最近的 10 条命令。通过命令 **history-command** max-size 可以设置历史命令缓冲区的大小
clock timezone *time-zone-name* { **add** | **minus** } *offset*	对本地时区信息进行设置。 参数解析如下所示： ① *time-zone-name* 参数用来指定时区名称。时区名称为字符串形式，需区分大小写，不支持空格，长度范围为 1~32。 ② **add** 关键词表示与 UTC 相比，该时区增加的时间偏移量。与 *offset* 参数结合在一起，就可以得到该时区所标识的时间。 ③ **minus** 关键词表示与 UTC 相比，该时区减少的时间偏移量。与 *offset* 参数结合在一起，就可以得到该时区所标识的时间。 ④ *offset* 参数指定了该时区与 UTC 的时间差。该参数的格式是 *HH:MM:SS*，参数解析如下所示。 a. *HH* 表示小时，在配合 **add** 关键词使用时，取值范围是 0~14；在配合 **minus** 关键词使用时，取值范围是 0~12。 b. *MM* 表示分，取值范围是 0~59。 c. *SS* 表示秒，取值范围是 0~59。 d. 当 *HH* 取最大值时，*MM* 和 *SS* 只能取值为 0
clock datetime *HH:MM:SS YYYY-MM-DD*	为设备设置当前的日期和时间。 参数解析如下所示： ① *HH:MM:SS* 参数指定了设备的当前时间。*HH* 表示小时，取值范围是 0~23；*MM* 表示分，取值范围是 0~59；*SS* 表示秒，取值范围是 0~59。 ② *YYYY-MM-DD* 参数指定了设备的当前日期。*YYYY* 表示年，取值范围是 2000~2037；*MM* 表示月，取值范围是 1~12；*DD* 表示日，取值范围是 1~31
sysname *host-name*	为设备设置主机名。 参数解析如下所示： *host-name* 参数指定了主机名，该参数为字符串形式，需区分大小写，支持空格，长度范围是 1~246。在缺省情况下，设备主机名为 Huawei
display this	根据用户所在的配置视图，显示相关的配置信息
user-interface [*ui-type*] *first-ui-number* [*last-ui-number*]	进入一个或多个用户界面视图。 参数解析如下所示。 ① *ui-type* 参数的取值可以是 **console** 或 **vty**： a. **console** 关键词用来指定控制口； b. **vty** 关键词用来指定 vty 接口。 ② *first-ui-umber* 参数指定了配置的第一个用户界面编号。当取值为 **console** 时，取值是 0；当取值为 **vty** 时，取值范围是 0 至配置的最大 **vty** 值。

实验 1　基础配置命令（续）

命令	命令解析
user-interface [*ui-type*] *first-ui-number* [*last-ui-number*]	③ *last-ui-number* 参数指定了配置的最后一个用户界面编号。添加了这个可选参数后，用户会同时进入多个用户界面视图。这个参数只有在 *ui-type* 的取值为 **vty** 时才有效，并且 *last-ui-number* 的取值要比 *first-ui-number* 的取值大
authentication-mode { **aaa** | **password** | **none** }	用来设置登录用户界面所使用的验证方式。在缺省情况下，通过 Console 口登录的验证方式为 AAA，其他登录方式缺省无配置，必须先配置用户界面的验证方式，否则用户无法成功登录设备。 参数解析如下所示。 ① **aaa** 关键词指定使用 AAA 方式进行登录验证。 ② **password** 关键词指定使用密码方式进行登录验证。密码为字符串形式，包括大小写字母、数字和特殊字符。 ③ **none** 关键词指定不进行登录验证
user privilege *level*	指定用户级别。在缺省情况下，Console 口用户界面下的用户级别是 15，其他用户界面下的用户级别是 0。 参数解析如下所示： *level* 参数指定了用户级别。该参数的取值范围是 0～15，取值越大，级别越高
aaa	从系统视图进入 AAA 视图，进行有关用户接入方面的安全配置，比如创建本地用户、设置用户级别、配置认证、授权方案等
local-user *user-name* { **password** { **cipher** | **irreversible-cipher** } *password* | **privilege level** *level* }	创建本地用户并设置本地用户的各项参数。 参数解析如下所示。 ① *user-name* 参数指定了本地用户的用户名。该参数为字符串形式，不区分大小写，不支持空格、星号、双引号和问号，其长度范围是 1～64。如果用户名中带有域名分隔符（比如@），那么@前面的部分是用户名，@后面的部分是域名。如果没有@，则整个字符串都是用户名，并且它所属的域为默认域。 ② **password** { **cipher** | **irreversible-cipher** } *password* 指定本地用户的登录密码。 参数解析如下所示。 a. **cipher** 关键词表示采用可逆算法对用户密码进行加密，安全性较低。 b. **irreversible-cipher** 关键词表示采用不可逆算法对用户密码进行加密，提供了更高的安全性。 c. *password* 参数指定了密码，该参数为字符串形式，区分大小写，长度范围是 8～128。为了提高安全性，用户输入的明文必须至少包括两种字符形式，例如大写字母、小写字母、数字和特殊字符，并且不能与用户名或用户名的倒写形式相同。 ③ **privilege level** *level* 指定了本地用户的级别。用户登录后只能使用等于或低于自己级别的命令。 参数解析如下所示： *level* 参数指定了级别，该参数的取值范围是 0～15，取值越大，用户的级别越高
local-user *user-name* **service-type** { **8021x** | **api** | **ftp** | **http** | **ppp** | **ssh** | **telnet** | **terminal** | **web** }	配置本地用户可使用的接入类型。在缺省情况下，本地用户未关联任何接入类型。 参数解析如下所示。 ① *user-name* 参数指定了本地用户的用户名。该参数为字符串形式，不区分大小写，不支持空格、星号、双引号和问号，其长度范围是 1～64。如果用户名中带有域名分隔符（比如@），那么@前面的部分是用户名，@后面的部分是域名。如果没有@，则整个字符串都是用户名，并且它所属的域为默认域。

实验 1　基础配置命令（续）

命令	命令解析
local-user *user-name* **service-type** { **8021x** | **api** | **ftp** | **http** | **ppp** | **ssh** | **telnet** | **terminal** | **web** }	② **8021x** 关键词指定用户类型为 802.1x 用户。 ③ **api** 关键词指定用户类型为 API 用户，通常用于 NETCONF 接入的用户。当用户类型为 API 用户时，用户名不能配置为 root。 ④ **ftp** 关键词指定用户类型为 FTP 用户。 ⑤ **http** 关键词指定用户类型为 HTTP 用户，通常用于 Web 网管登录。 ⑥ **ppp** 关键词指定用户类型为 PPP 用户。 ⑦ **ssh** 关键词指定用户类型为 SSH 用户。 ⑧ **telnet** 关键词指定用户类型为 Telnet 用户，通常指网络管理员。 ⑨ **terminal** 关键词指定用户类型为终端用户，通常指 Console 口用户。 ⑩ **web** 关键词指定用户类型为 Portal 认证用户
telnet server enable	使能 Telnet 服务器功能。在缺省情况下，Telnet 服务器功能处于去使能状态。只有当 Telnet 服务器的状态为打开时，才能连接到 Telnet 服务器。Telnet 服务器关闭后，用户只能通过 Console 口或 SSH 等其他方式登录设备
protocol inbound { **all** | **ssh** | **telnet** }	用户界面视图命令，指定 VTY 用户界面所支持的协议。在缺省情况下，系统支持 SSH。 参数解析如下所示： ① **all** 关键词指定了所有支持的协议，包括 SSH 和 Telnet； ② **ssh** 关键词指定了 SSH 协议； ③ **telnet** 关键词指定了 Telnet 协议
telnet *host-ip*	从当前设备使用 Telnet 协议登录其他设备。 参数解析如下所示： *host-ip* 参数指定了远端设备的 IPv4 地址或主机名（需要配合 DNS 使用）

实验 2　静态路由配置命令

命令	命令解析
interface *interface slot-number*	从系统视图进入接口视图。 参数解析如下所示： ① *interface* 参数应配置接口类型，比如 GigabitEthernet； ② *slot-number* 参数应配置接口编号，比如 0/0/0
ip address *ip-address* { *mask* | *mask-length* }	接口视图命令，配置接口的 IP 地址。在缺省情况下，接口上没有配置 IP 地址。 参数解析如下所示： ① *ip-address* 参数指定了接口的 IP 地址，该参数使用点分十进制格式； ② *mask* 参数指定了子网掩码，该参数使用点分十进制格式； ③ *mask-length* 参数指定了子网掩码长度，该参数的取值范围是 0～32
ip route-static *ip-address* { *mask* | *mask-length* } { *nexthop-address* | *interface-type interface-number* [*nexthop-address*] } [**preference** *preference*]	配置单播静态路由。在缺省情况下，系统中没有配置任何单播静态路由。 参数解析如下所示： ① *ip-address* 参数指定了目的 IP 地址，格式为点分十进制。 ② *mask* 参数指定了 IP 地址的掩码，格式为点分十进制。 ③ *mask-length* 参数指定了掩码长度。由于 32 位的掩码要求 "1" 是连续的，点分十进制格式的掩码可以用掩码长度代替。该参数的取值范围是 0～32。 ④ *nexthop-address* 参数指定了路由的下一跳 IP 地址，格式为点分十进制。 ⑤ *interface-type interface number* 参数指定了路由转发报文的接口类型和接口号。 ⑥ **preference** *preference* 指定了静态路由协议的优先级。优先级值越小，优先级越高。该参数的取值范围是 1～255，缺省值是 60

实验 2　静态路由配置命令（续）

命令	命令解析
display ip interface [*interface-type interface-number*]	查看接口上与 IP 相关的配置和统计信息，如接口上接收和发送的报文数、字节数和组播报文数，以及接口接收、发送、转发和丢弃的广播报文数。 参数解析如下所示： ① *interface-type* 参数应配置接口类型，比如 GigabitEthernet； ② *interface-number* 参数应配置接口编号，比如 0/0/0
display ip interface brief	查看接口上与 IP 相关的简要信息，如 IP 地址、子网掩码、物理链路和协议的 Up/Down 状态及处于不同状态的接口数量
tracert *host*	查看数据包从源到目的地的路径信息，以便检查网络连接是否可用。当网络出现连通性故障时，管理员可以使用这条命令来定位故障点
display ip routing-table	用来查看 IPv4 路由表的信息
display ip routing-table protocol *protocol*	查看指定路由协议的路由信息。 参数解析如下所示： *protocol* 参数指定了需要查看的路由协议，该参数可能取值为 **bgp**、**direct**、**isis**、**ospf**、**rip**、**static**、**unr**，具体以设备所支持的路由协议为准

实验 3　OSPF 配置命令

命令	命令解析
ospf [*process-id* | **router-id** *router-id*]	创建 OSPF 进程并进入 OSPF 路由协议视图。 参数解析如下所示： ① *process-id* 参数应配置 OSPF 进程号，取值范围是 1～65535，缺省值是 1； ② **router**-id *router-id* 指定了设备的 OSPF 路由器 ID，格式为点分十进制
area *area-id*	OSPF 视图配置命令，创建 OSPF 区域并进入 OSPF 区域视图。 参数解析如下所示： *area-id* 参数指定了区域的标识。其中 *area-id* 为 0 的区域称为骨干区域。该参数可以是十进制整数形式，也可以是点分十进制格式。采取整数形式时，取值范围是 0～4294967295
network *network-address wildcard-mask*	OSPF 区域视图配置命令，指定了运行 OSPF 协议的接口及接口所属区域。在缺省情况下，此接口不属于任何区域。 参数解析如下所示： ① *network-address* 参数指定了接口所在的网段地址，该参数的格式为点分十进制； ② *wildcard-mask* 参数指定了 IP 地址的通配符掩码，该参数的格式为点分十进制
ospf enable [*process-id*] **area** *area-id*	接口视图配置命令，在接口上使能 OSPF。 参数解析如下所示。 ① *process-id* 参数指定了接口所要参与的 OSPF 进程号，取值范围是 1～65535，缺省值是 1； ② **area** *area-id* 指定了接口所要参与的 OSPF 区域，可以是整数形式或 IP 地址格式。当采取整数形式时，取值范围是 0～4294967295
ospf cost *cost*	接口视图配置命令，配置接口上运行 OSPF 协议所需的开销。在缺省情况下，OSPF 根据接口的带宽自动计算其开销。 参数解析如下所示： *cost* 参数指定了运行 OSPF 协议所需的开销，该参数的取值范围是 1～65535

实验 3　OSPF 配置命令（续）

命令	命令解析
default-route-advertise [[**always**] [**cost** *cost*]]	OSPF 视图配置命令，将缺省路由通告到普通 OSPF 区域中。 参数解析如下所示： ① **always** 关键词令主机产生并发布一个描述缺省路由的 LSA，无论本机是否存在激活的非本 OSPF 进程的缺省路由。 a. 如果配置了 **always** 参数，设备不再计算来自其他设备的缺省路由； b. 如果没有配置 **always** 参数，本机路由表中必须有激活的非本 OSPF 进程的缺省路由时才生成缺省路由的 LSA。 ② **cost** *cost* 指定了该 ASE LSA 的开销值，该参数的取值范围是 0～16777214
authentication-mode { **md5** | **hmac-md5** | **hmac-sha256** } [*key-id* { **plain** *plain-text* | [**cipher**] *cipher-text* }]	接口视图/OSPF 区域视图配置命令，指定 OSPF 区域所使用的验证模式及验证密码。 参数解析如下所示： ① **md5** | **hmac-md5** | **hmac-sha256** 关键词指定相应的密文验证模式； ② *key-id* 参数指定了接口上密文验证的验证字标识符，两端的验证字标识符必须保持一致，该参数的取值范围是 1～255； ③ **plain** *plain-text* 指定了明文类型的密码，在查看配置文件时会看到明文形式的密码； ④ **cipher** *cipher-text* 指定了密文类型的密码，在查看配置文件时会看到密文形式的密码
display ospf [*process-id*] **peer** [[*interface-type interface-number*] [*neighbor-id*] | **brief** | **last-nbr-down**]	查看 OSPF 中各个区域邻居的信息。 参数解析如下所示： ① *process-id* 参数指定了 OSPF 进程号，该参数的取值范围是 1～65535； ② *interface-type interface-number* 参数指定了接口类型和接口号； ③ *neighbor-id* 参数指定了邻居的路由器 ID，该参数的格式是点分十进制； ④ **brief** 关键词用来显示 OSPF 各个区域中邻居的概要信息； ⑤ **last-nbr-down** 关键词用来显示 OSPF 区域中最后 Down 掉的邻居的概要信息
display ospf [*process-id*] **interface** [**all** | *interface-type interface-number*] [**verbose**]	查看 OSPF 的接口信息。 参数解析如下所示： ① *process-id* 参数指定了 OSPF 进程号，该参数的取值范围是 1～65535； ② **all** 关键词用来显示所有接口的 OSPF 详细信息； ③ *interface-type interface-number* 参数指定了接口类型和接口号； ④ **verbose** 关键词用来显示详细配置信息

实验 4　以太网配置命令

命令	命令解析
vlan *vlan-id*	创建 VLAN 并进入 VLAN 视图，如果 VLAN 已存在，直接进入该 VLAN 的视图。在缺省情况下，将所有接口都加入一个缺省的 VLAN 中，该 VLAN 标识为 1。 参数解析如下所示： *vlan-id* 参数指定了 VLAN ID，其取值范围是 1～4094
vlan batch { *vlan-id1* [**to** *vlan-id2*]}	批量创建 VLAN。 参数解析如下所示。 *vlan-id1* [**to** *vlan-id2*] 指定了批量创建的 VLAN ID，其中：

实验 4　以太网配置命令（续）

命令	命令解析
vlan batch { *vlan-id1* [**to** *vlan-id2*]}	① *vlan-id1* 标识第一个 VLAN 的编号，*vlan-id2* 标识最后一个 VLAN 的编号； ② *vlan-id2* 的取值必须大于或等于 *vlan-id1*； ③ 如果仅指定了 *vlan-id1*，则只创建 *vlan-id1* 所指定的 VLAN
port link-type { **access** | **hybrid** | **trunk** }	接口视图配置命令，用来配置接口的链路类型。在缺省情况下，接口的链路类型为 Hybrid。 参数解析如下所示： ① **access** 关键词将接口的链路类型设置为 Access； ② **hybrid** 关键词将接口的链路类型设置为 Hybrid； ③ **trunk** 关键词将接口的链路类型设置为 Trunk
port default vlan *vlan-id*	接口视图配置命令，用来配置接口的缺省 VLAN 并加入这个 VLAN。在缺省情况下，所有接口的 VLAN ID 为 1。 参数解析如下所示： *vlan-id* 参数设置了缺省 VLAN 的编号，取值范围是 1～4094
port trunk allow-pass vlan {{ *vlan-id1* [**to** *vlan-id2*]} | **all** }	接口视图配置命令，用来配置 Trunk 类型接口加入的 VLAN。在缺省情况下，Trunk 类型接口加入了 VLAN1。 参数解析如下所示。 ① *vlan-id1* [**to** *vlan-id2*]指定了 Trunk 类型接口加入的 VLAN，其中： a. *vlan-id1* 标识第一个 VLAN 的编号，*vlan-id2* 标识最后一个 VLAN 的编号； b. *vlan-id2* 的取值必须大于或等于 *vlan-id1*。 ② **all** 关键词指定了 Trunk 接口加入所有 VLAN
port hybrid untagged vlan {{ *vlan-id1* [**to** *vlan-id2*]} | **all** }	接口视图配置命令，用来配置 Hybrid 类型接口加入的 VLAN，并且这些 VLAN 的数据帧将以不打标签的方式通过接口。在缺省情况下，Hybrid 类型接口以不打标签方式加入了 VLAN1。 参数解析如下所示： ① *vlan-id1* [**to** *vlan-id2*]指定了 Hybrid 类型接口加入的 VLAN，其中： a. *vlan-id1* 标识第一个 VLAN 的编号，*vlan-id2* 标识最后一个 VLAN 的编号； b. *vlan-id2* 的取值必须大于或等于 *vlan-id1*。 ② **all** 关键词指定了 Hybrid 接口加入所有 VLAN
port hybrid tagged vlan {{ *vlan-id1* [**to** *vlan-id2*]} | **all** }	接口视图配置命令，用来配置 Hybrid 类型接口加入的 VLAN，并且这些 VLAN 的数据帧将以打标签的方式通过接口。在缺省情况下，Hybrid 类型接口以不打标签方式加入了 VLAN1。 参数解析如下所示： ① *vlan-id1* [**to** *vlan-id2*]指定了 Hybrid 类型接口加入的 VLAN，其中： a. *vlan-id1* 标识第一个 VLAN 的编号，*vlan-id2* 标识最后一个 VLAN 的编号； b. *vlan-id2* 的取值必须大于或等于 *vlan-id1*。 ② **all** 关键词指定了 Hybrid 接口加入所有 VLAN
mac-vlan mac-address *mac-address* [*mac-address-mask* | *mac-address-mask-length*] [**priority** *priority*]	将 MAC 地址与 VLAN 关联在一起。 参数解析如下所示： ① *mac-address* 参数设置了 MAC 地址，该参数的格式是 H-H-H，其中 H 是 4 位的十六进制数； ② *mac-address-mask* | *mac-address-mask-length* 参数指定了掩码、掩码长度； ③ **priority** *priority* 指定了 MAC 地址的 802.1p 优先级，取值范围是 0～7，值越大优先级越高，缺省值是 0

实验 4　以太网配置命令（续）

命令	命令解析
port-group *port-group-name*	创建并进入永久端口组视图
group-member *interface-type interface-number1* [**to** *interface-type interface-number2*]	将以太网端口添加到指定的永久端口组中。 参数解析如下所示： ① *interface-type interface-number1* 指定了第一个要加入端口组的端口； ② **to** *interface-type interface-number2* 指定了最后一个要加入端口组的端口
display vlan summary	查看系统中所有 VLAN 的汇总信息
display vlan	查看 VLAN 的相关信息
display mac-address dynamic [**vlan** *vlan-id* | *interface-type interface-number*]	查看动态 MAC 地址表项信息。如果不指定任何参数，将显示系统中所有动态 MAC 地址表项信息。 参数解析如下所示： ① **vlan** *vlan-id* 用来查看具体 VLAN 的动态 MAC 地址表项信息； ② *interface-type interface-number* 指定出接口为指定接口的动态 MAC 地址表项

实验 5　生成树配置命令

命令	命令解析
stp enable	使能 STP/RSTP 功能。在缺省情况下，设备的 STP/RSTP 功能处于启用状态
stp mode { **stp** | **rstp** }	配置交换设备的 STP/RSTP 工作模式。在缺省情况下，交换设备运行 MSTP 模式，MSTP 模式兼容 STP 和 RSTP 模式
stp root { **primary** | **secondary** }	将设备配置为根桥或备份根桥。 参数解析如下所示： ① **primary** 关键词将设备配置为根桥，配置后该设备优先级值自动为 0，并且不能更改设备优先级； ② **secondary** 关键词将设备配置为备份根桥，配置后该设备优先级值为 4096，并且不能更改设备优先级
stp priority *priority*	配置交换设备在系统中的优先级。 参数解析如下所示： *priority* 参数指定了优先级值。在缺省情况下，交换设备的优先级值为 32768
stp cost *cost*	接口视图配置命令，设置端口的路径开销值
stp timer forward-delay *forward-delay*	配置设备的 Forward Delay 时间。在缺省情况下，设备的 Forward Delay 时间是 15 s
stp timer hello *hello-time*	配置设备的 Hello Time 时间。在缺省情况下，设备的 Hello Time 时间是 2 s
stp timer max-age *max-age*	配置设备的 Max Age 时间。在缺省情况下，设备的 Max Age 时间是 20 s
stp bridge-diameter *diameter*	配置网络直径。在缺省情况下，网络直径为 7
stp root-protection	配置交换设备的 Root 保护功能。当端口的角色是指定端口时，配置的 Root 保护功能才生效。Root 保护功能一般只在根桥的端口上配置。配置了 Root 保护的端口，不可以配置环路保护
stp loop-protection	配置交换设备根端口或 Alternate 端口的环路保护功能。由于 Alternate 端口是根端口的备份端口，如果交换设备上有 Alternate 端口，需要在根端口和 Alternate 端口上同时配置环路保护。配置了根保护的端口，不可以配置环路保护

实验 5　生成树配置命令（续）

命令	命令解析
stp bpdu-protection	配置交换设备边缘端口的 BPDU 保护功能
stp edged-port enable	接口视图配置命令，将端口配置为边缘端口
error-down auto-recovery cause bpdu-protection interval *interval-value*	使能端口自动恢复为 Up 的功能，并设置端口自动恢复为 Up 的时延，使被关闭的端口经过时延后能够自动恢复。 参数解析如下所示： *interval-value* 参数设置了自动恢复的时延值
stp region-configuration	进入 MST 域视图。在缺省情况下，MST 域的 3 个参数均取缺省值： ① MST 域名为交换设备的 MAC 地址； ② MSTP 修订级别取值为 0； ③ 所有 VLAN 映射到 CIST 上
region-name *name*	配置交换设备的 MST 域名。在缺省情况下，MST 域名等于交换设备桥 MAC 的 MAC 地址。 参数解析如下所示： *name* 参数指定了交换设备的 MST 域名，该参数为字符串形式，不支持空格，区分大小写，长度为 1～32 字符
instance *instance-id* vlan { *vlan-id1* [to *vlan-id2*] }	将指定 VLAN 映射到指定的生成树实例上。在缺省情况下，所有 VLAN 均映射到 CIST，即实例 0 上。 参数解析如下所示： ① *instance-id* 参数指定了生成树实例的编号。该参数的取值范围是 0～4094，取值为 0 表示的是 CIST； ② *vlan-id1* 参数指定了起始的 VLAN 编号； ③ *vlan-id2* 参数指定了结束的 VLAN 编号
reversion-level *level*	配置交换设备的 MSTP 修订级别。在缺省情况下，交换设备 MST 域的修订级别为 0。 参数解析如下所示： *level* 参数指定了 MST 域的修订级别，取值范围是 0～65535
active region-configuration	激活 MST 域配置
display stp [interface *interface-type interface-number*] [brief]	查看生成树的状态和统计信息。 参数解析如下所示： ① **interface** *interface-type interface-number* 查看指定端口上的生成树状态和统计信息； ② **brief** 查看生成树的状态和统计信息摘要
display stp region-configuration	查看交换设备上当前生效的 MST 域配置信息，其中包括域名、域的修订级别、VLAN 与生成树实例的映射关系及配置的摘要

实验 6　链路聚合配置命令

命令	命令解析
interface eth-trunk *trunk-id*	创建并进入 Eth-Trunk 接口。 参数解析如下所示： *trunk-id* 参数指定了 Eth-Trunk 编号
trunkport *interface-type* { *interface-number1* [to *interface-number2*] }	Eth-Trunk 接口视图配置命令，在 Eth-Trunk 接口下增加成员接口。 参数解析如下所示： *interface-type* { *interface-number1* [to *interface-number2*] } 指定了接口类型和编号

实验 6　链路聚合配置命令（续）

命令	命令解析
eth-trunk *trunk-id*	接口视图配置命令，将当前接口加入指定的 Eth-Trunk 中。 参数解析如下所示： *trunk-id* 参数指定了 Eth-Trunk 的 ID
mode lacp-static	配置 Eth-Trunk 的工作模式为 LACP
least active-linknumber *link-number*	Eth-Trunk 接口视图配置命令，配置 Eth-Trunk 中活动接口数量的下限阈值。在缺省情况下，活动接口数的下限阈值为 1。 参数解析如下所示： *link-number* 参数指定了活动接口数量的下限阈值
max active-linknumber *link-number*	Eth-Trunk 接口视图配置命令，配置 Eth-Trunk 中活动接口数量的上限阈值。在缺省情况下，活动接口数的上限阈值为 8。 参数解析如下所示： *link-number* 参数指定了活动接口数量的上限阈值
lacp priority *priority*	系统视图、接口视图配置命令，设置系统 LACP 优先级或接口 LACP 优先级。在缺省情况下，系统 LACP 优先级和接口 LACP 优先级的值都是 32768。 参数解析如下所示： *priority* 参数指定了 LACP 优先级值。取值越小，LACP 优先级越高，取值范围是 0~65535
lacp preempt enable	Eth-Trunk 接口视图配置命令，使能 LACP 模式下的 LACP 优先级抢占功能
lacp preempt delay *delay-time*	Eth-Trunk 接口视图配置命令，配置抢占等待时间。在缺省情况下，LACP 抢占等待时间是 30 s。 参数解析如下所示： *delay-time* 参数指定了抢占等待时间，取值范围是 10~180，单位是 s
load-balance {dst-ip | dst-mac | src-ip | src-mac | src-dst-ip | src-dst-mac}	Eth-Trunk 接口视图配置命令，配置 Eth-Trunk 接口的负载分担模式。在缺省情况下，交换机上 Eth-Trunk 接口的负载分担模式为 src-dst-ip。 参数解析如下所示： ① **dst-ip** 关键词指定了基于目的 IP 地址进行负载分担； ② **dst-mac** 关键词指定了基于目的 MAC 地址进行负载分担； ③ **src-ip** 关键词指定了基于源 IP 地址进行负载分担； ④ **src-mac** 关键词指定了基于源 MAC 地址进行负载分担； ⑤ **src-dst-ip** 关键词指定了基于源和目的 IP 地址进行负载分担； ⑥ **src-dst-mac** 关键词指定了基于源和目的 MAC 地址进行负载分担
display eth-trunk { brief | *trunk-id* }	查看 Eth-Trunk 的信息。 参数解析如下所示： ① **brief** 关键词显示了 Eth-Trunk 的摘要信息； ② *trunk-id* 参数指定了要显示的 Eth-Trunk 编号
display trunkmembership eth-trunk *trunk-id*	查看 Eth-Trunk 接口及其成员信息。 参数解析如下所示： *trunk-id* 参数指定了 Eth-Trunk 接口的 ID

实验 7　VLAN 间路由配置命令

命令	命令解析
interface *interface-type interface-number.subinterface-number*	创建并进入子接口配置视图。 参数解析如下所示： ① **interface** *interface-type interface-number* 指定了接口的类型和编号； ② *subinterface-number* 参数指定了子接口的编号，取值范围是 1~4096
dot1q termination vid *low-pe-vid* [**to** *high-pe-vid*]	子接口视图命令，用来配置子接口 Dot1q 终结的单层 VLAN ID。 参数解析如下所示。 ① *low-pe-vid* 参数指定了用户报文中 Tag 的取值下限，取值范围是 2~4094。 ② *high-pe-vid* 参数指定了用户报文中 Tag 的取值上限，取值范围是 2~4094
arp broadcast enable	子接口视图命令，使能终结子接口的 ARP 广播功能。
ip interface vlanif *vlan-id*	创建 VLANIF 接口并进入 VLANIF 接口配置视图。 参数解析如下所示： *vlan-id* 参数指定了 VLANIF 接口的编号
display arp [**all**]	查看所有 ARP 表项
display mac-address	查看交换设备学习到的 MAC 地址与 VLAN 的对应关系

实验 8　ACL 配置命令

命令	命令解析
acl [**number**] *acl-number*	创建并进入编号 ACL 参数解析如下所示： ① **number** 关键词标识编号 ACL； ② *acl-number* 参数指定了 ACL 的编号，2000~2999 表示基本 ACL，3000~3999 表示高级 ACL
rule [*rule-id*] { **deny** | **permit** } [**source** { *source-address source-wildcard* | **any** }]	ACL 视图配置命令，用来配置 ACL 中的规则。 参数解析如下所示。 ① *rule-id* 参数指定了 ACL 规则的编号。 ② **deny** | **permit** 关键词指定了 ACL 规则的处理行为，**deny** 表示拒绝，**permit** 表示允许。 ③ **source** { *source-address source-wildcard* | **any** } 根据源 IP 地址匹配数据包，其中： a. *source-address* 指定了源 IP 地址； b. *source-wildcard* 指定了通配符掩码，0 表示必须匹配，1 表示不关心； c. **any** 匹配任意源 IP 地址
traffic-filter { **outbound** | **inbound** } **acl** *acl-number*	接口视图配置命令，在接口上应用 ACL 进行流量过滤。 参数解析如下所示。 ① **outbound** | **inbound** 关键词指定了在接口的哪个方向上配置报文过滤。**outbound** 表示出方向，**inbound** 表示入方向。 ② **acl** *acl-number* 指定了基于 IPv4 ACL 对报文进行过滤

实验 8　ACL 配置命令（续）

命令	命令解析
rule [*rule-id*] { **deny** | **permit** } { *protocol-number* | **tcp** } [**destination** { *destination-address destination-wildcard* | **any** } | **destination-port** { **eq** *port* | **gt** *port* | **lt** *port* | **range** *port-start port-end* } | **source** { *source-address source-wildcard* | **any** } | **source-port** { **eq** *port* | **gt** *port* | **lt** *port* | **range** *port-start port-end* } | **tcp-flag** { **ack** | **established** | **fin** | **psh** | **rst** | **syn** | **urg** }]	在高级 ACL 中过滤 TCP 流量。 参数解析如下所示。 ① *rule-id* 参数指定了 ACL 的规则 ID，取值范围是 0~4294967294。 ② **deny** | **permit** 设置了拒绝或允许行为，**deny** 表示拒绝，**permit** 表示允许。 ③ *protocol-number* 指定了 TCP 的协议号。 ④ **destination** { *destination-address destination-wildcard* | **any** } 指定了 ACL 规则匹配报文的目的地址信息。 ⑤ **destination-port** { **eq** *port* | **gt** *port* | **lt** *port* | **range** *port-start port-end* } 指定了 ACL 规则匹配报文的 TCP 目的端口，仅在报文协议是 TCP 或 UDP 时有效。其中： a. **eq** *port* 表示等于目的端口； b. **gt** *port* 表示大于目的端口； c. **lt** *port* 表示小于目的端口； d. **range** *port-start port-end* 表示端口范围。 ⑥ **source** { *source-address source-wildcard* | **any** } 指定 ACL 规则匹配报文的源地址信息。 ⑦ **source-port** { **eq** *port* | **gt** *port* | **lt** *port* | **range** *port-start port-end* } 指定 ACL 规则匹配报文的 TCP 源端口，仅在报文协议是 TCP 或 UDP 时有效。其中： a. **eq** *port* 表示等于目的端口； b. **gt** *port* 表示大于目的端口； c. **lt** *port* 表示小于目的端口； d. **range** *port-start port-end* 表示端口范围。 ⑧ **tcp-flag** { **ack** | **established** | **fin** | **psh** | **rst** | **syn** | **urg** } 指定了 ACL 规则匹配报文的 TCP 报文头部中的标记。其中： a. **ack** 指定 ACL 规则匹配报文的 TCP 报文头部中的 SYN Flag 类型为 ack（010000）； b. **established** 指定 ACL 规则匹配报文的 TCP 报文头部中的 SYN Flag 类型为 ack（010000）或 rst（000100）； c. **fin** 指定 ACL 规则匹配报文的 TCP 报文头部中的 SYN Flag 类型为 fin（000001）； d. **psh** 指定 ACL 规则匹配报文的 TCP 报文头部中的 SYN Flag 类型为 psh（001000）； e. **rst** 指定 ACL 规则匹配报文的 TCP 报文头部中的 SYN Flag 类型为 rst（000100）； f. **syn** 指定 ACL 规则匹配报文的 TCP 报文头部中的 SYN Flag 类型为 syn（000010）； g. **urg** 指定 ACL 规则匹配报文的 TCP 报文头部中的 SYN Flag 类型为 urg（100000）

实验 8　ACL 配置命令（续）

命令	命令解析
rule [*rule-id*] { **deny** | **permit** } { *protocol-number* | **udp** } [**destination** { *destination-address destination-wildcard* | **any** } | **destination-port** { **eq** *port* | **gt** *port* | **lt** *port* | **range** *port-start port-end* } | **source** { *source-address source-wildcard* | **any** } | **source-port** { **eq** *port* | **gt** *port* | **lt** *port* | **range** *port-start port-end* }]	在高级 ACL 中过滤 UDP 流量。 参数解析如下所示。 ① *rule-id* 参数指定了 ACL 的规则 ID，取值范围是 0～4294967294。 ② **deny** | **permit** 设置了拒绝或允许行为，**deny** 表示拒绝，**permit** 表示允许。 ③ *protocol-number* 指定了 UDP 的协议号。 ④ **destination** { *destination-address destination-wildcard* | **any** } 指定了 ACL 规则匹配报文的目的地址信息。 ⑤ **destination-port** { **eq** *port* | **gt** *port* | **lt** *port* | **range** *port-start port-end* } 指定了 ACL 规则匹配报文的 UDP 目的端口，仅在报文协议是 TCP 或 UDP 时有效。其中： a. **eq** *port* 表示等于目的端口； b. **gt** *port* 表示大于目的端口； c. **lt** *port* 表示小于目的端口； d. **range** *port-start port-end* 表示端口范围。 ⑥ **source** { *source-address source-wildcard* | **any** } 指定 ACL 规则匹配报文的源地址信息。 ⑦ **source-port** { **eq** *port* | **gt** *port* | **lt** *port* | **range** *port-start port-end* } 指定 ACL 规则匹配报文的 UDP 源端口，仅在报文协议是 TCP 或 UDP 时有效。其中： a. **eq** *port* 表示等于目的端口； b. **gt** *port* 表示大于目的端口； c. **lt** *port* 表示小于目的端口； d. **range** *port-start port-end* 表示端口范围
rule [*rule-id*] { **deny** | **permit** } { *protocol-number* | **icmp** } [**destination** { *destination-address destination-wildcard* | **any** } | **icmp-type** { *icmp-name* | *icmp-type* [*icmp-code*] } | **source** { *source-address source-wildcard* | **any** }]	在高级 ACL 中过滤 ICMP 流量。 参数解析如下所示： ① *rule-id* 参数指定了 ACL 的规则 ID，取值范围是 0～4294967294； ② **deny** | **permit** 设置了拒绝或允许行为，**deny** 表示拒绝，**permit** 表示允许； ③ *prtotocol-number* 指定了 ICMP 的协议号； ④ **destination** { *destination-address destination-wildcard* | **any** } 指定了 ACL 规则匹配报文的目的地址信息； ⑤ **icmp-type** { *icmp-name* | *icmp-type* [*icmp-code*] } 指定了 ACL 规则匹配报文的 ICMP 类型信息； ⑥ **source** { *source-address source-wildcard* | **any** } 指定 ACL 规则匹配报文的源地址信息
display acl { *acl-number* | **name** *acl-name* | **all** }	查看 ACL 的配置信息。 参数解析如下所示： ① *acl-number* 指定要查看的 ACL 编号； ② **name** *acl-name* 指定要查看的 ACL 名称； ③ **all** 指定查看所有 ACL

实验 9　AAA 配置命令

命令	命令解析
aaa	进入 AAA 视图
authentication-scheme *authentication-scheme-name*	AAA 视图配置命令，创建并进入认证方案。在缺省情况下，"default" 域使用名为 "radius" 的认证方案，"default_admin" 域使用名为 "default" 的认证方案，其他域使用名为 "radius" 的认证方案。 参数解析如下所示： *authentication-scheme-name* 指定了认证方案名称
authentication-mode { **hwtacacs** | **local** | **radius** } [**none**]	认证方案视图配置命令，配置当前认证方案适用的认证模式。在缺省情况下，认证模式为本地认证。 参数解析如下所示： ① **hwtacacs** 指定认证模式为 HWTACACS 认证。如果需要使用 HWTACACS 方式进行认证，则必须配置 HWTACACS 服务器模板中的认证服务器。 ② **local** 指定认证模式为本地认证。 ③ **radius** 指定认证模式为 RADIUS 认证。如果需要使用 RADIUS 方式进行认证，则必须配置 RADIUS 服务器模板中的认证服务器。 ④ **none** 指定认证模式为不进行认证，即直接让用户通过认证
authorization-scheme *authorization-scheme-name*	AAA 视图配置命令，创建并进入授权方案。在缺省情况下，系统中有一个名称为 "default" 的授权方案。用户可以修改 "default" 授权方案，但是不能删除。"default" 授权方案的策略为：授权模式采用本地授权，不启用按命令行授权。 参数解析如下所示： *authorization-scheme-name* 指定了授权方案名称
authorization-mode { **hwtacacs** | **if-authenticated** | **local** } [**none**]	授权方案视图配置命令，配置当前授权方案中的授权模式。在缺省情况下，授权模式为本地授权。 参数解析如下所示： ① **hwtacacs** 指定授权模式为 HWTACACS 授权。 ② **if-authenticated** 指定授权模式为 if-authenticated 授权。如果用户通过验证，则用户授权通过，否则授权不通过。在 RADIUS 认证方式下，配置 if-authenticated 授权不生效。 ③ **local** 指定授权模式为本地授权。 ④ **none** 指定授权模式为无须授权
service-scheme *service-scheme-name*	AAA 视图配置命令，创建并进入业务方案视图。在缺省情况下，设备中没有配置业务方案。 参数解析如下所示： *service-scheme-name* 指定了业务方案的名称
admin-user privilege level *level*	AAA 视图配置命令，指定用户可以作为管理员登录设备，并配置登录时的用户级别。 参数解析如下所示： *level* 指定了用户级别，取值范围是 0～15，取值越大，级别越高；不同级别的用户登录后，只能使用等于或低于自己级别的命令
domain *domain-name*	AAA 视图配置命令，创建并进入域视图。在缺省情况下，设备上存在名为 "default" 和 "default_admin" 两个域。可以修改这两个域下的配置，但是不能删除这两个域。 参数解析如下所示： *domain-name* 指定了域名，字符串形式，不区分大小写，长度范围是 1～64，不支持空格，不能仅配置为 "-" 或 "--"，且不能包含字符 "*"、"?"、""

实验 9　AAA 配置命令（续）

命令	命令解析
local-user *user-name* **password**	AAA 视图配置命令，配置本地账号的登录密码。在缺省情况下，本地账号的登录密码为空。 参数解析如下所示： ① *user-name* 指定了用户名； ② 本命令为交互式命令，输入命令并按下回车键后即可根据提示信息设置密码，输入的密码为字符串形式，区分大小写，长度范围是 8～128
local-user *user-name* **service-type** { **8021x** | **ftp** | **http** | **ppp** | **ssh** | **telnet** | **terminal** | **web** }	AAA 视图配置命令，配置本地用户的接入类型。在缺省情况下，本地用户关闭所有的接入类型。 参数解析如下所示。 ① *user-name* 指定了用户名。 ② **service-type** 指定了接入类型，其中： a. **8021x** 指定用户类型为 802.1x 用户； b. **ftp** 指定用户类型为 FTP 用户； c. **http** 指定用户类型为 HTTP 用户； d. **ppp** 指定用户类型为 PPP 用户； e. **ssh** 指定用户类型为 SSH 用户； f. **telnet** 指定用户类型为 Telnet 用户； g. **terminal** 指定用户类型为终端用户，通常指 Console 用户； h. **web** 指定用户类型为 Portal 认证用户
telnet server enable	使能 Telnet 服务器。在缺省情况下，Telnet 服务器功能处于去使能状态
user-interface [*ui-type*] *first-ui-number* [*last-ui-number*]	进入一个或多个用户界面视图。 参数解析如下所示： ① *ui-type* 指定用户界面的类型，取值可以是 **console** 或 **vty**； ② *first-ui-number* 指定配置的第一个用户界面编号； ③ *last-ui-number* 指定配置的最后一个用户界面编号
free user-interface *ui-type* *ui-number*	断开与指定用户界面的连接。 参数解析如下所示： ① *ui-type* 指定用户界面的类型，取值可以是 **console** 或 **vty**； ② *ui-number* 指定了用户界面的编号
display aaa configuration	查看 AAA 的概要信息，如域、认证方案、授权方案、计费方案的使用情况
display authentication-scheme [*authentication-scheme-name*]	查看认证方案的配置信息。 参数解析如下所示： *authentication-scheme-name* 指定查看的认证方案名称
display authorization-scheme [*authorization-scheme-name*]	查看授权方案的配置信息。 参数解析如下所示： *authorization-scheme-name* 指定查看的授权方案名称
display domain [**name** *domain-name*]	查看域的配置信息。 参数解析如下所示： *domain-name* 指定了查看的域名，如果不指定域名，则显示所有域的摘要信息

实验9 AAA 配置命令（续）

命令	命令解析
display local-user [**username** *user-name*]	查看本地用户的属性。 参数解析如下所示： *user-name* 查看特定用户名的本地用户属性
display telnet servers status	查看 Telnet 服务器的状态和配置信息
display aaa statistics offline-reason	查看用户下线原因
display access-user	查看在线用户（包括接入用户和管理员用户）的信息

实验10 NAT 配置命令

命令	命令解析
nat static global *global-address* **inside** *host-address*	接口视图配置命令，配置私网 IP 地址和公网 IP 地址的静态映射关系。 参数解析如下所示： ① **global** *global-address* 指定外部地址； ② **inside** *host-address* 指定内部地址
nat address-group *group-index start-address end-address*	配置 NAT 地址池。 参数解析如下所示： ① *group-index* 指定了 NAT 地址池索引号，不同设备有不同的取值范围，请查询文档或使用在线帮助； ② *start-address* 指定了地址池的起始地址； ③ *end-address* 指定了地址池的结束地址
nat outbound *acl-number* **address-group** *group-index* [**no-pat**]	将一个 ACL 和一个地址池关联起来，表示 ACL 中规定的地址可以使用地址池进行地址转换。 参数解析如下所示。 ① *acl-number* 指定了 ACL 编号，取值范围是 2000～3999。 ② **address-group** *group-index* 表示使用地址池的方式配置地址转换。如果不指定地址池，则直接使用该接口的 IP 地址作为转换后的地址，即 Easy IP 特性。设备型号不同，取值范围有所不同。 ③ **no-pat** 表示使用一对一的地址转换，只转换数据报文的地址而不转换端口信息
nat server protocol { **tcp** | **udp** } **global** *global-addres global-port* **inside** *host-address host-port*	定义一个内部服务器的映射表，外部用户可以通过地址和端口转换来访问内部服务器的某项服务。 参数解析如下所示。 ① *global-address* 提供给外部访问的 IP 地址（公网可路由 IP 地址）。 ② *global-port* 提供给外部访问的服务端口号。常用的端口号可以用关键词代替，例如 FTP 服务端口号为 21，同时可以使用 ftp 代替。如果不配置此参数，则表示是 **any** 的情况，即端口号为 0，任何类型的服务都提供。 ③ *host-address* 服务器的内部 IP 地址。 ④ *host-port* 服务器提供的服务端口号。如果不配置此参数，则与 *global-port* 端口号一致
display nat static	查看静态 NAT 转换规则

实验 10　NAT 配置命令（续）

命令	命令解析
display nat address-group [*group-index*] **verbose**	查看 NAT 地址池的配置信息。 参数解析如下所示： ① *group-index* 显示指定 NAT 地址池索引号信息； ② **verbose** 显示地址池的详细配置信息
display nat outbound [**acl** *acl-number* | **address-group** *group-index* | **interface** *interface-type interface-number*]	查看配置的 NAT Outbound 信息。 参数解析如下所示： ① **acl** *acl-number* 指定 ACL； ② **address-group** *group-index* 指定地址转换使用的地址池； ③ **interface** *interface-type interface-number* 指定查询接口的类型和编号
display nat session { **all** | **verbose** | **number** }	查看 NAT 映射表项。 参数解析如下所示： ① **all** 查看所有 NAT 映射表项； ② **verbose** 查看 NAT 映射表项的详细信息； ③ **number** 查看 NAT 映射表项数目
display nat server	查看 NAT Server 配置信息

实验 11　FTP 配置命令

命令	命令解析
dir [**/all**] [*filename* | *directory*]	用户视图配置命令，查看存储器中的文件和目录信息。 参数解析如下所示： ① **/all** 指定查看当前路径下的所有的文件和目录，其中包括已经删除至回收站的文件； ② *filename* 指定待查看的文件名称； ③ *directory* 指定待查看的目录路径
mkdir *directory*	用户视图配置命令，在当前存储器中创建目录。 参数解析如下所示： *directory* 指定需要创建的目录或路径
cd *directory*	用来修改用户当前的工作路径。 参数解析如下所示： *directory* 指定用户当前的工作路径
pwd	用户视图配置命令，显示当前工作路径
rmdir *directory*	用户视图配置命令，删除当前存储器中的指定目录。 参数解析如下所示： *directory* 指定需要删除的目录或路径
copy *source-filename destination-filename*	用户视图配置命令，复制文件。 参数解析如下所示： ① *source-filename* 指定被复制文件的路径及源文件名； ② *destination-filename* 指定目标文件的路径或目标文件名
move *source-filename destination-filename*	用户视图配置命令，将源文件从指定目录移动到目标目录中。 参数解析如下所示： ① *source-filename* 指定被移动的源文件的路径及源文件名； ② *destination-filename* 指定目标文件的路径或目标文件名

实验 11　FTP 配置命令（续）

命令	命令解析
rename *old-name new-name*	用户视图配置命令，对目录或文件进行重命名。 参数解析如下所示： ① *old-name* 指定当前的目录名或文件名； ② *new-name* 指定重命名后的目录名或文件名
delete [**/unreserved**] [**/force**] { *filename* | *devicename* }	用户视图配置命令，删除存储器中的指定文件。 参数解析如下所示： ① **/unreserved** 彻底删除指定文件，被删除的文件将不可恢复； ② **/force** 无须确认，直接删除文件； ③ *filename* 指定要删除的文件名； ④ *devicename* 指定要删除存储器中的所有文件
undelete { *filename* | *devicename* }	用户视图配置命令，恢复被删除至回收站的文件。 参数解析如下所示： ① *filename* 指定待恢复的文件名； ② *devicename* 指定存储器名
reset recycle-bin [*filename* | *devicename*]	用户视图配置命令，彻底删除指定路径下回收站中的文件。 参数解析如下所示： ① *filename* 指定要删除的文件名； ② *devicename* 指定存储器名
ftp server enable	开启设备的 FTP 服务器功能，允许 FTP 用户登录
local-user *user-name* **ftp-directory** *directory*	AAA 视图配置命令，设置允许 FTP 用户访问的目录。 参数解析如下所示： ① *user-name* 指定本地用户； ② *directory* 指定用户可访问的目录
put *local-filename* [*remote-filename*]	FTP 客户端视图配置命令，将本地的文件上传到远程 FTP 服务器。 参数解析如下所示： ① *local-filename* 指定 FTP 客户端的本地文件名； ② *remote-filename* 指定上传到远程 FTP 服务器上的文件名
get *remote-filename* [*local-filename*]	FTP 客户端视图配置命令，从远程 FTP 服务器下载文件并保存在本地。 参数解析如下所示： ① *remote-filename* 指定 FTP 服务器上需要下载的文件名； ② *local-filename* 指定下载后保存在本地的文件名
delete *remote-filename*	FTP 客户端视图配置命令，删除远程 FTP 服务器上的文件。 参数解析如下所示： *remote-filename* 指定待删除的文件名
bye	FTP 客户端视图配置命令，终止与远程 FTP 服务器的连接，并退回到用户视图

实验 12　DHCP 配置命令

命令	命令解析
ip pool *ip-pool-name*	创建全局地址池。 参数解析如下所示： *ip-pool-name* 指定地址池名称，不支持空格，长度范围是 1～64 字符，可以包含数字、字母和特殊字符

实验 12　DHCP 配置命令（续）

命令	命令解析
network *ip-address* [**mask** { *mask* | *mask-length* }]	IP 地址池视图配置命令，配置全局地址池下可分配的网段地址。 参数解析如下所示： ① *ip-address* 指定了网络地址段； ② *mask* 指定了 IP 地址池的网络掩码，不指定该参数时，使用自然掩码； ③ *mask-length* 指定了网络的掩码长度
excluded-ip-address *start-ip-address* [*end-ip-address*]	IP 地址池视图配置命令，配置 IP 地址池中不参与自动分配的 IP 地址范围。 参数解析如下所示。 ① *start-ip-address* 指定了不参与自动分配的IP 地址段的起始IP 地址。 ② *end-ip-address* 指定了不参与自动分配的 IP 地址段的结束 IP 地址。如果不指定该参数，表示只有一个 IP 地址
gateway-list *ip-address*	IP 地址池视图配置命令，为 DHCP 客户端配置出口网关地址。 参数解析如下所示： *ip-address* 指定了出口网关的 IP 地址
lease { **day** *day* [**hour** *hour* [**minute** *minute*]] | **unlimited** }	IP 地址池视图配置命令，配置地址池下的地址租期。 参数解析如下所示。 ① **day** *day* 指定了租用有效期的天数，取值范围是 0~999，单位是天，缺省值是 1； ② **hour** *hour* 指定了租用有效期的小时数，取值范围是 0~23，单位是小时，缺省值是 0； ③ **minute** *minute* 指定了租用有效期的分钟数，取值范围是 0~59，单位是分，缺省值是 0； ④ **unlimited** 指定租期为无限制
dns-list *ip-address*	IP 地址池视图配置命令，为 DHCP 客户端配置 DNS 服务器地址。 参数解析如下所示： *ip-address* 指定了 DNS 服务器的地址，最多可以配置 8 个 DNS 服务器的 IP 地址，各地址之间以空格分隔
dhcp enable	用来开启 DHCP 功能。
dhcp select global	接口视图配置命令，用来开启接口采用全局地址池的 DHCP 服务器功能
dhcp select interface	接口视图配置命令，用来开启接口采用接口地址池的 DHCP 服务器功能
dhcp server excluded-ip-address *start-ip-address* [*end-ip-address*]	接口视图配置命令，配置接口地址池中不参与自动分配的 IP 地址范围。 参数解析如下所示。 ① *start-ip-address* 指定了不参与自动分配的 IP 地址段的起始IP 地址。 ② *end-ip-address* 指定了不参与自动分配的 IP 地址段的结束 IP 地址。如果不指定该参数，表示只有一个 IP 地址

实验 12　DHCP 配置命令（续）

命令	命令解析
dhcp server lease {day *day* [hour *hour* [minute *minute*]] | unlimited}	接口视图配置命令，配置地址池下的地址租期。 参数解析如下所示。 ① **day** *day* 指定了租用有效期的天数。该参数的取值范围是 0～999，单位是天，缺省值是 1。 ② **hour** *hour* 指定了租用有效期的小时数。该参数的取值范围是 0～23，单位是小时，缺省值是 0。 ③ **minute** *minute* 指定了租用有效期的分钟数，取值范围是 0～59，单位是分，缺省值是 0。 ④ **unlimited** 指定租期为无限制
ip address dhcp-alloc	接口视图配置命令，开启接口下的 DHCP 客户端功能
dhcp select relay	接口视图配置命令，使能 DHCP 中继功能
dhcp relay server-ip *ip-address*	接口视图配置命令，配置 DHCP 中继所代理的 DHCP 服务器的 IP 地址。 参数解析如下所示： *ip-address* 指定了 DHCP 服务器的 IP 地址
display ip pool name *ip-pool-name*	查看已配置的 IP 地址池信息。 参数解析如下所示： **name** *ip-pool-name* 指定查看特定全局地址池的配置信息

实验 13　WLAN 配置命令

命令	命令解析
wlan	从系统视图进入 WLAN 视图
ap-group name *group-name*	WLAN 视图配置命令，创建 AP 组，并进入 AP 组视图，若 AP 组已存在则直接进入 AP 组视图。在缺省情况下，系统上存在名为 default 的 AP 组。 参数解析如下所示： *group-name* 指定了 AP 组的名称，长度范围是 1～35 字符，不能包含问号和空格，双引号不能出现在字符串的首尾
regulatory-domain-profile *profile-name*	WLAN 视图配置命令，将指定的域管理模板引用到 AP 或 AP 组。在缺省情况下，AP 组引用名为 default 的域管理模板，AP 未引用域管理模板。 参数解析如下所示： *profile-name* 指定了域管理模板的名称
country-code *country-code*	域管理模板视图配置命令，配置设备的国家码标识。在缺省情况下，设备的国家码标识为 cn。 参数解析如下所示： *country-code* 指定了设备的国家码标识
capwap source interface { loopback *loopback-number* | vlanif *vlan-id* }	配置 AC 与接入设备建立 CAPWAP 隧道的源接口。 参数解析如下所示： ① **loopback** *loopback-number* 指定了 Loopback 接口为源接口，取值范围是 0～1023； ② **vlanif** *vlan-id* 指定了 VLANIF 接口为源接口，取值范围是 1～4094

实验 13　WLAN 配置命令（续）

命令	命令解析
ap auth-mode { **mac-auth** | **no-auth** | **sn-auth** }	WLAN 视图配置命令，配置 AP 认证模式。在缺省情况下，AP 认证模式为 MAC 地址认证。 参数解析如下所示： ① **mac-auth** 指定 AP 认证模式为 MAC 地址认证； ② **no-auth** 指定 AP 认证模式为不认证； ③ **sn-auth** 指定 AP 认证模式为序列号认证
ap-id *ap-id* { **ap-mac** *ap-mac* | **ap-sn** *ap-sn* | **ap-mac** *ap-mac* **ap-sn** *ap-sn* }	WLAN 视图配置命令，离线增加 AP 设备或进入 AP 视图。 参数解析如下所示： ① **ap-mac** *ap-mac* 指定了 AP 的 MAC 地址，格式为 H-H-H，其中 H 为 4 位的十六进制数； ② **ap-sn** *ap-sn* 指定了 AP 的序列号，长度范围是 1～31，只能包含字母和数字
ap-name *ap-name*	AP 视图配置命令，用来配置单个 AP 的名称。 参数解析如下所示： *ap-name* 指定了 AP 名称，区分大小写，长度范围是 1～31
ap-group *ap-group*	AP 视图配置命令，配置 AP 所加入的组。 参数解析如下所示： *ap-group* 指定了 AP 加入的组
ssid-profile name *profile-name*	WLAN 视图配置命令，创建 SSID 模板，并进入模板视图，若模板已存在则直接进入模板视图。在缺省情况下，系统上存在名为 default 的 SSID 模板。 参数解析如下所示： *profile-name* 指定了 SSID 模板的名称，不区分大小写，长度范围是 1～35
ssid *ssid*	SSID 模板视图配置命令，配置当前 SSID 模板中的服务组合识别码。在缺省情况下，SSID 模板中的 SSID 为 HUAWEI-WLAN。 参数解析如下所示： *ssid* 指定了 SSID 的名称，区分大小写，长度范围是 1～32，支持中文字符，也支持中英文字符混合（说明：仅支持使用 UTF-8 编码格式的命令行编辑器编辑中文字符。若 STA 不支持 UTF-8 编码格式，则 STA 在搜索到含中文的 SSID 后无法正常显示 SSID 名称。）
security-profile name *profile-name*	WLAN 视图配置命令，用来创建安全模板并进入安全模板视图。在缺省情况下，系统中已有名为 default、default-wds 和 default-mesh 的安全模板
security { **wpa** | **wpa2** | **wpa-wpa2** } **psk** { **pass-phrase** | **hex** } *key-value* { **aes** | **tkip** | **aes-tkip** }	安全视图配置命令，用来配置 WPA/WPA2 预共享密钥认证和加密。 参数解析如下所示： *key-value* 指定了密码
vap-profile name *profile-name*	WLAN 视图配置命令，创建 VAP 模板，并进入模板视图，若模板已存在则直接进入模板视图。在缺省情况下，系统上存在名为 default 的 VAP 模板。 参数解析如下所示： *profile-name* 指定了 VAP 模板的名称，不区分大小写，长度范围是 1～35

实验 13　WLAN 配置命令（续）

命令	命令解析
forward-mode { **direct-forward** | **tunnel** }	VAP 模板视图配置命令，配置 VAP 模板下的数据转发方式。在缺省情况下，VAP 模板下的数据转发方式为直接转发。 参数解析如下所示： ① **direct-forward** 指定数据转发方式为直接转发； ② **tunnel** 指定数据转发方式为隧道转发
service-vlan vlan-id *vlan-id*	VAP 模板视图配置命令，配置 VAP 的业务 VLAN。在缺省情况下，VAP 的业务 VLAN 为 VLAN1。 参数解析如下所示： *vlan-id* 指定 VAP 的业务 VLAN 为单个 VLAN
ssid-profile *profile-name*	VAP 模板视图配置命令，用来应用 SSID 模板，在缺省情况下，VAP 模板中应用的是名为 default 的 SSID 模板。只有在 VAP 模板中应用了 SSID 模板后，SSID 模板中的配置才会对使用了该 VAP 模板的所有 AP 生效。 参数解析如下所示： *profile-name* 指定了 SSID 模板的名称
security-profile *profile-name*	VAP 模板视图配置命令，用来应用安全模板，在缺省情况下，VAP 模板中应用的是名为 default 的安全模板。 参数解析如下所示： *profile-name* 指定了安全模板的名称
vap-profile *profile-name* **wlan** *wlan-id* **radio** { *radio-id* | **all** }	AP 组视图配置命令，用来将指定的 VAP 模板引用到射频。 参数解析如下所示： ① *profile-name* 指定了 VAP 模板的名称； ② **wlan** *wlan-id* 指定了 VAP 的 ID，取值范围是 1～16； ③ **radio** 指定了射频； ④ *radio-id* 指定了射频的 ID； ⑤ **all** 指定了所有的射频
display ap { **all** | **ap-group** *ap-group* }	查看 AP 信息。 参数解析如下所示： ① **all** 查看所有已添加的 AP 信息； ② **ap-group** *ap-group* 指定 AP 所属的 AP 组名称
display ssid-profile { **all** | **name** *profile-name* }	查看 SSID 模板的配置信息和引用信息。 参数解析如下所示： ① **all** 查看所有 SSID 模板的信息； ② **name** *profile-name* 查看指定 SSID 模板的信息
display security-profile { **all** | **name** *profile-name* }	查看安全模板的配置信息和引用信息。 参数解析如下所示： ① **all** 查看所有安全模板的信息； ② **name** *profile-name* 查看指定安全模板的信息
display vap-profile { **all** | **name** *profile-name* }	查看 VAP 模板的配置信息和引用信息。 参数解析如下所示： ① **all** 查看所有 VAP 模板的信息； ② **name** *profile-name* 查看指定 VAP 模板的信息

实验 13　WLAN 配置命令（续）

命令	命令解析
display vap { all | ssid *ssid* }	查看 VAP 的相关信息。 参数解析如下所示： ① **all** 查看所有 VAP 的相关信息； ② **ssid** *ssid* 查看指定 SSID 的 VAP 相关信息
display station ssid *ssid*	查看在线的无线用户信息。 参数解析如下所示： **ssid** *ssid* 查看指定 SSID 的 STA 接入信息

实验 14　IPv6 配置命令

命令	命令解析
ipv6	使能设备转发 IPv6 单播报文，其中包括本地 IPv6 报文的发送与接收
ipv6 enable	接口视图配置命令，用来在接口上使能 IPv6 功能
ipv6 address auto link-local	接口视图配置命令，为接口配置自动生成的链路本地地址
ipv6 address { *ipv6-address prefix-length* | *ipv6-address* | *prefix-length* }	接口视图配置命令，配置接口的全球单播地址
dhcpv6 pool *pool-name*	创建 IPv6 地址池或进入 IPv6 地址池视图。 参数解析如下所示： *pool-name* 指定了 IPv6 地址池名称，区分大小写，长度范围是 1～31
address prefix *ipv6-prefix/ipv6-prefix-length*	IPv6 地址池视图配置命令，配置 IPv6 网络前缀。 参数解析如下所示。 *ipv6-prefix/ipv6-prefix-length* 指定了 IPv6 网络前缀和前缀长度，其中： a. *ipv6-prefix* 总长度为 128 位，通常分为 8 组，每组为 4 个十六进制数的形式，格式为 X:X:X:X:X:X:X:X； b. *ipv6-prefix-length* 为整数形式，取值范围是 1～128
dhcp enable	使能 DHCP 功能
dhcpv6 server *pool-name*	接口视图配置命令，启用 DHCPv6 服务器功能，并指定自动分配 IPv6 地址所使用的 IPv6 地址池。 参数解析如下所示： *pool-name* 指定了使用的 IPv6 地址池
ipv6 address auto dhcp	接口视图配置命令，在接口下启用 DHCPv6 客户端功能
undo ipv6 nd ra halt	接口视图配置命令，启用设备发送 RA 报文的功能。在缺省情况下，华为路由器不会发送 RA 报文
ipv6 nd autoconfig managed-address-flag	接口视图配置命令，设置 RA 报文中有状态自动配置地址的标志位（也就是将 M flag 设置为 1），令客户端使用服务器分配的 IPv6 地址。在缺省情况下，RA 报文中有状态自动配置地址的标志位为 0
ipv6 nd autoconfig other-flag	接口视图配置命令，设置 RA 报文中有状态自动配置其他信息的标志位（也就是将 O flag 设置为 1）
ipv6 address auto global default	接口视图配置命令，使设备请求默认网关信息，在收到 RA 报文生成 IPv6 地址的同时，还可以学习 RA 报文中的源 IPv6 地址，并且把它作为 IPv6 缺省路由的下一跳地址

实验 14 IPv6 配置命令（续）

命令	命令解析
ipv6 address auto global	接口视图配置命令，用来无状态自动生成 IPv6 全球单播地址
ipv6 route-static *dest-ipv6-address* *ipv6-prefix-length* *nexthop-ipv6-address*	配置 IPv6 静态路由。 参数解析如下所示： ① *dest-ipv6-address* 指定了目的 IPv6 地址； ② *ipv6-prefix-length* 指定了 IPv6 前缀长度； ③ *nexthop-ipv6-address* 指定了下一跳 IPv6 地址
display ipv6 interface [*interface-type interface-number* | **brief**]	查看接口的 IPv6 信息。 参数解析如下所示： ① *interface-type interface-number* 查看指定接口的 IPv6 信息； ② **brief** 查看接口的摘要信息
display ipv6 routing-table	查看 IPv6 路由表
display dhcpv6 server	查看设备上 DHCPv6 服务器的配置
ping ipv6 *ipv6-address* **-i** *interface-type interface-number*	针对 IPv6 链路本地地址进行 ping 测试